Lecture Notes in Computer Science   12944

More information about this subseries at http://www.springer.com/series/7407

Thierry Lecroq · Hélène Touzet (Eds.)

String Processing and Information Retrieval

28th International Symposium, SPIRE 2021
Lille, France, October 4–6, 2021
Proceedings

Editors
Thierry Lecroq (ID)
Université de Rouen Normandie
Mont-St-Aignan, France

Hélène Touzet (ID)
CNRS, CRIStAL
Villeneuve d'Ascq, France

ISSN 0302-9743   ISSN 1611-3349   (electronic)
Lecture Notes in Computer Science
ISBN 978-3-030-86691-4   ISBN 978-3-030-86692-1   (eBook)
https://doi.org/10.1007/978-3-030-86692-1

LNCS Sublibrary: SL1 – Theoretical Computer Science and General Issues

This Springer imprint is published by the registered company Springer Nature Switzerland AG
The registered company address is: Gewerbestrasse 11, 6330 Cham, Switzerland

Preface

This volume contains the papers presented at the International Symposium on String Processing and Information Retrieval (SPIRE 2021), which was supposed to be held in Lille, France, but was instead held virtually during October 4–6, 2021.

SPIRE started in 1993 as the South American Workshop on String Processing, therefore it was held in Latin America until 2000 when SPIRE traveled to Europe. From then on, SPIRE meetings have been held in Australia, Japan, the UK, Spain, Italy, Finland, Portugal, Israel, Brazil, Chile, Colombia, Mexico, Argentina, Bolivia, Peru and the USA. In this edition, we continued the long and well-established tradition of encouraging high-quality research at the broad nexus of algorithms and data structures for sequences and graphs, data compression, databases, data mining, information retrieval, and computational biology. As per usual, SPIRE 2021 continues to provide an opportunity to bring together specialists and young researchers working in these areas.

Using different email lists, the SPIRE 2021 call for papers was distributed around the world, resulting in 30 submissions. The EasyChair system was used to facilitate management of submissions and reviewing. Each submission was reviewed by at least 3, and on average 3.9, Program Committee (PC) members. The PC decided to accept 18 papers (14 long papers and 4 short papers). The program also includes three invited talks:

- Christina Boucher (University of Florida, USA)
- Daniel Lemire (University of Québec, Canada)
- Nicola Prezza (Ca' Foscari University of Venice, Italy)

These proceedings contain all 18 presented papers, together with two extended versions of the invited talks.

We thank the authors for their valuable combinatorial contributions and the reviewers for their thorough, constructive, and enlightening commentaries on the manuscripts. A special thanks goes also to the SPIRE Steering Committee members for their availability, help, and advice. Lastly, we are grateful to the organizing committee, chaired by Stéphane Janot and Antoine Limasset, and to GDR Bioinformatique Moléculaire (CNRS) and CRIStAL, whose financial support made it possible to make SPIRE 2021 a free event for all attendees.

We are happy to welcome you at SPIRE 2021.

July 2021

Thierry Lecroq
Hélène Touzet

Organization

Steering Committee

Ricardo Baeza-Yates	Northeastern University, USA, Universitat Pompeu Fabra, Spain, and University of Chile, Chile
Christina Boucher	University of Florida, USA
Nieves R. Brisaboa	University of A Coruña, Spain
Travis Gagie	Dalhousie University, Canada
Alistair Moffat	University of Melbourne, Australia
Gonzalo Navarro	University of Chile, Chile
Berthier Ribeiro-Neto	Federal University of Minas Gerais, Brazil
Simon J. Puglisi	University of Helsinki, Finland
Sharma Thankachan	University of Central Florida, USA
Nivio Ziviani	Universidade Federal Minas Gerais, Brazil

Program Committee Chairs

Thierry Lecroq	University of Rouen, France
Hélène Touzet	CNRS, CRIStAL, Lille, France

Program Committee

Lorraine Ayad	Brunel University London, UK
Golnaz Badkobeh	University of Warwick, UK
Ricardo Baeza-Yates	Northeastern University, USA, NTENT and Universitat Pompeu Fabra, Spain, and Universidad de Chile, Chile
Djamal Belazzougui	Research Centre for Scientific and Technical Information, Alger
Philip Bille	Technical University of Denmark, Denmark
Iovka Boneva	University of Lille, France
Broňa Brejová	Comenius University in Bratislava, Slovakia
Nieves R. Brisaboa	Universidade da Coruña, Spain
Ayelet Butman	Holon Institute of Technology, Israel
Nadia El-Mabrouk	University of Montreal, Canada
Simone Faro	University of Catania, Italy
Gabriele Fici	Università di Palermo, Italy
Travis Gagie	Diego Portales University, Chile
Arnab Ganguly	University of Wisconsin-Whitewater, USA
Cecilia Hernandez	University of Concepción, Chile
Tomohiro I.	Kyushu Institute of Technology, Japan
Shunsuke Inenaga	Kyushu University, Japan

Giuseppe F. Italiano	University of Rome "Tor Vergata", Italy
Jaap Kamps	University of Amsterdam, The Netherlands
Dominik Kempa	Johns Hopkins University, USA
Tomasz Kociumaka	University of California, Berkeley, USA
Tsvi Kopelowitz	University of Michigan, USA
M. Oguzhan Kulekci	Istanbul Technical University, Turkey
Susana Ladra	University of A Coruña, Spain
Thierry Lecroq	University of Rouen, France
Moshe Lewenstein	Bar Ilan University, Israel
Zsuzsanna Liptak	University of Verona, Italy
Giovanni Manzini	University of Pisa, Italy
Juan Mendivelso	Universidad Nacional de Colombia, Colombia
Laurent Mouchard	University of Rouen, France
Veli Mäkinen	University of Helsinki, Finland
Gonzalo Navarro	University of Chile, Chile
Yakov Nekrich	Michigan Technological University, USA
Kunsoo Park	Seoul National University, South Korea
Nadia Pisanti	University of Pisa, Italy
Solon Pissis	Centrum Wiskunde & Informatica, The Netherlands
Cinzia Pizzi	University of Padua, Italy
Jakub Radoszewski	University of Warsaw, Poland
Giovanna Rosone	Università di Pisa, Italy
Leena Salmela	University of Helsinki, Finland
Srinivasa Rao Satti	Norwegian University of Science and Technology, Norway
Marinella Sciortino	University of Palermo, Italy
Blerina Sinaimeri	Inria, France
Jouni Sirén	University of California, Santa Cruz, USA
Jens Stoye	Bielefeld University, Germany
Yasuo Tabei	RIKEN Center for Advanced Intelligence Project, Japan
Lynda Tamine	IRIT, France
Hélène Touzet	CNRS, CRIStAL, Lille, France
Bojian Xu	Eastern Washington University, USA
Binhai Zhu	Montana State University, USA

Organizing Committee

Rayan Chikhi	Institut Pasteur, Paris, France
Areski Flissi	CNRS, Lille, France
Stéphane Janot (Co-chair)	Université de Lille, France
Antoine Limasset (Co-chair)	CNRS, Lille, France
Camille Marchet	CNRS, Lille, France
Mikaël Salson	Université de Lille, France

Additional Reviewers

Akagi, Tooru
Bernardini, Giulia
Brandon, Adrian G.
Bulteau, Laurent
Charalampopoulos, Panagiotis
Fariña, Antonio
Funakoshi, Mitsuru
Gibney, Daniel
Girgis, Hani
Guerrini, Veronica
Jo, Seungbum
Kanda, Shunsuke
Leinonen, Miika

Mantaci, Sabrina
Mieno, Takuya
Nishimoto, Takaaki
Oco, Nathaniel
Paramá, José R.
Romana, Giuseppe
Rosenfeld, Matthieu
Rossi, Massimiliano
Sweering, Michelle
Urbina, Cristian
Walen, Tomasz
Wittler, Roland
Zuba, Wiktor

Contents

Data Structures

Repeats

Information Retrieval

Pattern Matching

Invited Papers

r-Indexing the eBWT

Christina Boucher[1]([⊠]) [iD], Davide Cenzato[2] [iD], Zsuzsanna Lipták[2] [iD],
Massimiliano Rossi[1] [iD], and Marinella Sciortino[3] [iD]

[1] Department of Computer and Information Science and Engineering,
University of Florida, Gainesville, FL, USA
{cboucher,rossi.m}@cise.ufl.edu

[2] Department of Computer Science, University of Verona, Verona, Italy
{davide.cenzato,zsuzsanna.liptak}@univr.it

[3] Department of Computer Science, University of Palermo, Palermo, Italy
marinella.sciortino@unipa.it

Abstract. The extended Burrows Wheeler Transform (eBWT) was
introduced by Mantaci et al. [TCS 2007] to extend the definition of the
BWT to a collection of strings. In our prior work [SPIRE 2021], we
give a linear-time algorithm for the eBWT that preserves the fundamental property of the original definition (i.e., the independence from the
input order). The algorithm combines a modification of the Suffix Array
Induced Sorting (SAIS) algorithm [IEEE Trans Comput 2011] with Prefix Free Parsing [AMB 2019; JCB 2020]. In this paper, we show how this
construction algorithm leads to r-indexing the eBWT, i.e., run-length
encoded eBWT and SA samples of Gagie et al. [SODA 2018] can be constructed efficiently from the components of the PFP. Moreover, we show
that finding maximal exact matches (MEMs) between a query string and
the r-index of the eBWT can be efficiently supported.

1 Introduction

There exists a number of large sequencing projects that aim to identify the biological variation of individuals of a given species. For example, the 100K Human
Genome Project [29], the 1001 Arabidopsis Project [28], and the 3,000 Rice
Genomes Project (3K RGP) [27]. Although biological variation is common and
necessary among cultivars or individuals of these sequencing projects, a large
portion of sequencing data is shared—leading to repetition within the dataset.
Given the thousands of individuals within these sequencing projects, indexing
them in a manner that allows variation to be identified and compared is challenging. The FM-index [18] has been the cornerstone of this indexing as most
standard read alignment algorithms (e.g., BWA [16] and Bowtie [15]) build and
use the FM-index of the set of one or more reference genomes. More specifically,
these read alignment algorithms build the Burrows Wheeler Transform (BWT)
and suffix array (SA) or SA samples to find alignments between sequence reads
and the database of genomes. However, as the number of genomes increases,
there has been an aim to build the BWT and SA samples in space that is linear
in number of runs in the BWT which is typically denoted as r.

© Springer Nature Switzerland AG 2021
T. Lecroq and H. Touzet (Eds.): SPIRE 2021, LNCS 12944, pp. 3–12, 2021.
https://doi.org/10.1007/978-3-030-86692-1_1

Mäkinen and Navarro may have been the first to pose the search from a FM-index that can be constructed in $\mathcal{O}(r)$ space [17]. They introduced the *run-length Burrows Wheeler Transform* (RL BWT) and showed how to locate all occurrences of a query string $P[1..p]$ in string $T[1..n]$ in $\mathcal{O}\big((p+s+occ)\big(\frac{\log \sigma}{\log \log r}+(\log \log n)^2\big)\big)$-time, where occ is the number of occurrences, σ is the alphabet size, and s is a parameter. The downside is that they require $\mathcal{O}(r+n/s)$-space for the SA samples. Given a query string P and the RL BWT of a string T, Policriti and Prezza [25] showed how to find a single SA sample in the interval in RL BWT containing P in $\mathcal{O}(r)$-space. Then in 2018, Gagie et al. [8] showed how to fully support locate queries, i.e., locate *all occ* SA samples in $\mathcal{O}(r)$-space. The resulting data structure is referred to as the *r-index*. We note that this result defined the r-index but did not give an algorithm to construct it. The algorithm to construct the r-index was described later by Boucher et al. [5] and Kuhnle et al. [14]—both are based on a preprocessing technique called *Prefix Free Parsing* (PFP). PFP produces two temporary structures called the *dictionary* and *parse*. From these components the r-index can be built. This was a significant achievement but was not fully set up to accomplish read alignment since finding short exact matches between a read and an index was not defined for the r-index. To accomplish this latter task Rossi et al. [26] augmented PFP to construct an auxiliary data structure, called *thresholds*, in addition to the r-index. The addition of thresholds allows for finding *maximal exact matches (MEMs)* between a query string (e.g., sequence read) and an index (e.g., genomes), where a MEM is defined as an exact match that cannot be extended to the left or to the right.

In this paper, we consider the *extended Burrows Wheeler Transform* (eBWT), which extends the definition of the BWT to a collection of strings. Previously, we showed that the eBWT can be constructed by combining a modified version of the Suffix Array Induced Sorting (SAIS) algorithm of Nong et al. [24] with PFP. Here, we show that it follows from this construction that the run-length encoding of the eBWT and the SA samples of Gagie et al. [8] can be constructed in linear time in the size of the input, and linear space in the size of the dictionary and parse. Similarly, the thresholds of Rossi et al. [26] can be constructed for the eBWT and thus, be used to find MEMs.

2   Preliminaries

Basic Definitions. A string $T = T[1..n]$ is a sequence of characters $T[1] \cdots T[n]$ drawn from an ordered alphabet Σ of size σ. We denote by $|T|$ the length n of T. Given a multiset of m strings $\mathcal{M} = \{T_1, T_2, \ldots, T_m\}$, we denote the total length of the strings in $\mathcal{M}$ as $\|\mathcal{M}\|$, i.e., $\|\mathcal{M}\| = |T_1| + \ldots + |T_m|$.

Given two integers $1 \le i, j \le n$ where $i \le j$, the *substring* $T[i] \cdots T[j]$ is denoted by $T[i..j]$, the j-th *prefix* $T[1..j]$ is denoted by $T[..j]$, and the i-th *suffix* $T[i..n]$ by $T[i..]$. A substring S of T is called *proper* if $S \ne T$.

Given two strings S and T, we denote by $\mathtt{lcp}(S,T)$ the length of the *longest common prefix* (LCP) of S and T, i.e., $\mathtt{lcp}(S,T) = \max\{i \mid S[1..i] = T[1..i]\}$.

Given a string $T = T[1..n]$ and an integer k, we denote by T^k the kn-length string $TT\cdots T$ (k-fold concatenation of T), and by T^ω the infinite string $TT\cdots$ obtained by concatenating an infinite number of copies of T. A string T is called *primitive* if $T = S^k$ implies $T = S$ and $k = 1$. For any string T, there exists a unique primitive word S and a unique integer k such that $T = S^k$. We refer to $S = T[1..\frac{n}{k}]$ as root(T) and to k as exp(T). Thus, $T = \text{root}(T)^{\exp(T)}$.

Suffix Array. We denote by $<_{\text{lex}}$ the lexicographic order: for two strings $S[1..m]$ and $T[1..n]$, $S <_{\text{lex}} T$ if S is a proper prefix of T, or there exists an index $1 \leq i \leq n,m$ such that $S[1..i-1] = T[1..i-1]$ and $S[i] < T[i]$. Given a string $T[1..n]$, the *suffix array* [20], denoted by SA = SA$_T$, is the permutation of $\{1,\ldots,n\}$ such that $T[\text{SA}[i]..]$ is the i-th lexicographically smallest suffix of T.

Given a string $T[1..n]$ and the SA of T, we denote the *inverse suffix array* as ISA, and define it as ISA[SA[i]] = i for all $i = 1,\ldots,n$.

Definition of ϕ. Kärkkäinen et al. [12] introduced the permutation ϕ. It is defined as follows: $\phi(i) = \text{SA}[\text{ISA}[i] - 1]$ if ISA[i] > 1; and $\phi(i) = \text{SA}[n]$ otherwise. We can rewrite this as: $\phi(\text{SA}[j]) = \text{SA}[j-1]$, for all $j > 1$.

ω-order. We denote by $\prec_\omega$ the ω-order [10,21], defined as follows: for two strings S and T, $S \prec_\omega T$ if root(S) = root(T) and exp(S) < exp(T), or $S^\omega <_{\text{lex}} T^\omega$ (this implies root(S) $\neq$ root(T)). One can verify that the ω-order relation is different from the lexicographic one. For instance, $CG <_{\text{lex}} CGA$ but $CGA \prec_\omega CG$.

Conjugate Array. The string S is a *conjugate* of the string T if $S = T[i..n]T[1..i-1]$, for some $i \in \{1,\ldots,n\}$ (also called the *i-th rotation* of T). The conjugate S is also denoted conj$_i(T)$. For a string T, the *conjugate array*[1] CA = CA$_T$ of T is the permutation of $\{1,\ldots,n\}$ such that CA[i] = j if conj$_j(T)$ is the i-th conjugate of T with respect to the lexicographic order, with ties broken according to string order, i.e. if CA[i] = j and CA[i'] = j' for some $i < i'$, then either conj$_j(T) <_{\text{lex}}$ conj$_{j'}(T)$, or conj$_j(T) =$ conj$_{j'}(T)$ and $j < j'$.

Given a string T, U is a *circular* or *cyclic substring* of T if it is a substring of TT of length at most $|T|$, or equivalently, if it is the prefix of some conjugate of T. For instance, ATA is a cyclic substring of $AGCAT$. It is sometimes also convenient to regard a given string $T[1..n]$ itself as *circular* (or *cyclic*); in this case we set $T[0] = T[n]$ and $T[n+1] = T[1]$.

Burrows Wheeler Transform. Given a string T, BWT(T) [6] is a permutation of the letters of T which equals the last column of the matrix of the lexicographically sorted conjugates of T. The construction is reversible, allowing the original string T to be computed in linear time [6]. The BWT itself can be computed from the conjugate array, since for all $i = 1,\ldots,n$, BWT(T)[i] = $T[\text{CA}[i] - 1]$, where T is considered to be cyclic.

It should be noted that in many applications, it is assumed that an end-of-string-character (usually denoted \$), which is not element of Σ, is appended to the string; this character is assumed to be smaller than all characters from

---

[1] Our conjugate array CA is called *circular suffix array* and denoted SA$_\circ$ in [2,11], and *BW-array* in [13], but in both cases defined for primitive strings only.

Σ. Computing the conjugate array becomes equivalent to computing the suffix array, since $CA_{T\$}[i] = SA_{T\$}[i]$. Thus, applying one of the linear-time suffix array computation algorithms [22] leads to linear-time computation of the BWT.

LF-*mapping*. Last-to-first-mapping (LF-mapping) is the mapping of the lexico-graphical rank of a conjugate of T, $conj_i(T)$ to the lexicographical rank of the conjugate $conj_{i-1}(T)$. In particular, the LF-mapping maps characters from the last column of the matrix of the lexicographically sorted conjugates of T (commonly referred to as L) to the corresponding occurrence of the character in the first column of the same matrix (commonly referred to as F); hence the name last-to-first. It is the fundamental operation behind backward search, and what allows the BWT to be reversed.

r-index. Given a text $T[1..n]$ whose BWT has r runs, and a pattern $P[1..p]$, the r-index [9] is a data structure that supports *count* queries, i.e. comput-ing the number *occ* of occurrences of P in T, in $\mathcal{O}(p \log \log_w(\sigma + n/r))$ time and $\mathcal{O}(r)$ words of space, where w is the machine word size. It also sup-ports *locate* queries, i.e., return the *occ* positions in T where P occurs, in $\mathcal{O}(p \log \log_w(\sigma + n/r) + occ \log \log_w(n/r))$ time and $\mathcal{O}(r)$ space [9, Theorem 3.6]. Recently, Nishimoto and Tabei [23] improved the previous running times for counting to $\mathcal{O}(p \log \log_w \sigma)$ and locating to $\mathcal{O}(p \log \log_w \sigma + occ)$.

The r-index is made of three main components, which are: (1) a data struc-ture that stores the run-length encoded BWT supporting LF-mapping queries, (2) a SA sample for each of the r runs, and (3) a data structure supporting ϕ operations. In particular, in [9], (1) builds on the RLFM-index of Mäkinen et al. [19] combined with the data structures of Belazzougui and Navarro [3], while (2) is an array storing the SA samples at the end of each run. Lastly, (3) is imple-mented as a predecessor data structure built on SA samples at the beginning of each run, with the corresponding SA sample at the end of the previous run as a satellite information.

3   r-Indexing the eBWT

In this section, we show how to construct and use the r-index for the eBWT. The eBWT [21] (extended BWT) is a generalization of the BWT to a multiset $\mathcal{M} = \{T_1, \ldots, T_m\}$ of strings that is independent of the order in which the strings appear in the multiset. Similarly to the BWT, eBWT($\mathcal{M}$) is a permutation of the characters of the strings in $\mathcal{M}$ which is, however, based on the ω-order between the conjugates of the strings in $\mathcal{M}$, rather than on lexicographic order. It uses the *generalized conjugate array* GCA$_\mathcal{M}$, which is an array whose k-th entry GCA$_\mathcal{M}[k]$ equals (j, d) if $conj_d(T_j)$ is the k-th conjugate in ω-order. In [4], we showed how to compute the eBWT and the generalized conjugate array GCA$_\mathcal{M}$ in time linear in the total size of $\mathcal{M}$.

Remark 1. In [4], we showed that given a multiset of strings $\mathcal{M} = \{T_1, \ldots, T_m\}$, we can compute the $\text{GCA}_{\mathcal{M}}$ from the GCA of the roots of the strings in $\mathcal{M}$. Therefore, we assume that the multiset of strings $\mathcal{M} = \{T_1, \ldots, T_m\}$ consists of m primitive strings. Moreover, we assume that ties between same roots in $\mathcal{M}$ in the eBWT are broken by string index.

Given a pattern $P[1..p]$, we say that P occurs in $\mathcal{M} = \{T_1, T_2, \ldots, T_m\}$ if P occurs as a substring of any of $T_1^\omega, T_2^\omega, \ldots, T_m^\omega$. Formally, we define an *occurrence* of P in $\mathcal{M}$ as a pair (j, d) such that for all $i = 1, \ldots, p$, $T_d[j + i - 1] = P[i]$, where T_d is considered as circular.

It follows from the definition of the r-index that we need three main components to build an r-index on the eBWT, namely (1) a data structure that stores the run-length encoded eBWT supporting LF-mapping queries, (2) the samples of the GCA at the end of each run, and (3) a data structure supporting ϕ operations, extended to the GCA. Here, we note that we make the assumption throughout this paper that no two strings have the same set of conjugates. For the eBWT, we need to store an auxiliary data structure marking the first conjugate of all the strings in $\mathcal{M}$.

We now show that all the components can be stored in $\mathcal{O}(r)$-space, building upon the results of Gagie et al. [9].

For (1), the data structure storing the eBWT supporting LF-mapping, we can use the same data structure used by the r-index. We summarize this in the next corollary, which follows directly from Gagie et al. [9].

Corollary 1. *Given a multiset of strings $\mathcal{M} = \{T_1, \ldots, T_m\}$, of total length N and pattern $P[1..p]$, we can build an index of $\mathcal{O}(r)$ words such that we can count all occurrences of P in $\mathcal{M}$ in $\mathcal{O}(p \log \log_w(\sigma + N/r))$-time.*

In the following proposition, we describe an analogue of a known property of the BWT for the eBWT. It states that the indices corresponding to characters within the same run are mapped contiguously by the LF-mapping. It follows from the fact that the LF-property holds for the eBWT [21].

Proposition 1. *Let $\mathcal{M} = \{T_1, \ldots, T_m\}$ be a multiset of strings of total length N and $\text{GCA}_{\mathcal{M}} = [(j_1, i_1), (j_2, i_2), \ldots, (j_N, i_N)]$ its conjugate array, i.e. $\text{GCA}_{\mathcal{M}}[\nu] = (j_\nu, i_\nu)$ for $\nu = 1, \ldots, N$. If, for some integers k and h, $\text{eBWT}[k] = \text{eBWT}[k-1]$ and $\text{GCA}[h] = (j_k - 1, i_k)$, then $\text{GCA}_{\mathcal{M}}[h-1] = (j_{k-1} - 1, i_{k-1})$, where the strings are considered cyclic, i.e. $T[0] = T[|T|]$ and $T[|T| + 1] = T[1]$.*

Next, we show that for a multiset of strings $\mathcal{M}$ where no two strings have the same set of conjugates, at least one of the characters of each string in the set appears at the beginning of a run, and at least one of the characters of each string appears at the end of a run.

Proposition 2. *Given a multiset of strings $\mathcal{M} = \{T_1, \ldots, T_m\}$ of total length N and where no two strings have the same set of conjugates, with $\text{GCA}_{\mathcal{M}} = [(j_1, i_1), (j_2, i_2), \ldots, (j_N, i_N)]$. Then, for all $1 \leq k \leq m$, there exist two integers*

h and h' such that $i_h = i_{h'} = k$, and the following are true: (1) either $h = 1$ or $\text{eBWT}_\mathcal{M}[h-1] \neq \text{eBWT}_\mathcal{M}[h]$, and (2) either $h' = N$ or $\text{eBWT}_\mathcal{M}[h'+1] \neq \text{eBWT}_\mathcal{M}[h']$.

The following corollary follows directly from the previous proposition.

Corollary 2. *Given a multiset of strings $\mathcal{M} = \{T_1, \dots, T_m\}$ of total length N and where no two strings have the same set of conjugates, we have $m \leq r$.*

From the previous corollary, it follows that we can store the $\mathcal{O}(m)$ positions of the starting rotations of each string in $\mathcal{O}(r)$-space. Moreover, the following result immediately follows from the previous corollary. It says that, for example, the compression ratio for a set of reads of length 150 cannot be better than 150, when using either the BWT or the eBWT.

Corollary 3. *Given a multiset of ℓ-length strings $\mathcal{M} = \{T_1, \dots, T_m\}$ such that no two strings have the same set of conjugates, we have that $N/r \leq \ell$.*

We can extend the definition of ϕ to the $\text{GCA}_\mathcal{M}$, such that for each $h > 1$ ϕ associates to the h-th pair (j_h, i_h) in $\text{GCA}_\mathcal{M}$ the pair (j_{h-1}, i_{h-1}). However, we have one additional constraint when we extend the description of ϕ to the eBWT—namely, we need to ensure that each string in the set $\mathcal{M}$ has at least one conjugate appearing at the beginning and at least one conjugate appearing at the end of a run. This follows from Proposition 2. Therefore, by storing the samples at the beginning and the end of each run, we are guaranteed to cover all strings. Hence, the number of samples is $\mathcal{O}(r)$. We can then build a predecessor data structure for each string T_i in $\mathcal{M}$ with the GCA samples at the beginning of each run corresponding to samples of T_i, and associating the GCA sample at the end of the preceding run. Note that we need the predecessor search to be circular, i.e., the predecessor of the first element is the last element. Therefore, using Proposition 1, we can prove the following result.

Proposition 3. *Given a multiset of strings $\mathcal{M} = \{T_1, \dots, T_m\}$, of total length N and where no two strings have the same set of conjugates, we evaluate ϕ in $\mathcal{O}(r)$ words of space and $\mathcal{O}(\log\log_w(N/r))$-time, where w is the machine word size.*

We can summarize our results in the following theorem.

Theorem 1. *Given a multiset of strings $\mathcal{M} = \{T_1, \dots, T_m\}$ of total length N and where no two strings have the same set of conjugates, we can build an index of $\mathcal{O}(r)$ words such that given pattern $P[1..p]$, we can count the occ occurrences of P in $\mathcal{M}$ in $\mathcal{O}(p\log\log_w(\sigma + N/r))$-time and we can locate the occ positions in $\mathcal{M}$, in additional $\mathcal{O}(occ\log\log_w(N/r))$-time.*

4   Constructing the r-Index of the eBWT

In [4], we demonstrated how to construct the eBWT efficiently and in a manner that preserves the original definition of Mantaci et al. [21]. In particular, we combined a modified version of the Suffix Array Induced Sorting (SAIS) algorithm of Nong et al. [24] with PFP to develop a novel construction algorithm. Hence, we produce a dictionary and parse of the eBWT as a result of this algorithm. It follows from Kuhnle et al. [14] that we can produce the run-length encoding of the eBWT and the GCA samples from the dictionary and parse in linear time and space in their size.

Corollary 4. *Given a multiset $\mathcal{M}$ of input strings, we can build the run-length encoding of the* eBWT *and the* GCA *samples in* $\mathcal{O}(||D|| + ||\mathcal{P}||)$*-space, where D and $\mathcal{P}$ are the dictionary and parse defined by the* PFP*.*

As previously noted, the main components of the r-index—namely the RL BWT and the SA samples—are not enough to support efficient MEM-finding. In order to accomplish this, Bannai et al. [1] showed that MEM-finding can be supported by computing *matching statistics* for a query string P, which is defined for each position in P as the length of the longest substring starting at that position that occurs in the indexed string T, and the starting position in T of one of its occurrences. From the matching statistics for P, one can compute the occurrence of a MEM using a two-pass process: first, working right to left, for each suffix of P until it finds a suffix of the text that matches for as long as possible; then, working left to right, it uses random access to T to determine the length of those matches. Thus, to compute the matching statistics, Bannai et al. described the addition of a small data structure to the r-index that is referred to as *thresholds*. A threshold between a consecutive pair of runs can be defined as the position of the minimum LCP value in the interval between them. Rossi et al. [26] showed how to compute all thresholds efficiently by modifying the PFP construction of the r-index. Here, we see that such a modification can also be made for the eBWT construction, allowing the thresholds for the eBWT to be constructed along with the GCA samples.

Corollary 5. *Given a set of input strings $\mathcal{M}$, we can construct the thresholds in addition to the run-length encoding of the* eBWT *and the* GCA *samples in* $\mathcal{O}(||D|| + ||\mathcal{P}||)$*-space, where D and $\mathcal{P}$ are the dictionary and parse defined by the* PFP*.*

It follows directly from Bannai et al. [2] and Rossi et al. [26] that given a query string $P[1..p]$ that has *occ* occurrences of a MEM in $\mathcal{M}$, we can find a single MEM in $\mathcal{O}(p(\log\log_w(\sigma + N/r) + t_{RA}))$-time and $\mathcal{O}(r + s_{RA})$ words of space, where t_{RA} and s_{RA} are the time and space of any data structure that is able to provide random access to the string. Moreover, we can extend this search to find all *occ* occurrences in additional $\mathcal{O}(occ \log\log_w(N/r))$-time.

Figure 1 depicts an example of matching statistics query of the pattern $P =$ CABAA against the collection of strings $\mathcal{M} = \{$AAC, AACAC, AABACAAC$\}$.

$$\mathcal{M} = \{AAC, AACAC, AABACAAC\}$$

P	Prefix	POS	$GCA_\mathcal{M}$	A	B	C	$eBWT_\mathcal{M}$	
			(1,3)	*	*	*	C	AABACAAC
			(6,3)				C	AACAABAC
$P = CABAA$			(1,1)				C	AAC
			(1,2)				C	AACAC
C	CA	(3,2)	(2,3)			*	A	ABACAACA
A	ABA	(2,3)	(7,3)				A	ACAABACA
B	BA	(3,3)	(4,3)	*			B	ACAACAAB
A	AA	(1,2)	(2,1)				A	ACA
A	A	(2,2)	(4,2)				C	ACAAC
			(2,2)		*		A	ACACA
			(3,3)				A	BACAACAA
			(8,3)				A	CAABACAA
			(5,3)				A	CAACAABA
			(3,1)				A	CAA
			(5,2)				A	CAACA
			(3,2)				A	CACAA

Fig. 1. An illustration of the thresholds for calculating the matching statistics of a query string $P[1..p]$ in a set of strings $\mathcal{M}$. Shown on the left is P, the longest *Prefix* of the suffix of that occurs in $\mathcal{M}$, and the position of the corresponding prefix in the text. Shown on the right, continuing from left to right, is the $GCA_\mathcal{M}$ of $\mathcal{M}$, the thresholds $THR_\mathcal{M}$ for the characters A, B, and C, the $eBWT_\mathcal{M}$, and all conjugates of all strings in $\mathcal{M}$, with the samples of the $GCA_\mathcal{M}$ highlighted in red. The arrows illustrate the position in the $GCA_\mathcal{M}$ which corresponds to the prefix on the left.

5   Conclusions

In this paper, we describe how the fundamentals of the r-index can be transferred to the context of the eBWT. We note that the eBWT has an advantage over other BWT-based data structures for string collections, which is that it is independent of the order of the input strings. An r-index based on the eBWT inherits this important property. Yet, we note that the applicability of this data structure has not been fully explored. Thus, we think that implementing this r-index of the eBWT and evaluating the efficiency of its construction on large datasets is warranted. From a more theoretical point of view, recasting some of the more recent results—including the results of Nishimoto and Tabei [23], Bannai et al. [1], and Cobas et al. [7]—of the r-index to the context of eBWT merits attention.

Acknowledgements. We thank Travis Gagie for comments on preliminary versions of this manuscript. CB and MR are funded by National Science Foundation NSF IIBR (Grant No. 2029552), NSF SCH (Grant No. 2013998), National Institutes of Health (NIH) NIAID (Grant No. HG011392) and NIH NIAID (Grant No. R01AI141810).

References

1. Bannai, H., Gagie, T., Tomohiro, I.: Refining the r-index. Theor. Comput. Sci. **812**, 96–108 (2020)
2. Bannai, H., Kärkkäinen, J., Köppl, D., Piatkowski, M.: Constructing the bijective and the extended burrows-wheeler-transform in linear time. In: Proceedings of the 32nd Annual Symposium on Combinatorial Pattern Matching, CPM 2021. LIPIcs, vol. 191, pp. 7:1–7:16 (2021)
3. Belazzougui, D., Navarro, G.: Optimal lower and upper bounds for representing sequences. ACM Trans. Algorithms **11**(4), 31:1-31:21 (2015)
4. Boucher, C., Cenzato, D., Lipták, Zs., Rossi, M., Sciortino, M.: Computing the original eBWT faster, simpler, and with less memory. In: Lecroq, T., Touzet, H. (eds.) SPIRE 2021. LNCS, vol. 12944, pp. 129–142. Springer, Cham (2021)
5. Boucher, C., Gagie, T., Kuhnle, A., Langmead, B., Manzini, G., Mun, T.: Prefix-free parsing for building big BWTs. Algorithms Mol. Biol. **14**(1), 13:1-13:15 (2019)
6. Burrows, M., Wheeler, D.J.: A block sorting lossless data compression algorithm. Technical report 124, Digital Equipment Corporation (1994)
7. Cobas, D., Gagie, T., Navarro, G.: A fast and small subsampled r-index. In: Proceedings of the 32nd Annual Symposium on Combinatorial Pattern Matching, CPM 2021. LIPIcs, vol. 191, pp. 13:1–13:16 (2021)
8. Gagie, T., Navarro, G., Prezza, N.: Optimal-time text indexing in BWT-runs bounded space. In: Proceedings of the Twenty-Ninth Annual ACM-SIAM Symposium on Discrete Algorithms, SODA 2018, pp. 1459–1477 (2018)
9. Gagie, T., Navarro, G., Prezza, N.: Fully functional suffix trees and optimal text searching in BWT-runs bounded space. J. ACM **67**(1), 2:1-2:54 (2020)
10. Gessel, I.M., Reutenauer, C.: Counting permutations with given cycle structure and descent set. J. Combin. Theory Ser. A **64**(2), 189–215 (1993)
11. Hon, W.-K., Ku, T.-H., Lu, C.-H., Shah, R., Thankachan, S.V.: Efficient algorithm for circular Burrows-Wheeler Transform. In: Kärkkäinen, J., Stoye, J. (eds.) CPM 2012. LNCS, vol. 7354, pp. 257–268. Springer, Heidelberg (2012). https://doi.org/10.1007/978-3-642-31265-6_21
12. Kärkkäinen, J., Manzini, G., Puglisi, S.J.: Permuted longest-common-prefix array. In: Kucherov, G., Ukkonen, E. (eds.) CPM 2009. LNCS, vol. 5577, pp. 181–192. Springer, Heidelberg (2009). https://doi.org/10.1007/978-3-642-02441-2_17
13. Kucherov, G., Tóthmérész, L., Vialette, S.: On the combinatorics of suffix arrays. Inf. Process. Lett. **113**(22–24), 915–920 (2013)
14. Kuhnle, A., Mun, T., Boucher, C., Gagie, T., Langmead, B., Manzini, G.: Efficient construction of a complete index for pan-genomics read alignment. J. Comput. Biol. **27**(4), 500–513 (2020)
15. Langmead, B., Trapnell, C., Pop, M., Salzberg, S.: Ultrafast and memory-efficient alignment of short DNA sequences to the human genome. Genome Biol. **10**, R25 (2009)
16. Li, H., Durbin, R.: Fast and accurate short read alignment with Burrows-Wheeler Transform. Bioinformatics **25**(14), 1754–1760 (2009)
17. Mäkinen, V., Navarro, G.: Succinct suffix arrays based on run-length encoding. Nord J. Comput. **12**, 40–66 (2005)
18. Mäkinen, V., Välimäki, N., Laaksonen, A., Katainen, A.: Algorithms and Applications. Springer, Heidelberg (2010)
19. Mäkinen, V., Navarro, G., Sirén, J., Välimäki, N.: Storage and retrieval of highly repetitive sequence collections. J. Comput. Biol. **17**(3), 281–308 (2010)

20. Manber, U., Myers, G.W.: Suffix arrays: a new method for on-line string searches. SIAM J. Comput. **22**(5), 935–948 (1993)
21. Mantaci, S., Restivo, A., Rosone, G., Sciortino, M.: An extension of the Burrows-Wheeler Transform. Theor. Comput. Sci. **387**(3), 298–312 (2007)
22. Navarro, G.: Compact Data Structures: A Practical Approach. Cambridge University Press, Cambridge (2016)
23. Nishimoto, T., Tabei, Y.: Optimal-time queries on BWT-runs compressed indexes. In: Proceedings of the 48th International Colloquium on Automata, Languages, and Programming, ICALP 2021. LIPIcs, vol. 198, pp. 101:1–101:15 (2021)
24. Nong, G., Zhang, S., Chan, W.H.: Two efficient algorithms for linear time suffix array construction. IEEE Trans. Comput. **60**(10), 1471–1484 (2011)
25. Policriti, A., Prezza, N.: LZ77 computation based on the run-length encoded BWT. Algorithmica **80**, 1986–2011 (2017)
26. Rossi, M., Oliva, M., Langmead, B., Gagie, T., Boucher, C.: MONI: a pangenomics index for finding MEMs. In: Proceedings of the 25th Annual International Conference on Research in Computational Molecular Biology, RECOMB 2021 (2021)
27. Sun, C., et al.: RPAN: rice pan-genome browser for 3000 rice genomes. Nucleic Acids Res. **45**(2), 597–605 (2017)
28. The 1001 Genomes Consortium. Epigenomic diversity in a global collection of arabidopsis thaliana accessions. Cell **166**(2), 492–505 (2016)
29. Turnbull, C., et al.: The 100,000 genomes project: bringing whole genome sequencing to the NHS. Br. Med. J. **361** (2018)

Unicode at Gigabytes per Second

Daniel Lemire[✉]

DOT-Lab Research Center, University of Quebec (TELUQ), Montreal, Canada
`daniel.lemire@teluq.ca`

Abstract. We often represent text using Unicode formats (UTF-8 and UTF-16). The UTF-8 format is increasingly popular, especially on the web (XML, HTML, JSON, Rust, Go, Swift, Ruby). The UTF-16 format is most common in Java, .NET, and inside operating systems such as Windows.

Software systems frequently have to convert text from one Unicode format to the other. While recent disks have bandwidths of 5 GiB/s or more, conventional approaches transcode non-ASCII text at a fraction of a gigabyte per second.

We show that we can validate and transcode Unicode text at gigabytes per second on current systems (x64 and ARM) without sacrificing safety. Our open-source library can be ten times faster than the popular ICU library on non-ASCII strings and even faster on ASCII strings.

Keywords: Unicode · Vectorization · Internationalization

1 Introduction

From the early days of computing, programmers have had to represent characters in software. They needed to agree to standards so that different software written by different vendors would be interoperable. One of the earliest such standards is ASCII—first specified in the early 1960s. The ASCII standard is still popular today: it uses one byte per character—with the most significant bit set to zero. Unfortunately, ASCII could only ever represent up to 128 characters—far less than needed.

Thus many diverging standards emerged for representing characters in software. The existence of multiple incompatible formats made the production of interoperable localized software difficult. Conversion between some formats could sometimes be lossy or ambiguous.

Unicode arose in the late 1980s as an attempt to provide a single agreed-upon standard. Initially, it was believed that using 16 bits per character would be sufficient. However, engineers realized over time that a wider range of characters should be supported—if the standard was to be universal. Thus the Unicode

This manuscript is based on a forthcoming long-form article written with Wojciech Muła, *Transcoding Billions of Unicode Characters per Second with SIMD Instructions*.

T. Lecroq and H. Touzet (Eds.): SPIRE 2021, LNCS 12944, pp. 13–18, 2021.
https://doi.org/10.1007/978-3-030-86692-1_2

standard was extended to potentially include up to 1114112 characters. Characters are sometimes called code points and represented as integer values between 0 and 1114112. In practice, only a small fraction of all possible code points have been assigned, but more are assigned over time with each Unicode revision. The Unicode standard is an extension of the ASCII standard in the sense that the first 128 Unicode code-point values match the ASCII characters.

There are several ways to represent Unicode characters in bytes. Due to the original expectation that Unicode would fit in 16-bit space, a format based on 16-bit words (UTF-16) format was published in 1996 and formalized in 2000 [3]. It may use either 16-bit or 32-bit per character. The UTF-16 format was adopted by programming languages such as Java, and became a default under Windows.

Unfortunately, UTF-16 is not backward compatible with ASCII at a byte level. Thus an ASCII compatible format was proposed and formalized in 2003: UTF-8 [8]. Over time, it became widely used. Text interchange formats such as JSON, HTML or XML are expected to be in the UTF-8 format. Programming languages such as Go, Rust and Swift use UTF-8 by default. When used as part of data interchange documents, UTF-8 is commonly more concise due to its ability to use one byte per character to represent ASCII text.

Though UTF-8 dominates in many applications, it does not make UTF-16 obsolete. The UTF-16 format has advantages. Indeed, most text represented in the UTF-16 format has exactly 2-byte per character, except for the occasional special character (e.g., an emoji). Having a flat 2-byte per character makes some operations faster. Both formats require validation: not all arrays of bytes are valid. However, the UTF-8 format is more expensive to validate. In some cases, UTF-16 may be even more concise as well (e.g., when representing Chinese text).

Thus, for the foreseeable future, we need to validate both UTF-16 and UTF-8 strings, and to convert (transcode) text between the two formats.

We should expect these operations to be fast, and they are. However, speed and efficiency are relative. Cloud vendors offer high bandwidth between node instances (e.g., 3.3 GiB/s) and the bandwidth of disks is rising fast (e.g., 5 GiB/s with PCIe 4.0) [1]. We believe that we should be able to match such high speeds inside the processor when transcoding text.

For fast processing, we should seek to make the best possible use out of our processors. Commodity processors support single-instruction-multiple-data (SIMD) instructions. These instructions operate on several words at once unlike regular instructions. Starting with the Haswell microarchitecture (2013), Intel and AMD processors support the AVX2 instruction set and 256-bit vector registers. Most mobile phones tablets have 64-bit ARM processors (aarch64) with NEON instructions (128-bit registers). Hence, on recent x64 processors, we can compare two strings of 32 characters in a single instruction. Algorithms designed for SIMD instructions typically require fewer instructions per byte. Software that retires fewer instructions tends to use less power and to be faster.

2 Related Work

There are many ways to validate and transcode Unicode text. One may use series of branches or finite-state approaches [4]. We are most interested in the fastest techniques. Keiser and Lemire [6] describe a fast SIMD-based UTF-8 validation directly on byte streams. We make use of their approach (see Sect. 4).

Cameron [2] proposed that we transform text using *bit streams*. Given byte arrays, the bit-stream approach creates eight bit arrays, each one corresponding to a bit position within a byte. There is one bit stream for the least significant bit, one for the second least significant bit, and so forth. We must first transform the input data into such bit streams and then convert the data back from bit streams to an array of bytes. Cameron applied this strategy to UTF-8 to UTF-16 transcoding.

Independently, Inoue et al. [5] proposed a UTF-8 to UTF-16 SIMD-accelerated transcoder, but it does not validate its inputs. The authors did not make their implementation (for PowerPC processors) available.

We are not aware of any further work in the scientific literature regarding the application of SIMD instruction to the validation or transcoding of Unicode text. Except for fast ASCII paths, we do not know of any widespread use of SIMD instructions for validating or transcoding Unicode text.

3 The Unicode Formats

ASCII characters require one byte with UTF-8 and two bytes with UTF-16. UTF-16 can represent all characters—except for the supplemental characters such as emojis—using two bytes. The UTF-8 format uses two bytes for Latin, Hebrew and Arabic alphabets. Asiatic characters (including Chinese and Japanese) require three UTF-8 bytes. Both UTF-8 and UTF-16 require 4 bytes for the supplemental characters. We often represent Unicode characters using its integer value in hexadecimal as, for example, U+7F (for 127).

UTF-8 encodes values in sequences of one to four bytes. We refer to the first byte of a sequence as a leading byte; the most significant bits of the leading byte indicate the length of the sequence:

- If the most significant bit is zero, we have a sequence of one byte (ASCII).
- If the three most significant bits are 110, we have a two-byte sequence.
- If the four most significant bits are 1110, we have a three-byte sequence.
- Finally, if the five most significant bits are 11110, we have a four-byte sequence.

All bytes following the leading byte in a sequence are continuation bytes, and they must have their two most significant bits as 10. Except for the required most significant bit sequences, other bits (from 7 bits to 21 bits) provide the code-point value. The most significant bits of the code-point value are in the leading byte, followed by lesser significant bits in the second byte and so forth, with the least significant bits in the last byte of the sequence. Valid UTF-8 sequences must follow the following exhaustive rules:

1. The five most significant bits of any byte cannot be all ones.
2. The leading byte must be followed by the right number of continuation bytes.
3. A continuation byte must be preceded by a leading byte.
4. The decoded character must be larger than U+7F for two-byte sequences, larger than U+7FF for three-byte sequences, and larger than U+FFFF for four-byte sequences.
5. The decoded code-point value must be less than 1114112.
6. The code-point value must not be in the range U+D800...DFFF.

In the UTF-16 format, characters in U+0000...D7FF and U+E000...FFFF are stored as 16-bit values—using two bytes. The characters in the range U+010000...10FFFF require two 16-bit words (a surrogate pair). The first word in the pair is in 0xD800...DBFF whereas the second word is in 0xDC00...DFFF. The code-point value is made of the 10 least significant bits of the two words—using the second word as least significant—adding 0x10000 to the result. During validation, only the possible surrogate pairs require attention.

4   Algorithms

All commodity software with SIMD instructions (e.g., x64, ARM, POWER) have fast instructions to permute bytes within a SIMD register according to a sequence of indexes. Our transcoding techniques depend critically on this feature: we code in a table the necessary parameters—including the indexes (sometimes called shuffle masks)—necessary to process a variety of incoming characters.

Our accelerated UTF-8 to UTF-16 transcoding algorithm processes up to 12 input UTF-8 bytes at a time. From the input bytes, we can quickly determine the leading bytes and thus the end beginning of each character. We use a 12-bit word as a key in a 1024-entry table. Each entry in the table contains the number of UTF-8 bytes that will be consumed and an index into another table where we find shuffle masks. The value of the index into the other table also determines one of three possible code paths. The first 64 index values indicate that we have 6 characters spanning between one and two bytes. Index values in [64, 145) indicate that we have 4 characters spanning between one and three bytes. The remaining indexes represent the general case: 3 characters spanning between one and four bytes. The shuffle mask can then be applied to the 12 input bytes to form a vector register that can be transformed efficiently. We use this 12-byte routine inside 64-byte blocks. After loading a 64-byte block, we apply the Keiser-Lemire validation routine [6]. Afterward, we identify the leading bytes, and then process the block in multiple iterations, using up to 12 bytes each time. In the special case where all 64 bytes are ASCII, we use a fast path. For even greater efficiency, we have three other fast paths within the 12-byte routine: we check whether the next 16 bytes are ASCII bytes, whether they are all two-byte characters, or all three-byte characters.

Our UTF-16 to UTF-8 algorithm iteratively reads a block of input bytes in a SIMD register. If all 16-bit words in the loaded SIMD register are in the range U+0000...007f, we use a fast routine to convert the 16 input bytes into eight

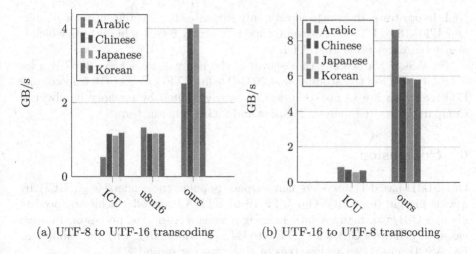

(a) UTF-8 to UTF-16 transcoding   (b) UTF-16 to UTF-8 transcoding

Fig. 1. Transcoding speeds for various test files.

equivalent ASCII bytes. If all 16-bit words are in the range U+0000...07ff, then we use a fast routine to produce sequences of one-byte or two-byte UTF-8 characters. Given an 8-bit bitset which indicates which 16-bit words are ASCII, we load a byte value from a table indicating how many bytes will be written, and a 16-byte shuffle mask. If all 16-bit words are in the ranges U+0000...d777, U+e000...ffff, we use another similar specialized routine to produce sequences of one-byte, two-byte and three-byte UTF-8 characters. Otherwise, when we detect that the input register contains at least one part of a surrogate pair, we fall back to a conventional code path.

5   Experiments

We make available our software as a portable open-source C++ library.[1] As a benchmarking system, we use a recent AMD processor (AMD EPYC 7262, Zen 2 microarchitecture, 3.39 GHz) and GCC 10. We compare against a popular library: International Components for Unicode (UCI) [7] (version 67.1). We also use the u8u16 library [2]. Unlike UCI and our own work, the u8u16 library only provides UTF-8 to UTF-16 transcoding. For our experiments, we use lipsum text in various languages.[2] All of our transcoding tests include validation. To measure the speed, we record the time by repeating the task 2000 times. We compare the average time with the minimal time and find that we have an accuracy of at least 1%. We divide the input volume by the time required for the transcoding. Figure 1 shows our results. Our UTF-8 to UTF-16 transcoding speed exceeds 4 GiB/s for Chinese and Japanese texts, which is about four times faster than

---

[1] https://github.com/simdutf/simdutf.
[2] https://github.com/rusticstuff/simdutf8.

UCI. In our tests, the u8u16 library only surpasses ICU significantly for Arabic. Our UTF-16 to UTF-8 transcoding speed is nearly 6 GiB/s in all tests which is nearly ten times faster than UCI.

For ASCII transcoding (not shown in the figures), we achieve 36 GiB/s for UTF-16 to UTF-8 transcoding, and 20 GiB/s for UTF-8 to UTF-16 transcoding. Effectively, we are so fast that we are nearly limited by memory bandwidth. Comparatively, UCI delivers 2 GiB/s and 1 GiB/s in our tests.

6 Conclusion

Our SIMD-based transcoders can surpass popular transcoders (e.g., UCI) by a wide margin (e.g., 4×). Our UTF-16 to UTF-8 transcoder achieves speed of about 6 GiB/s for many Asiatic languages using a recent x64 processor. In some cases, we achieve 4 GiB/s for UTF-8 to UTF-16 transcoding with full validation. For ASCII inputs, we achieve tens of gigabytes per second.

References

1. Barr, J.: The floodgates are open - increased network bandwidth for EC2 instances. https://aws.amazon.com/blogs/aws/the-floodgates-are-open-increased-network-bandwidth-for-ec2-instances/
2. Cameron, R.D.: A case study in SIMD text processing with parallel bit streams: UTF-8 to UTF-16 transcoding. In: Proceedings of the 13th ACM SIGPLAN Symposium on Principles and practice of parallel programming, pp. 91–98. ACM (2008)
3. Hoffman, P., Yergeau, F.: UTF-16, an encoding of ISO 10646 (2000). https://tools.ietf.org/html/rfc2781. Accessed July 2021. Internet Engineering Task Force, Request for Comments: 3629
4. Höhrmann, B.: Flexible and economical UTF-8 decoder (2010). http://bjoern.hoehrmann.de/utf-8/decoder/dfa/. Accessed July 2021
5. Inoue, H., Komatsu, H., Nakatani, T.: Accelerating UTF-8 decoding using SIMD instructions. Inf. Process. Soc. Jpn. Trans. Program. 1(2), 1–8 (2008). (in Japanese)
6. Keiser, J., Lemire, D.: Validating UTF-8 in less than one instruction per byte. Softw. Pract. Exp. 51(5), 950–964 (2021)
7. International Components for Unicode (UCI). http://site.icu-project.org
8. Yergeau, F.: UTF-8, a transformation format of ISO 10646 (2003). https://tools.ietf.org/html/rfc3629. Accessed July 2021. Internet Engineering Task Force, Request for Comments: 3629

Combinatorics

Longest Common Rollercoasters

Kosuke Fujita[1], Yuto Nakashima[1] [iD], Shunsuke Inenaga[1,2] [iD],
Hideo Bannai[3(✉)] [iD], and Masayuki Takeda[1] [iD]

[1] Department of Informatics, Kyushu University, Fukuoka, Japan
`fujita.kosuke.040@s.kyushu-u.ac.jp`,
`{yuto.nakashima,inenaga,takeda}@inf.kyushu-u.ac.jp`
[2] PRESTO, Japan Science and Technology Agency, Kawaguchi, Japan
[3] M&D Data Science Center, Tokyo Medical and Dental University, Tokyo, Japan
`hdbn.dsc@tmd.ac.jp`

Abstract. For an integer $k \geq 3$, a k-rollercoaster is a numeric string
such that any maximal strictly non-increasing or non-decreasing substring has length at least k. We consider the problem of computing the
longest common k-rollercoaster between two integer strings S and T,
i.e., the longest k-rollercoaster that is a subsequence common to both S
and T. We give two algorithms that solve this problem; The first runs
in $O(nmk)$ time and space, where n, m are respectively the lengths of S
and T. The second runs in $O(rk \log^3 m \log \log m)$ time and $O(rk)$ space,
where $r = O(mn)$ is the number of pairs (i, j) of matching points such
that $S[i] = T[j]$, assuming that $m \leq n$ and that S, T only consist of
characters which occur in both strings. The second algorithm is faster
than the first one when r is sub-linear in $nm/\log^3 m \log \log m$.

Keywords: k-rollercoaster · Dynamic programming · Dynamic range
query

1 Introduction

A subsequence of a string S is a string that can be obtained by removing 0
or more characters from S. Given a string of length n, the problem of finding a longest subsequence that satisfies certain properties has been studied for
various settings. For strings over an ordered alphabet, the longest increasing
subsequence problem is well known, and is known to have a $O(n \log n)$ time
solution [5]. The longest palindromic subsequence problem, where the objective
is to find the longest subsequence that is a palindrome, has a simple $O(n^2)$
dynamic programming solution. The longest square subsequence problem (originally referred to as the longest scattered subsequence problem [19]), where the
objective is to find the longest subsequence that is a square, is also known
to be solvable in $O(n^2)$ time [19], and later improved to $O(n^2 \log \log n / \log n)$
time [22]. Algorithms running in $O(\ell \min\{n, r\} \log \frac{n}{\ell} + r \log n + n)$ time [16], or
$O(\ell \min\{n, r\}(1 + \log(\min\{\ell, \frac{r}{\ell}\})) + n\ell + r)$ time [21], have also been proposed,
where r is the number of pairs (i, j) of matching points such that $S[i] = T[j]$, and

© Springer Nature Switzerland AG 2021
T. Lecroq and H. Touzet (Eds.): SPIRE 2021, LNCS 12944, pp. 21–32, 2021.
https://doi.org/10.1007/978-3-030-86692-1_3

ℓ is the length of the solution. Notice that while $r = O(n^2)$ in the worst case, r can be much smaller than $O(n^2)$ when the matching points are sparse. Thus, the aforementioned algorithms, and the ones to be mentioned later, whose complexities are dependent on r can be faster than their basic dynamic programming counterparts in the case where the matching points are sparse.

These problems can be naturally extended to the case where two strings S, T of respective lengths n, m ($n \geq m$) are given, and the objective is to compute a longest such sequence that is common to both S and T. The longest common palindromic subsequence problem was considered by Chowdhury et al. [11,12], where they gave an $O(n^4)$ time algorithm, as well as an $O(r^2 \log^2 n \log \log n)$ time algorithm. This was later improved to $O(\sigma r^2 + n)$ time by Inenaga and Hyyrö [15], where σ is the number of distinct characters occurring in both strings, and to $O(r^2 + n)$ time by Bae and Lee [5]. The longest common square subsequence problem was considered by Inoue et al. [17] where they gave several algorithms running in $O(n^6)$, $O(rn^4)$, or $O(\sigma r^3 + n)$ time. The longest common increasing subsequence problem was considered by Yang et al. [23], where they gave an $O(mn)$ time algorithm. This was improved by Chan et al. [9] to $O(\min\{r \log \ell, n\ell + r\} \log \log n + Sort)$ time, where $Sort$ is the time to sort the elements of the input strings. Also, Kutz et al. [20] showed an $O((n + m\ell) \log \log \sigma + Sort)$ time algorithm. More recently, Duraj [13] proposed an $O(n^2(\log \log n)^2 / \log^{1/6} n)$ time algorithm. Agrawal and Gawrychowski [2] further give an $O(n^2 \log \log n / \sqrt{\log n})$ time algorithm.

In this paper, we focus on characteristic subsequences called *rollercoasters* [8]. For an integer $k \geq 3$, a k-rollercoaster is a string over an ordered alphabet such that any maximal strictly non-increasing or non-decreasing substring has length at least k. Biedl et al. [6,7] gave an $O(n \log n)$ time solution, as well as an $O(n \log \log n)$ time solution when the string is a permutation of $\{1, \ldots, n\}$. Later, Gawrychowski et al. [14] showed an $O(nk^2)$ time algorithm.

As with the other properties mentioned above, we consider the following extension of this problem: Given two strings S, T of respective lengths n, m, and a non-negative integer k, compute a longest k-rollercoaster that is common to both S and T. We show two algorithms: the first runs in $O(nmk)$ time and space. The second runs in $O(rk \log^3 m \log \log m)$ time and $O(rk)$ space, assuming that $m \leq n$ and that S, T only consist of characters which occur in both strings. This can be translated to $O(rk \log^3 m \log \log m + Sort)$ time and $O(rk+n)$ space without the latter assumption.

1.1   Related Work

The Longest Property-Preserved Common Factor problem which asks to find a longest common *substring* that satisfy certain properties, was considered by Ayad et al. [3,4]. They consider three settings and properties, and show that: 1) A given string x can be preprocessed in linear time so that for any query string y, a longest common square-free factor between x and y can be computed in linear time. 2) Given a set of k strings and integer $1 < k' \leq k$, a longest periodic substring common to at least $1 < k' \leq k$ of the strings can be computed in linear time. 3) Given two strings, a longest palindromic factor common to the two strings can be computed in linear time.

Kai et al. [18] further generalized and unified these settings. For the properties: square-free, square, periodic, palindromic, and Lyndon, they showed that a given set X of k strings can be preprocessed in linear time so that for any string y and $1 < k' \leq k$, the longest property preserved substring common to y and at least k' of the strings in X can be computed in linear time.

2 Preliminaries

Let Σ be an ordered alphabet. A string is an element of Σ^*. The length of a string $X \in \Sigma^*$ is denoted by $|X|$. For any integer $1 \leq i \leq |X|$, $X[i]$ denotes the ith character of X, i.e., $X = X[1] \cdots X[|X|]$. Also, for any $1 \leq i \leq j \leq |X|$, we denote $X[i..j] = X[i] \cdots X[j]$. A (positioned) substring $X[i..j]$ of X is a *run* in X, if it is a maximal strictly increasing (+-run) or a maximal strictly decreasing (--run) substring, i.e., $i = 1$ or $X[i-1..j]$ is neither strictly increasing nor strictly decreasing, and $j = |X|$ or $X[i..j+1]$ is neither strictly increasing nor strictly decreasing. X is a k-rollercoaster if any run in X has length at least k.

For an integer string X, let $X_1, \ldots, X_x$ be the sequence of runs in X ordered by their occurrence in X. Notice that for any $i \in [1, x-1]$, $X_i[|x|] = X_{i+1}[1]$. We use the notion of $(k, h)_w$-rollercoasters defined by Biedl et al. [6,7]. For $w \in \{+,-\}$ and integer $h \in [1, k]$, X is a $(k, h)_w$-rollercoaster if $X_1, \ldots, X_x$ satisfies the following:

1. The last run X_x is a w-run.
2. $|X_i| \geq k$ for all $i \in [1, |x|-1]$.
3. If $h \in [1, k-1]$, $|X_x| = h$, and $|X_x| \geq k$ otherwise, i.e., $h = k$.

In other words, X is a $(k, h)_w$-rollercoaster if all but the last run has length at least k, and the last run is a w-run of length exactly h if $h < k$, or at least k if $h = k$. Note that a $(k, k)_w$-rollercoaster is a k-rollercoaster, while a $(k, h)_w$-rollercoaster for $h < k$ is not.

For example, for $X = (8, 5, 4, 1, 6, 7, 9, 7, 5)$, the runs of X are $(8, 5, 4, 1)$, $(1, 6, 7, 9)$, $(9, 7, 5)$. Thus, X is a 3-rollercoaster but not a 4-rollercoaster, and is also a $(3, 3)_+$-rollercoaster and a $(4, 3)_+$-rollercoaster as well, but not a $(3, 2)_+$-rollercoaster or a $(4, 2)_+$-rollercoaster.

A string s is a *subsequence* of a string t, if s can be obtained by removing 0 or more symbols from s. Previous work focused on the problem of finding the longest k-rollercoaster that is a subsequence of a given string. We consider the following problem:

Problem 1 (Longest Common k-Rollercoaster). Given two integer strings $S[1..n]$, $T[1..m]$, and integer $k \geq 3$, find a longest k-rollercoaster that is a subsequence of both S and T.

We will further assume $n \geq m$ and that S and T only consists of characters which occur in both S and T. Notice that for a linearly sortable alphabet, the problem can be reduced, with extra $O(n)$ time and space, to finding the k-rollercoaster between S', T' which are respectively subsequences of S, T, that satisfy this condition. For general ordered alphabets, this can be done with extra $O(n \log m)$ time and $O(n)$ space.

3   Algorithms

3.1 $O(nmk)$ Time and Space Algorithm

Unless noted otherwise, we will use S, T to denote integer strings with respective lengths n, m. Also, we will assume $w \in \{+, -\}$, $h \in [1, k]$, $i \in [1, n]$, and $j \in [1, m]$. Also, let $\overline{w} = -$ if $w = +$ and $\overline{w} = +$ if $w = -$.

Let $R_w^{k,h}(S, T, i, j)$ denote the set of longest $(k, h)_w$-rollercoasters that is common to $S[1..i]$ and $T[1..j]$ and ends with $T[j]$. Let $L_w^{k,h}[i, j]$ denote the length of elements in $R_w^{k,h}(S, T, i, j)$. Then, it is clear that the length of a longest common k-rollercoaster between S and T is $\max\{L_w^{k,k}[n, j] \mid w \in \{+, -\}, j \in [1, m]\}$.

For example, if $S = (1, 2, 7, 4, 3, 8, 3, 4, 2, 1)$ and $T = (3, 1, 4, 2, 8, 7, 1, 4, 3, 2)$, $R_-^{3,2}(S, T, 8, 9) = \{(1, 2, 7, 3), (1, 2, 8, 3), (1, 4, 8, 3)\}$.

Finally, we define $M_w^{k,h}[i, j]$ as the length of the longest common $(k, h)_w$-rollercoaster between $S[1..i]$ and $T[1..j-1]$ that ends with an element that is less than $T[j]$ when $w = +$ and greater than $T[j]$ when $w = -$. More formally,

$$M_w^{k,h}[i, j] = \begin{cases} \max(\{0\} \cup \{L_+^{k,h}[i, j'] \mid 1 \le j' < j, T[j'] < T[j]\}) & \text{if } w = + \\ \max(\{0\} \cup \{L_-^{k,h}[i, j'] \mid 1 \le j' < j, T[j'] > T[j]\}) & \text{if } w = - \end{cases}$$

For example, for $S = (1, 2, 4, 8, 5, 1, 4, 2)$ and $T = (8, 1, 4, 2, 2, 8, 1)$, we have $L_+^{3,2}[5, 2] = 0$, $L_+^{3,2}2[5, 3] = L_+^{3,2}[5, 4] = L_+^{3,2}[5, 5] = 2$, and therefore, $M_+^{3,2}[5, 6] = 2$. Intuitively, $M_w^{k,h}[i, j]$ denotes the length of the longest $(k, h)_w$-rollercoaster that might be extended by $T[j]$.

In the following, we claim a recurrence formula for computing $L_w^k[i, j]$. As the conditions of each recurrence is somewhat involved, these will be denoted as shown below:

- (C_1): $i = 0$ or $j = 0$
- (C_2): $S[i] = T[j]$
- (C_3): $L_{\overline{w}}^{k,k}[i, j] = 0$
- (C_4): $M_w^{k,h-1}[i-1, j] = 0$
- (C_5): $\max\{M_w^{k,k-1}[i-1, j], M_w^{k,k}[i-1, j]\} = 0$

Lemma 1. *The following recurrences for $L_w^{k,h}[i, j]$ hold:*
For $h = 1$,

$$L_w^{k,1}[i, j] = \begin{cases} 0 & \text{if } C_1, \\ 1 & \text{if } \neg C_1 \wedge C_2 \wedge C_3, \\ L_{\overline{w}}^{k,k}[i, j] & \text{if } \neg C_1 \wedge C_2 \wedge \neg C_3, \\ L_w^{k,1}[i-1, j] & \text{if } \neg C_1 \wedge \neg C_2. \end{cases} \tag{1}$$

For $h \in [2, k-1]$,

$$L_w^{k,h}[i,j] = \begin{cases} 0 & \text{if } C_1 \\ 0 & \text{if } \neg C_1 \wedge C_2 \wedge C_4 \\ M_w^{k,h-1}[i-1,j]+1 & \text{if } \neg C_1 \wedge C_2 \wedge \neg C_4 \\ L_w^{k,h}[i-1,j] & \text{if } \neg C_1 \wedge \neg C_2. \end{cases} \tag{2}$$

Finally, for $h = k$,

$$L_w^{k,k}[i,j] = \begin{cases} 0 & \text{if } C_1 \\ 0 & \text{if } \neg C_1 \wedge C_2 \wedge C_5 \\ \max\{M_w^{k,k-1}[i-1,j], M_w^{k,k}[i-1,j]\}+1 & \text{if } \neg C_1 \wedge C_2 \wedge \neg C_5 \\ L_w^{k,k}[i-1,j] & \text{if } \neg C_1 \wedge \neg C_2. \end{cases} \tag{3}$$

Proof. We show the case for $w = +$. The case for $w = -$ can be shown in a symmetric fashion.

For any $h \in [1, k]$, if case C_1 holds, then it is clear that $L_+^{k,h}[i,j] = 0$ holds since at least one of $S[1..i]$ or $T[1..j]$ is empty. If $\neg C_1 \wedge \neg C_2$, let $U \in R_+^{k,h}(S,T,i,j)$, i.e., U is a longest common $(k,h)_+$-rollercoaster between $S[1..i]$ and $T[1..j]$ that ends with $T[j]$. Then, from $S[i] \neq T[j]$ ($\neg C_2$), it holds that the last element of U cannot be $S[i]$. Therefore, $R_+^{k,h}(S,T,i,j) = R_+^{k,h}(S,T,i-1,j)$ which implies $L_+^{k,h}[i,j] = L_+^{k,h}[i-1,j]$.

In what follows, we consider the cases where $\neg C_1 \wedge C_2$ holds, i.e., $i,j > 0$ and $S[i] = T[j]$.

Case $h = 1$: Since $S[i] = T[j]$ holds, $L_+^{k,1}[i,j]$ is at least 1. Notice that a $(k,1)_+$-rollercoaster is either a sequence of length 1, or a $(k,k)_-$-rollercoaster.

- C_3: $L_-^{k,k}[i,j] = 0$. This implies that the length of the longest $(k,k)_-$-rollercoaster between $S[1..i]$ and $T[1..j]$ that ends with $T[j]$ is 0, and thus the longest common $(k,1)_+$-rollercoaster between $S[1..i]$ and $T[1..j]$ that ends with $T[j]$ is of length 1, i.e., $L_+^{k,1}[i,j] = 1$.

- $\neg C_3$: $L_-^{k,k}[i,j] > 0$. Since $k \geq 3$, this implies that $L_-^{k,k}[i,j] \geq 3 > 1$, and therefore $L_+^{k,1}[i,j] = L_-^{k,k}[i,j]$.

Case $h \in [2, k-1]$: Recall that $M_+^{k,h-1}[i-1,j]$ is the length of the longest common $(k,h-1)_+$-rollercoaster between $S[1..i-1]$ and $T[1..j-1]$ that ends with en element less than $T[j]$.

- C_4: $M_+^{k,h-1}[i-1,j] = 0$. If we can show $R_+^{k,h}(S,T,i,j) = \emptyset$, then $L_+^{k,h}[i,j] = 0$ follows. Suppose to the contrary that there exists some $U \in R_+^{k,h}(S,T,i,j)$. Then, since $|U| \geq h \geq 2$ and $S[i] = T[j]$, $U[1..|U|-1]$ is a common $(k,h-1)_+$-rollercoaster between $S[1..i-1]$ and $T[1..j-1]$ that ends with an element smaller than $T[j]$. Therefore, by definition, $M_+^{k,h-1}[i-1,j] \geq |U|-1$. Since $M_+^{k,h-1}[i-1,j] = 0$, this implies $|U| \leq 1$, but this contradicts $|U| \geq 2$.

$\neg C_4$: $M_+^{k,h-1}[i-1,j] > 0$. Therefore, it must be that there is some $j^* = \arg\max_{j'}\{L_+^{k,h-1}[i-1,j'] \mid j' < j, T[j'] < T[j]\}$ such that $M_+^{k,h-1}[i-1,j] = L_+^{k,h-1}[i-1,j^*] > 0$. This implies $R_+^{k,h-1}(S,T,i-1,j^*) \neq \emptyset$. Let $U' \in R_+^{k,h-1}(S,T,i-1,j^*)$. Then, since $j^* < j$, $T[j^*] < T[j]$, and $S[i] = T[j]$, the sequence obtained by appending $T[j]$ to U' is a common $(k,h)_+$-rollercoaster between $S[1..i]$ and $T[1..j]$ that ends with $T[j]$. Therefore, $L_+^{k,h}[i,j] \geq |U'| + 1 = M_+^{k,h-1}[i-1,j] + 1$.

Since $L_+^{k,h}[i,j] > 0$, there exists some $U \in R_+^{k,h}(S,T,i,j)$. Then, $U[1..|U| - 1]$ is a common $(k,h-1)_+$-rollercoaster between $S[1..i-1]$ and $T[1..j-1]$ that ends with an element smaller than $T[j]$. Therefore, $L_+^{k,h}[i,j] - 1 = |U| - 1 \leq M_+^{k,h-1}[i-1,j]$.

Thus, it holds that $L_+^{k,h}[i,j] = M_+^{k,h-1}[i-1,j] + 1$.

Case $h = k$: The proof is essentially the same to the case $h \in [2, k-1]$.

C_5: $\max\{M_+^{k,k-1}[i-1,j], M_+^{k,k}[i-1,j]\} = 0$. If we can show $R_+^{k,k}(S,T,i,j) = \emptyset$, then $L_+^{k,k}[i,j] = 0$ follows. Suppose to the contrary that there exists some $U \in R_+^{k,k}(S,T,i,j)$. Then, since $|U| \geq k \geq 3$ and $S[i] = T[j]$, $U[1..|U| - 1]$ is either a common $(k,k)_+$-rollercoaster if the last run of U is longer than k, or a common $(k,k-1)_+$-rollercoaster if the last run of U is exactly k, between $S[1..i-1]$ and $T[1..j-1]$ that ends with an element smaller than $T[j]$. Therefore, by definition, either $M_+^{k,k}[i-1,j] \geq |U| - 1$ or $M_+^{k,k-1}[i-1,j] \geq |U| - 1$. Since $M_+^{k,k}[i-1,j] = M_+^{k,k-1}[i-1,j] = 0$, this implies $|U| \leq 1$, but this contradicts $|U| \geq 3$.

$\neg C_5$: $\max\{M_+^{k,k-1}[i-1,j], M_+^{k,k}[i-1,j]\} > 0$. Let $k' \in \{k-1,k\}$ be such that $M_+^{k,k'}[i-1,j] = \max\{M_+^{k,k-1}[i-1,j], M_+^{k,k}[i-1,j]\}$. It must be that there is some $j^* = \arg\max_{j'}\{L_+^{k,k'}[i-1,j'] \mid j' < j, T[j'] < T[j]\}$ such that $M_+^{k,k'}[i-1,j] = L_+^{k,k'}[i-1,j^*] > 0$. This implies $R_+^{k,k'}(S,T,i-1,j^*) \neq \emptyset$. Let $U' \in R_+^{k,k'}(S,T,i-1,j^*)$. Then, since $j^* < j$, $T[j^*] < T[j]$, and $S[i] = T[j]$, the sequence obtained by appending $T[j]$ to U' is a common $(k,k)_+$-rollercoaster between $S[1..i]$ and $T[1..j]$ that ends with $T[j]$. Therefore, $L_+^{k,k}[i,j] \geq |U'| + 1 = M_+^{k,k'}[i-1,j] + 1 = \max\{M_+^{k,k-1}[i-1,j], M_+^{k,k}[i-1,j]\} + 1$.

Since $L_+^{k,k}[i,j] > 0$, there exists some $U \in R_+^{k,k}(S,T,i,j)$. Then, $U[1..|U| - 1]$ is either a common $(k,k)_+$-rollercoaster if the last run of U is longer than k, or a common $(k,k-1)_+$-rollercoaster if the last run of U is exactly k, between $S[1..i-1]$ and $T[1..j-1]$ that ends with an element smaller than $T[j]$. Therefore, $L_+^{k,k}[i,j] - 1 = |U| - 1 \leq \max\{M_+^{k,k-1}[i-1,j], M_+^{k,k}[i-1,j]\}$.

Thus, it holds that $L_+^{k,k}[i,j] = \max\{M_w^{k,k-1}[i-1,j], M_w^{k,k}[i-1,j]\} + 1$. $\qquad \square$

Notice that when computing $L_w^{k,h}[i,j]$ for any i,j and any $2 \leq h \leq k$, the value $M_w^{k,h-1}[i-1,j] = \max(\{0\} \cup \{L_w^{k,h-1}[i-1,j'] \mid 1 \leq j' < j, T[j'] < T[j]\})$ is required only when $S[i] = T[j]$ (from C_2), and depends on values $L_w^{k,h-1}[i-1,j']$

where $1 \leq j' < j$ and $T[j'] < S[i] = T[j]$. Thus, the required values can be determined in increasing order of j for fixed i and h. Overall, for $h = 1$, $L_w^{k,1}[i,j]$ may depend on $L_{\overline{w}}^{k,k}[i,j]$. However, for $2 \leq h \leq k$, the value of $L_w^{k,h}[i,j]$ depends on $L_w^{k,h'}[i',j']$ such that $h' \leq h$, $j' \leq j$, and $i' < i$. Thus, the dependency is acyclic, and these values can be computed by dynamic programming in a suitable order.

Algorithm 1 shows pseudo-code for computing $L_w^{k,h}$ and the length of the longest k-rollercoaster. In the code, θ represents the value for M computed for increasing j for fixed i and h. Let w^*, j^* be such that $L_{w^*}^{k,k}[n,j^*] = \max\{L_w^{k,k}[n,j] \mid w \in \{\texttt{+},\texttt{-}\}, j \in [1,m]\}$. P and ϕ are used for holding which values were used in the recurrence, so that a longest k-rollercoaster may be retrieved after computing its length, by backtracking. This can be done by calling Reconstruct1(w^*, k, n, j^*) of Algorithm 2.

Thus, we obtain the following theorem.

Theorem 1. *Given integer strings $S[1..n]$, $T[1..m]$, and integer $k \geq 3$, a longest common k-rollercoaster between S and T can be computed in $O(nmk)$ time and space.*

3.2 $O(rk \log^3 m \log \log m)$ Time and $O(rk)$ Space Algorithm

Notice that in the recurrences of Lemma 1, the value of $L_w^{k,h}[i,j]$ is not $L_w^{k,h}[i-1,j]$ only when $S[i] = T[j]$, or when $i = 0$ or $j = 0$ in which case the value of $L_w^{k,h}[i,j]$ is 0. This implies that we essentially need only the values of $L_w^{k,h}[i,j]$ for pairs i,j such that $S[i] = T[j]$.

It is easy to see that we can modify the recurrences of Lemma 1 for i,j restricted to the case where $S[i] = T[j]$, as follows.

Lemma 2. *The recurrences of Lemma 1 for pairs i,j where $S[i] = T[j]$ can be modified as follows: For $h = 1$,*

$$L_w^{k,1}[i,j] = \begin{cases} L_{\overline{w}}^{k,k}[i,j] & \text{if } L_{\overline{w}}^{k,k}[i,j] > 0 \\ 1 & \text{otherwise} \end{cases} \tag{4}$$

For $h \in [2, k-1]$,

$$L_w^{k,h}[i,j] = \begin{cases} \mathcal{M}_+^{k,h-1}[i,j] + 1 & \text{if } \mathcal{M}_+^{k,h-1}[i,j] > 0 \\ 0 & \text{otherwise} \end{cases} \tag{5}$$

Finally, for $h = k$,

$$L_w^{k,k}[i,j] = \begin{cases} \max\{\mathcal{M}_+^{k,k-1}[i,j], \\ \qquad \mathcal{M}_+^{k,k}[i,j]\} \end{cases} + 1 \quad if \begin{array}{l} \max\{\mathcal{M}_+^{k,k-1}[i,j], \\ \qquad \mathcal{M}_+^{k,k}[i,j]\} \end{array} > 0 \\ \quad 0 \qquad\qquad\qquad\qquad\quad otherwise \tag{6}$$

where

$$\mathcal{M}_+^{k,h}[i,j] = \max(\{0\} \cup \{L_+^{k,h}[i',j'] \mid 1 \leq i' < i, 1 \leq j' < j, S[i'] = T[j'] < T[j]\}).$$

Algorithm 1: Computing the longest common k-rollercoaster in $O(nmk)$ time and space.

Input: Integer strings $S[1..n], T[1..m]$, integer $k \geq 3$
Output: Length of longest common k-rollearcoaster between $S[1..n]$ and $T[1..m]$

1 Initialize $L_+^{k,1}, \cdots, L_+^{k,k}, L_-^{k,1}, \cdots, L_-^{k,k}$ to 0.;
2 Initialize $P_+^1, \cdots, P_+^k, P_-^1, \cdots, P_-^k$ to $(0,0,0,0)$.;
3 **for** $i = 1$ **to** n **do**
4   **for** $h = 2$ **to** $k - 1$ **do**
5          $\theta_+ \leftarrow 0$;
6          $\phi_+ \leftarrow (0,0,0,0)$;
7   **for** $j = 1$ **to** m **do**
8   **if** $S[i] = T[j]$ **then**
9   **if** $\theta_+ > 0$ **then** $L_+^{k,h}[i,j] \leftarrow \theta_+ + 1$;
10   **else** $L_+^{k,h}[i,j] \leftarrow 0$;
11                  $P_+^h[i,j] \leftarrow \phi_+$;
12   **else**
13                  $L_+^{k,h}[i,j] \leftarrow L_+^{k,h}[i-1,j]$, $P_+^h[i,j] \leftarrow P_+^h[i-1,j]$;
14   **if** $S[i] > T[j]$ **and** $\theta_+ < L_+^{k,h-1}[i-1,j]$ **then**
15                      $\theta_+ \leftarrow L_+^{k,h-1}[i-1,j]$;
16                      $\phi_+ \leftarrow (+, h-1, i-1, j)$;
17   Compute $L_-^{k,h}[i], P_-^h[i]$ similarly. ;
18   Compute $L_+^{k,k}, L_-^{k,k}, P_+^k, P_-^k$ similarly ;
19   **for** $j = 1$ **to** m **do**
20   **if** $S[i] = T[j]$ **then**
21   **if** $L_-^{k,k}[i,j] = 0$ **then**
22                  $L_+^{k,1}[i,j] \leftarrow 1$;
23                  $P_+^1[i,j] \leftarrow (0,0,0,0)$;
24   **else** $L_+^{k,1}[i,j] \leftarrow L_-^{k,k}[i,j]$, $P_+^1[i,j] \leftarrow P_-^k[i,j]$;
25   **else** $L_+^{k,1}[i,j] \leftarrow L_+^{k,1}[i-1,j]$, $P_+^1[i,j] \leftarrow P_+^1[i-1,j]$;
26   Compute $L_-^{k,1}, P_-^1$ similarly ;
27 **return** $\max\{L_w^{k,k}[n,j] \mid w \in \{+,-\}, j \in [1,m]\}$;

Proof. Considering the condition $S[i] = T[j]$ and the recurrences in Lemma 1, we only have to prove $\mathcal{M}_w^{k,h}[i,j] = M_w^{k,h}[i-1,j]$. We consider the case for $w = +$. The case for $w = -$ is symmetric.

Recall that

$$M_+^{k,h}[i-1,j] = \max(\{0\} \cup \{L_+^{k,h}[i-1,j'] \mid 1 \leq j' < j, T[j'] < T[j]\}).$$

Suppose for some $1 \leq i' < i$ and $1 \leq j' < j$, there is a common $(k,h)_+$-rollercoaster between $S[1..i']$ and $T[1..j']$ of length $\mathcal{M}_+^{k,h}[i,j] = L_+^{k,h}[i',j']$ that ends with an element $S[i'] = T[j'] < T[j]$. This implies $M_+^{k,h}[i-1,j] \geq \mathcal{M}_+^{k,h}[i,j]$.

Algorithm 2: Reconstruct1(w, h, i, j)

Input: w, h, i, j

Output: Reconstructing a longest common k-rollercoaster between $S[1..n]$
and $T[1..m]$

1 **if** $P_w^h[i, j] = (0, 0, 0, 0)$ **then return**;

2 **else**

3      $(w', h', i', j') \leftarrow P_w^h[i, j]$;

4   Reconstruct1(w', h', i', j');

5   **Print** $T[j]$;

6   **return**

On the other hand, suppose for some j', there is a common $(k, h)_+$-rollercoaster of length $M_+^{k,h}[i-1, j] = L_+^{k,h}[i-1, j']$ between $S[1..i-1]$ and $T[1..j']$ that ends with $T[j'] < T[j]$. Let i' be the largest value less than i such that $S[i'] = T[j']$. Since $T[j']$ does not occur in $S[i'+1..i-1]$, it must be that $L_+^{k,h}[i', j'] = L_+^{k,h}[i-1, j']$, which implies that $\mathcal{M}_+^{k,h}[i, j] \geq M_+^{k,h}[i-1, j]$.   □

We use the following property to compute the recurrence formulas in Lemma 2 efficiently.

Property 1. For any $1 \leq i' < i$ and $1 \leq j \leq m$ such that $S[i'] = S[i] = T[j]$, $L_w^{k,h}[i', j] \leq L_w^{k,h}[i, j]$.

Proof. A common $(k, h)_w$-rollercoaster of length $L_w^{k,h}[i', j]$ between $S[1..i']$ and $T[1..j]$ that ends with element $S[i'] = T[j]$, is also a $(k, h)_w$-rollercoaster between $S[1..i]$ and $T[1..j]$ that ends with element $S[i] = T[j]$.

We reorganize the computation as follows, and consider the relevant values of $L_w^{k,h}[i, j]$ to be computed as weights on a point on a 2-dimensional grid fixing i and h. More precisely, we consider $2k$ 2-dimensional grids, each corresponding to a distinct pair $(w, h) \in \{+, -\} \times [1, k]$. For each position j of T, we consider a point at $(j, T[j])$ on the grid. The points will initially have weight 0, and the weights will be maintained incrementally for $i = 1, \ldots, n$, to represent $L_w^{k,h}[i, j]$, by updating them for all points $\{(j, T[j]) \mid j \in [1, m], T[j] = S[i]\}$.

From Property 1, the points in the range $[1..j-1] \times [1..T[j]-1]$ for some i have weights which represent $\{L_+^{k,h}[i', j'] \mid 1 \leq i' < i, 1 \leq j' < j, S[i'] = T[j'] < T[j]\}$. Thus, $\mathcal{M}_w^{k,h}[i, j]$ can be computed by a range maximum query on the weighted grid constructed for $i - 1$, in the range $[1..j-1] \times [1..T[j]-1]$. This can be computed efficiently using the following result by Chazelle [10].

Lemma 3 (Theorem 3 of [10]). *There is an $O(n)$ space data structure that maintains n weighted points that supports 2D range maximum queries, as well as insertion and deletion of a point, in $O(\log^3 n \log \log n)$ time.*

Note that a k-rollercoaster can be reconstructed in a similar fashion to the first algorithm. Thus, we obtain the following result.

Theorem 2. *For two integer strings $S[1..n]$, $T[1..m]$ and integer $k \geq 3$, a longest k-rollercoaster between S and T can be computed in $O(rk \log^3 m \log\log m)$ time and $O(rk)$ space, where r is the number of pairs (i,j) such that $S[i] = T[j]$.*

4   Conclusion

We considered the problem of finding a longest common k-rollercoaster between two strings. Notice that for $k = 1$, any non-empty string is a k-rollercoaster, and thus, the problem is equivalent to the classic longest common subsequence problem which has $O(n^2/polylog(n))$ time solutions, but believed to be unlikely to have $O(n^{2-\varepsilon})$ time solutions [1]. It is an open problem whether $O(n^2/polylog(n))$ time algorithms exist for general k.

Acknowledgments. This work was supported by JSPS KAKENHI Grant Numbers JP18K18002 (YN), JP21K17705 (YN), JP17H01697 (SI), JP20H04141 (HB), JP18H04098 (MT), and JST PRESTO Grant Number JPMJPR1922 (SI).

References

1. Abboud, A., Backurs, A., Williams, V.V.: Tight hardness results for LCS and other sequence similarity measures. In: Guruswami, V. (ed.) IEEE 56th Annual Symposium on Foundations of Computer Science, FOCS 2015, Berkeley, CA, USA, 17–20 October 2015, pp. 59–78. IEEE Computer Society (2015). https://doi.org/10.1109/FOCS.2015.14
2. Agrawal, A., Gawrychowski, P.: A faster subquadratic algorithm for the longest common increasing subsequence problem. In: Cao, Y., Cheng, S., Li, M. (eds.) 31st International Symposium on Algorithms and Computation, ISAAC 2020, 14–18 December 2020, Hong Kong, China (Virtual Conference). LIPIcs, vol. 181, pp. 4:1–4:12. Schloss Dagstuhl - Leibniz-Zentrum für Informatik (2020). https://doi.org/10.4230/LIPIcs.ISAAC.2020.4
3. Ayad, L.A.K., et al.: Longest property-preserved common factor. In: Gagie, T., Moffat, A., Navarro, G., Cuadros-Vargas, E. (eds.) SPIRE 2018. LNCS, vol. 11147, pp. 42–49. Springer, Cham (2018). https://doi.org/10.1007/978-3-030-00479-8_4
4. Ayad, L.A.K., et al.: Longest property-preserved common factor: a new string-processing framework. Theor. Comput. Sci. **812**, 244–251 (2020). https://doi.org/10.1016/j.tcs.2020.02.012
5. Bae, S.W., Lee, I.: On finding a longest common palindromic subsequence. Theor. Comput. Sci. **710**, 29–34 (2018). https://doi.org/10.1016/j.tcs.2017.02.018
6. Biedl, T.C., et al.: Rollercoasters and caterpillars. In: Chatzigiannakis, I., Kaklamanis, C., Marx, D., Sannella, D. (eds.) 45th International Colloquium on Automata, Languages, and Programming, ICALP 2018, 9–13 July 2018, Prague, Czech Republic. LIPIcs, vol. 107, pp. 18:1–18:15. Schloss Dagstuhl - Leibniz-Zentrum für Informatik (2018). https://doi.org/10.4230/LIPIcs.ICALP.2018.18
7. Biedl, T.C., et al.: Rollercoasters: long sequences without short runs. SIAM J. Discret. Math. **33**(2), 845–861 (2019). https://doi.org/10.1137/18M1192226

8. Biedl, T., Chan, T.M., Derka, M., Jain, K., Lubiw, A.: Improved bounds for drawing trees on fixed points with L-shaped edges. In: Frati, F., Ma, K.-L. (eds.) GD 2017. LNCS, vol. 10692, pp. 305–317. Springer, Cham (2018). https://doi.org/10.1007/978-3-319-73915-1_24
9. Chan, W., Zhang, Y., Fung, S.P.Y., Ye, D., Zhu, H.: Efficient algorithms for finding a longest common increasing subsequence. J. Comb. Optim. **13**(3), 277–288 (2007). https://doi.org/10.1007/s10878-006-9031-7
10. Chazelle, B.: A functional approach to data structures and its use in multidimensional searching. SIAM J. Comput. **17**(3), 427–462 (1988). https://doi.org/10.1137/0217026
11. Chowdhury, S.R., Hasan, M.M., Iqbal, S., Rahman, M.S.: Computing a longest common palindromic subsequence. In: Arumugam, S., Smyth, W.F. (eds.) IWOCA 2012. LNCS, vol. 7643, pp. 219–223. Springer, Heidelberg (2012). https://doi.org/10.1007/978-3-642-35926-2_24
12. Chowdhury, S.R., Hasan, M.M., Iqbal, S., Rahman, M.S.: Computing a longest common palindromic subsequence. Fundam. Inform. **129**(4), 329–340 (2014). https://doi.org/10.3233/FI-2014-974
13. Duraj, L.: A sub-quadratic algorithm for the longest common increasing subsequence problem. In: Paul, C., Bläser, M. (eds.) 37th International Symposium on Theoretical Aspects of Computer Science, STACS 2020, 10–13 March 2020, Montpellier, France. LIPIcs, vol. 154, pp. 41:1–41:18. Schloss Dagstuhl - Leibniz-Zentrum für Informatik (2020). https://doi.org/10.4230/LIPIcs.STACS.2020.41
14. Gawrychowski, P., Manea, F., Serafin, R.: Fast and longest rollercoasters. In: Niedermeier, R., Paul, C. (eds.) 36th International Symposium on Theoretical Aspects of Computer Science, STACS 2019, 13–16 March 2019, Berlin, Germany. LIPIcs, vol. 126, pp. 30:1–30:17. Schloss Dagstuhl - Leibniz-Zentrum für Informatik (2019). https://doi.org/10.4230/LIPIcs.STACS.2019.30
15. Inenaga, S., Hyyrö, H.: A hardness result and new algorithm for the longest common palindromic subsequence problem. Inf. Process. Lett. **129**, 11–15 (2018). https://doi.org/10.1016/j.ipl.2017.08.006
16. Inoue, T., Inenaga, S., Bannai, H.: Longest square subsequence problem revisited. In: Boucher, C., Thankachan, S.V. (eds.) SPIRE 2020. LNCS, vol. 12303, pp. 147–154. Springer, Cham (2020). https://doi.org/10.1007/978-3-030-59212-7_11
17. Inoue, T., Inenaga, S., Hyyrö, H., Bannai, H., Takeda, M.: Computing longest common square subsequences. In: Navarro, G., Sankoff, D., Zhu, B. (eds.) Annual Symposium on Combinatorial Pattern Matching, CPM 2018, 2–4 July 2018, Qingdao, China. LIPIcs, vol. 105, pp. 15:1–15:13. Schloss Dagstuhl - Leibniz-Zentrum für Informatik (2018). https://doi.org/10.4230/LIPIcs.CPM.2018.15
18. Kai, K., Nakashima, Y., Inenaga, S., Bannai, H., Takeda, M., Kociumaka, T.: On longest common property preserved substring queries. In: Brisaboa, N.R., Puglisi, S.J. (eds.) SPIRE 2019. LNCS, vol. 11811, pp. 162–174. Springer, Cham (2019). https://doi.org/10.1007/978-3-030-32686-9_12
19. Kosowski, A.: An efficient algorithm for the longest tandem scattered subsequence problem. In: Apostolico, A., Melucci, M. (eds.) SPIRE 2004. LNCS, vol. 3246, pp. 93–100. Springer, Heidelberg (2004). https://doi.org/10.1007/978-3-540-30213-1_13
20. Kutz, M., Brodal, G.S., Kaligosi, K., Katriel, I.: Faster algorithms for computing longest common increasing subsequences. J. Discrete Algorithms **9**(4), 314–325 (2011). https://doi.org/10.1016/j.jda.2011.03.013
21. Russo, L.M.S., Francisco, A.P.: Small longest tandem scattered subsequences. CoRR abs/2006.14029 (2020). https://arxiv.org/abs/2006.14029

22. Tiskin, A.: Semi-local string comparison: algorithmic techniques and applications. Math. Comput. Sci. **1**(4), 571–603 (2008). https://doi.org/10.1007/s11786-007-0033-3
23. Yang, I., Huang, C., Chao, K.: A fast algorithm for computing a longest common increasing subsequence. Inf. Process. Lett. **93**(5), 249–253 (2005). https://doi.org/10.1016/j.ipl.2004.10.014

Minimal Unique Palindromic Substrings After Single-Character Substitution

Mitsuru Funakoshi[1,2]([✉]) [iD] and Takuya Mieno[1,2] [iD]

[1] Department of Informatics, Kyushu University, Fukuoka, Japan
{mitsuru.funakoshi,takuya.mieno}@inf.kyushu-u.ac.jp
[2] Japan Society for the Promotion of Science, Tokyo, Japan

Abstract. A *palindrome* is a string that reads the same forward and backward. A palindromic substring w of a string T is called a *minimal unique palindromic substring* (*MUPS*) of T if w occurs only once in T and any proper palindromic substring of w occurs at least twice in T. MUPSs are utilized for answering the *shortest unique palindromic substring problem*, which is motivated by molecular biology [Inoue et al., 2018]. Given a string T of length n, all MUPSs of T can be computed in $O(n)$ time. In this paper, we study the problem of updating the set of MUPSs when a character in the input string T is substituted by another character. We first analyze the number d of changes of MUPSs when a character is substituted, and show that d is in $O(\log n)$. Further, we present an algorithm that uses $O(n)$ time and space for preprocessing, and updates the set of MUPSs in $O(\log \sigma + (\log \log n)^2 + d)$ time where σ is the alphabet size. We also propose a variant of the algorithm, which runs in optimal $O(1 + d)$ time when the alphabet size is constant.

Keywords: String algorithm · Palindrome · Edit operation

1 Introduction

Palindromes are strings that read the same forward and backward. Finding palindromic structures has important applications to analyze biological data such as DNA, RNA, and proteins, and thus algorithms and combinatorial properties on palindromic structures have been heavily studied (e.g., see [6,9,13,16,17,19,20] and references therein). In this paper, we treat a notion of palindromic structures called *minimal unique palindromic substring* (*MUPS*) that is introduced in [14]. A palindromic substring $T[i..j]$ of a string T is called a MUPS of T if $T[i..j]$ occurs exactly once in T and $T[i+1..j-1]$ occurs at least twice in T. MUPSs are utilized for solving the *shortest unique palindromic substring* (*SUPS*) problem proposed by Inoue et al. [14], which is motivated by an application in molecular biology. They showed that there are no more than n MUPSs in any length-n

M. Funakoshi—Supported by JSPS KAKENHI Grant Number JP20J21147.
T. Mieno—Supported by JSPS KAKENHI Grant Number JP20J11983.

T. Lecroq and H. Touzet (Eds.): SPIRE 2021, LNCS 12944, pp. 33–46, 2021.
https://doi.org/10.1007/978-3-030-86692-1_4

string, and proposed an $O(n)$-time algorithm to compute all MUPSs of a given string of length n over an integer alphabet of size $n^{O(1)}$. After that, Watanabe et al. [23] considered the problem of computing MUPSs in an *run-length encoded (RLE)* string. They showed that there are no more than m MUPSs in a string whose RLE size is m. Also, they proposed an $O(m \log \sigma_R)$-time and $O(m)$-space algorithm to compute all MUPSs of a string given in RLE, where σ_R is the number of distinct single-character runs in the RLE string. Recently, Mieno et al. [18] considered the problems of computing palindromic structures in the *sliding window* model. They showed that the set of MUPSs in a sliding window can be maintained in a total of $O(n \log \sigma_W)$ time and $O(D)$ space while a window of size D shifts over a string of length n from the left-end to the right-end, where σ_W is the maximum number of distinct characters in the windows. This result can be rephrased as follows: The set of MUPSs in a string of length D can be updated in amortized $O(\log \sigma_W)$ time using $O(D)$ space after deleting the first character or inserting a character to the right-end.

To the best of our knowledge, there is no efficient algorithm for updating the set of MUPSs after editing a character at *any* position so far. Now, we consider the problem of updating the set of MUPSs in a string after substituting a character at any position. Formally, we tackle the following problem: Given a string T of length n over an integer alphabet of size $n^{O(1)}$ to preprocess, and then given a query of single-character substitution. Afterwards, we return the set of MUPSs of the edited string. In this paper, we first show that the number d of changes of MUPSs after a single-character substitution is $O(\log n)$. In addition, we present an algorithm that uses $O(n)$ time and space for preprocessing, and updates the set of MUPSs in $O(\log \sigma + (\log \log n)^2 + d) \subset O(\log n)$ time. We also propose a variant of the algorithm, which runs in optimal $O(1 + d)$ time when the alphabet size is constant.

Related Work. There are some results for the problem of computing string regularities, including palindromes, on dynamic strings [2,3,5,8]. In this work, we consider the problem of computing MUPSs after a single edit operation as a first step toward a fully dynamic setting. This line of research was initiated by Amir et al. [4], who tackled the problem of computing the longest common factor after one edit. After that, other notions of string regularities are treated in a similar setting [1,12,22]. In particular, regarding palindromic structures, Funakoshi et al. [12] proposed algorithms for computing the longest palindromic substring after single-character or block-wise edit operations.

2   Preliminaries

2.1   Notations

Strings. Let Σ be an *alphabet* of size σ. An element of Σ is called a *character*. An element of Σ^* is called a *string*. The length of a string T is denoted by $|T|$. The *empty string* ε is the string of length 0. For a string $T = xyz$, then $x, y,$ and

z are called a *prefix*, *substring*, and *suffix* of T, respectively. They are called a *proper prefix*, *proper substring*, *proper suffix* of T if $x \neq T$, $y \neq T$, and $z \neq T$, respectively. For each $1 \leq i \leq |T|$, $T[i]$ denotes the i-th character of T. For each $1 \leq i \leq j \leq |T|$, $T[i..j]$ denotes the substring of T starting at position i and ending at position j. For convenience, let $T[i'..j'] = \varepsilon$ for any $i' > j'$. A positive integer p is said to be a *period* of a string T if $T[i] = T[i+p]$ for all $1 \leq i \leq |T|-p$. For strings X and Y, let $lcp(X,Y)$ denotes the length of the *longest common prefix* (in short, lcp) of X and Y, i.e., $lcp(X,Y) = \max\{\ell \mid X[1..\ell] = Y[1..\ell]\}$. For a string T and two integers $1 \leq i \leq j \leq |T|$, let $lce_T(i,j)$ denotes the length of the *longest common extension* (in short, lce) of i and j in T, i.e., $lce_T(i,j) = lcp(T[i..|T|], T[j..|T|])$. For non-empty strings T and w, $beg_T(w)$ denotes the set of beginning positions of occurrences of w in T. Also, for a text position i in T, $inbeg_{T,i}(w)$ denotes the set of beginning positions of occurrences of w in T where each occurrence covers position i. Namely, $beg_T(w) = \{b \mid T[b..e] = w\}$ and $inbeg_{T,i}(w) = \{b \mid T[b..e] = w \text{ and } i \in [b,e]\}$. Further, let $xbeg_{T,i}(w) = beg_T(w) \backslash inbeg_{T,i}(w)$. For convenience, $|beg_T(\varepsilon)| = |inbeg_{T,i}(\varepsilon)| = |xbeg_{T,i}(\varepsilon)| = |T| + 1$ for any T and i. We say that w is *unique* in T if $|beg_T(w)| = 1$, and that w is *repeating* in T if $|beg_T(w)| \geq 2$. Note that the empty string is repeating in any other string. Since every unique substring $u = T[i..j]$ of T occurs exactly once in T, we will sometimes identify u with its corresponding interval $[i,j]$. In what follows, we consider an arbitrarily fixed string T of length $n \geq 1$ over an alphabet Σ of size $\sigma = n^{O(1)}$.

Palindromes. For a string w, w^R denotes the reversed string of w. A string w is called a *palindrome* if $w = w^R$. A palindrome w is called an *even-palindrome* (resp. *odd-palindrome*) if its length is even (resp. odd). For a palindrome w, its length-$\lfloor |w|/2 \rfloor$ prefix (resp. length-$\lfloor |w|/2 \rfloor$ suffix) is called the *left arm* (resp. *right arm*) of w, and is denoted by larm_w (resp. rarm_w). Also, we call $\mathsf{Larm}_w = \mathsf{larm}_w \cdot s_w$ (resp. $\mathsf{Rarm}_w = s_w \cdot \mathsf{rarm}_w$) the *extended left arm* (resp. *extended right arm*) of w where s_w is the character at the center of w if w is an odd-palindrome, and s_w is empty otherwise. Note that when w is an even-palindrome, $\mathsf{Rarm}_w = \mathsf{rarm}_w$ and $\mathsf{Larm}_w = \mathsf{larm}_w$. For a non-empty palindromic substring $w = T[i..j]$ of a string T, the center of w is $\frac{i+j}{2}$ and is denoted by $center(w)$. A non-empty palindromic substring $T[i..j]$ of a string T is said to be *maximal* if $i = 1$, $j = n$, or $T[i-1] \neq T[j+1]$. For a non-empty palindromic substring $w = T[i..j]$ of a string T and a non-negative integer ℓ, $v = T[i-\ell..j+\ell]$ is said to be an *extension* of w if $1 \leq i - \ell \leq j + \ell \leq n$ and v is a palindrome. Also, $T[i + \ell..j - \ell]$ is said to be a *shrink* of w. A non-empty string w is called a *1-mismatch palindrome* if there is exactly one mismatched position between $w[1..\lfloor |w|/2 \rfloor]$ and $w[\lceil |w|/2 \rceil + 1..|w|]^R$. Informally, a 1-mismatch palindrome is a pseudo palindrome with a mismatch position between their arms. As in the case of palindromes, a 1-mismatch palindromic substring $T[i..j]$ of a string T is said to be *maximal* if $i = 1$, $j = n$ or $T[i-1] \neq T[j+1]$.

A palindromic substring $T[i..j]$ of a string T is called a *minimal unique palindromic substring* (*MUPS*) of T if $T[i..j]$ is unique in T and $T[i+1..j-1]$

is repeating in T. We denote by MUPS(T) the set of intervals corresponding to MUPSs of a string T. A MUPS cannot be a substring of another palindrome with a different center. Also, it is known that the number of MUPSs of T is at most n, and set MUPS(T) can be computed in $O(n)$ time for a given string T over an integer alphabet [14]. The following lemma states that the total sum of occurrences of strings which are extended arms of MUPSs is $O(n)$:

Lemma 1. *The total sum of occurrences of the extended right arms of all MUPSs in a string T is at most $2n$. Similarly, the total sum of occurrences of the extended left arms of all MUPSs in T is at most $2n$.*

Proof. It suffices to prove the former statement for the extended *right* arms since the latter can be proved symmetrically. Let w_1 and w_2 be distinct odd-length MUPSs of T with $|w_1| \leq |w_2|$. For the sake of contradiction, we assume that $T[j..j + |\mathsf{Rarm}_{w_1}| - 1] = \mathsf{Rarm}_{w_1}$ and $T[j..j + |\mathsf{Rarm}_{w_2}| - 1] = \mathsf{Rarm}_{w_2}$ for some position j in T. Namely, Rarm_{w_1} is a prefix of Rarm_{w_2}. Then, larm_{w_1} is a suffix of larm_{w_2} by palindromic symmetry. This means that w_1 is a substring of w_2. This contradicts that w_2 is a MUPS of T. Thus, all occurrences of the extended right arms of all odd-length MUPSs are different, i.e., the total number of the occurrences is at most n. Similarly, the total number of all occurrences of the right arms of all even-length MUPSs is also at most n. □

2.2 Tools

This subsection lists some data structures used in our algorithm. Our model of computation is a standard word RAM model with machine word size $\Omega(\log n)$.

Suffix Trees. The *suffix tree* of T is the compacted trie for all suffixes of T [24]. We denote by $\mathsf{STree}(T)$ the suffix tree of T. If a given string T is over an integer alphabet of size $n^{O(1)}$, $\mathsf{STree}(T)$ can be constructed in $O(n)$ time [10]. Not all substrings of T correspond to nodes in $\mathsf{STree}(T)$. However, the loci of such substrings can be made explicit in linear time:

Lemma 2 (Corollary 8.1 in [15]). *Given m substrings of T, represented by intervals in T, we can compute the locus of each substring in $\mathsf{STree}(T)$ in $O(n + m)$ total time. Moreover, the loci of all the substrings in $\mathsf{STree}(T)$ can be made explicit in $O(n + m)$ extra time.*

Also, this lemma implies the following corollary:

Corollary 1. *Given m substrings of T, represented by intervals in T, we can sort them in $O(n + m)$ time.*

LCE Queries. An *LCE query* on a string T is, given two indices i, j of T, to compute $lce_T(i, j)$. Using $\mathsf{STree}(T\$)$ enhanced with a lowest common ancestor data structure, we can answer any LCE query on T in constant time where $\$$ is a special character with $\$ \notin \Sigma$. In the same way, we can compute the lcp value between any two suffixes of T or T^R in constant time by using $\mathsf{STree}(T\$T^R\#)$ where $\#$ is another special character with $\# \notin \Sigma$.

Eertrees. The *eertree* (a.k.a. palindromic tree) of T is a pair of rooted edge-labeled trees $\mathcal{T}_{\text{odd}}$ and $\mathcal{T}_{\text{even}}$ representing all distinct palindromes in T [19]. The roots of $\mathcal{T}_{\text{odd}}$ and $\mathcal{T}_{\text{even}}$ represent ε. Each non-root node of $\mathcal{T}_{\text{odd}}$ (resp. $\mathcal{T}_{\text{even}}$) represents an odd-palindrome (resp. even-palindrome) which occurs in T. Let $pal(v)$ be the palindrome represented by a node v. For the root r_{odd} of $\mathcal{T}_{\text{odd}}$, there is an edge (r_{odd}, u) labeled by $a \in \Sigma$ if there is a node u with $pal(u) = a$. For any node v in the eertree except for r_{odd}, there is an edge (v, w) labeled by $a \in \Sigma$ if there is a node w with $pal(w) = a \cdot pal(v) \cdot a$. We denote by $\text{EERTREE}(T)$ the eertree of T. We will sometimes identify a node u in $\text{EERTREE}(T)$ with its corresponding palindrome $pal(u)$. Also, the path from a node u to a node v in $\text{EERTREE}(T)$ is denoted by $pal(u) \rightsquigarrow pal(v)$. If a given string T is over an integer alphabet of size $n^{O(1)}$, $\text{EERTREE}(T)$ can be constructed in $O(n)$ time [19].

Path-Tree LCE Queries. A *path-tree LCE query* is a generalized LCE query on a rooted edge-labeled tree $\mathcal{T}$ [7]: Given three nodes u, v, and w in $\mathcal{T}$ where u is an ancestor of v, to compute the lcp between the path-string from u to v and any path-string from w to a descendant leaf. The following result is known:

Theorem 1 (Theorem 2 of [7]). *For a tree $\mathcal{T}$ with N nodes, a data structure of size $O(N)$ can be constructed in $O(N)$ time to answer any pathtree LCE query in $O((\log \log N)^2)$ time.*

We will use later path-tree LCE queries on the eertree of the input string.

Stabbing Queries. Let $\mathcal{I}$ be a set of n intervals, each of which is a subinterval of the universe $U = [1, O(n)]$. An *interval stabbing query* on $\mathcal{I}$ is, given a query point $q \in U$, to report all intervals $I \in \mathcal{I}$ such that I is *stabbed* by q, i.e., $q \in I$. We can answer such a query in $O(1 + k)$ time after $O(n)$-time preprocessing, where k is the number of intervals to report [21].

3 Changes of MUPSs After Single Character Substitution

In the following, we fix the original string T of length n, the text position i in T to be substituted, and the string T' after the substitution. Namely, $T[i] \neq T'[i]$ and $T[j] = T'[j]$ for each j with $1 \leq j \leq n$ and $j \neq i$. This section analyzes the changes of the set of MUPSs when $T[i]$ is substituted by $T'[i]$. For palindromes covering editing position i, Lemma 3 holds. All the proofs omitted due to lack of space can be found in the full version [11].

Lemma 3. *For a palindrome w, if $inbeg_{T,i}(w) \neq \emptyset$, then $inbeg_{T',i}(w) = \emptyset$.*

For a position i, let $\mathcal{W}_i$ be the set of palindromes w such that $|inbeg_{T,i}(w)| \geq 1$, $|xbeg_{T,i}(w)| = 1$, and w is minimal, i.e., $|inbeg_{T,i}(v)| = 0$ or $|xbeg_{T,i}(v)| \geq 2$ where $v = w[2..|w| - 1]$. This set $\mathcal{W}_i$ is useful for analyzing the number of changes of MUPSs in the proof of Theorem 2.

Lemma 4. *For any position i in T, $|\mathcal{W}_i| \in O(\log n)$.*

Lemma 5. *For each position i in T, the number of MUPSs covering i is $O(\log n)$.*

By using Lemmas 3, 4, and 5, we show the following theorem:

Theorem 2. $|\mathsf{MUPS}(T) \triangle \mathsf{MUPS}(T')| \in O(\log n)$ *always holds.*

Proof. In the following, we consider the number of MUPSs to be removed.

First, at most one interval can be a MUPS of T centered at i. Also, any other interval in $\mathsf{MUPS}(T)$ covering position i cannot be an element of $\mathsf{MUPS}(T')$ since its corresponding string in T' is no longer a palindrome. By Lemma 5, the number of such MUPSs is $O(\log n)$.

Next, let us consider MUPSs not covering position i. When a MUPS w of T not covering i is no longer a MUPS of T', then either (A) w is repeating in T' or (B) w is unique in T' but is not minimal.

Let w_1 be a MUPS of the case (A). Since w_1 does not cover i, is unique in T, and is repeating in T', $|inbeg_{T',i}(w_1)| \geq 1$ and $|xbeg_{T',i}(w_1)| = 1$. Let v_1 be the minimal shrink of w_1 such that $|inbeg_{T',i}(v_1)| \geq 1$ and $|xbeg_{T',i}(v_1)| = 1$. Contrary, w_1 is the only MUPS of the case (A) which is an extension of v_1 since $|xbeg_{T',i}(v_1)| = 1$. Namely, there is a one-to-one relation between w_1 and v_1. By Lemma 4, the number of palindromes that satisfy the above conditions of v_1 is $O(\log n)$. Thus, the number of MUPSs of the case (A) is also $O(\log n)$.

Let w_2 be a MUPS of the case (B). In T', w_2 covers some MUPS as a proper substring since it is not a MUPS and is unique in T'. Let v_2 be the MUPS of T', which is a proper substring of w_2. While v_2 is unique in T', it is repeating in T since w_2 is a MUPS of T. Namely, $|inbeg_{T,i}(v_2)| \geq 1$ and $|xbeg_{T,i}(v_2)| = 1$ hold. Also, v_2 is actually minimal: Let $u_2 = v_2[2..|v_2| - 1]$. If we assume that $|inbeg_{T,i}(u_2)| \geq 1$ and $|xbeg_{T,i}(u_2)| = 1$, then u_2 becomes unique in T', and this contradicts that v_2 is a MUPS of T'. Furthermore, similar to the above discussions, there is a one-to-one relation between w_2 and v_2. Again by Lemma 4, the number of palindromes that satisfy the above conditions of v_2 is $O(\log n)$. Thus, the number of MUPSs of the case (B) is also $O(\log n)$.

Therefore, $|\mathsf{MUPS}(T) \backslash \mathsf{MUPS}(T')| \in O(\log n)$ holds. Also, $|\mathsf{MUPS}(T') \backslash \mathsf{MUPS}(T)| \in O(\log n)$ holds by symmetry.

To summarize, $|\mathsf{MUPS}(T) \triangle \mathsf{MUPS}(T')| = |\mathsf{MUPS}(T) \backslash \mathsf{MUPS}(T') \cup \mathsf{MUPS}(T') \backslash \mathsf{MUPS}(T)| = |\mathsf{MUPS}(T) \backslash \mathsf{MUPS}(T')| + |\mathsf{MUPS}(T') \backslash \mathsf{MUPS}(T)| \in O(\log n)$.

4   Algorithms for Updating Set of MUPSs

In this section, we propose an algorithm for updating the set of MUPSs when a single-character in the original string is substituted by another character. We denote by $sub(i, s)$ the substitution query, that is, to substitute $T[i]$ by another character s. First, we define a sub-problem that will be used in our algorithm:

Problem 1. Given a substitution query $sub(i, s)$ on T, compute the longest odd-palindromic substring v of T' such that $center(v) = i$ and v occurs in T if it exists. Also, if such v exists, determine whether v is unique in T or not. Furthermore, if v is unique in T, compute the shrink of v that is a MUPS of T.

We show the following lemma:

Lemma 6. *After $O(n)$-time preprocessing, we can answer Problem 1 in $O(\delta(n, \sigma) + (\log \log n)^2)$ time where $\delta(n, \sigma)$ denote the time to retrieve any child of the root of the odd-tree of* EERTREE(T).

When $\sigma \in O(n)$, we can easily achieve $\delta(n, \sigma) \in O(1)$ with linear space, by using an array of size σ. Otherwise, we achieve $\delta(n, \sigma) \in O(\log \sigma)$ for a general ordered alphabet by using a binary search tree.

In the rest of this paper, we propose an algorithm to compute the changes in MUPSs after a single-character substitution. Our strategy is basically to pre-compute changes in MUPSs for some queries as much as possible within linear time. The other changes will be detected on the fly by using some data structures.

4.1 Computing MUPSs to Be Removed

We categorize MUPSs to be removed into three types as follows:

R1) A MUPS of T that covers i.
R2) A MUPS of T that does not cover i and is repeating in T'.
R3) A MUPS of T that does not cover i and is unique but not minimal in T'.

In the following, we describe how to compute all MUPSs for each type separately.

Type R1. All MUPSs covering editing position i are always removed. Thus, we can detect them in $O(1 + \alpha_{rem})$ time after a simple linear time preprocessing (e.g., using stabbing queries), where α_{rem} is the number of MUPSs of Type R1.

Type R2. Before describing our algorithm, we give a few observations about MUPSs of Type R2. Let w be a MUPS of Type R2. Since w is unique in T and is repeating in T', $|inbeg_{T', i}(w)| \geq 1$. When w occurs in T' centered at editing position i, we retrieve such w by applying Problem 1. If it is not the case, we can utilize the following observations: Consider the starting position j of an occurrence of w in T' such that $T'[j..j + |w| - 1] = w$ and $i \in [j, j + |w| - 1]$. If position i is covered in the right arm of $T'[j..j + |w| - 1]$, then Larm_w occurs at position j in both T and T'. Further, the Hamming distance between $T[j + |\mathsf{Larm}_w|..j + |w| - 1]$ and $w[|\mathsf{Larm}_w| + 1..|w|] = \mathsf{rarm}_w$ equals 1. Namely, for each occurrence at position k of string Larm_w in T, w can occur at k in T' only if the Hamming distance between $T[k + |\mathsf{Larm}_w|..k + |w| - 1]$ and $w[|\mathsf{Larm}_w| + 1..|w|]$ equals 1. In other words, if the Hamming distance is greater than 1, w cannot occur at k in T'.

In the preprocessing phase, we first apply the $O(n)$-time preprocessing of Lemma 6 for Problem 1. Next, we initialize the set $\mathcal{A}_{R2} = \emptyset$. The set $\mathcal{A}_{R2}$ will become an *index* of MUPSs of Type R2 when the preprocessing is finished. For each MUPS $w = T[b..e]$ of T, we process the followings: For the beginning position $j \neq b$ of each occurrence of Larm_w in T, we first compute the lcp value between $T[j + |\mathsf{Larm}_w|..|T|]\$$ and rarm_w with allowing one mismatch. Note that $T[j + |\mathsf{Larm}_w|..|T|]\$$ must have at least one mismatch with rarm_w, since $T[j..j + |\mathsf{Larm}_w| - 1] = \mathsf{Larm}_w$, $j \neq b$, and $T[b..e]$ is unique in T. If there are two mismatch positions between them, do nothing for this occurrence since w cannot occur at j after any substitution. We can check this by querying LCE at most twice. Otherwise, let $q = j + |\mathsf{Larm}_w| - 1 + d$ be the mismatched position in T. When the q-th character of T is substituted by the character $\mathsf{rarm}_w[d]$, $w = T[b..e]$ occurs at $j \neq b$, i.e., it is a MUPS of Type R2. So we add MUPS $w = T[b..e]$ into $\mathcal{A}_{R2}$ with the pair of index and character $(q, \mathsf{rarm}_w[d])$ as the key. In addition, symmetrically, we update $\mathcal{A}_{R2}$ for each occurrence of Rarm_w in T. After finishing the above processes for every MUPS of T, we sort the elements of $\mathcal{A}_{R2}$ by radix sort on the keys. If there are multiple identical elements with the same key, we unify them into a single element. Also, if there are multiple elements with the same key, we store them in a linear list. By Lemma 1, the total number of occurrences of arms of MUPSs is $O(n)$, and hence, the total preprocessing time is $O(n)$.

Given a query $sub(i, s)$, we query Problem 1 with the same pair (i, s) as the input. Then, we complete checking whether there exists a MUPS of Type R2 centered at i. Next, consider the existence of the remaining MUPSs of Type R2. First, an element in $\mathcal{A}_{R2}$ corresponding to the key (i, s) can be detected in $O(\log \sigma_i)$ time by using random access on indices and binary search on characters, where σ_i is the number of characters s_i such that the key (i, s_i) exists in $\mathcal{A}_{R2}$. After that, we can enumerate all the other elements with the key by scanning the corresponding linear list. Thus, the total query time is $O(\delta(n, \sigma) + (\log \log n)^2 + \log \sigma_i + \beta_{rem})$ where β_{rem} is the number of MUPSs of Type R2. Finally, we show $\sigma_i \in O(\min\{\sigma, \log n\})$. Let us consider palindromes in T' whose right arm covers position i. Those whose left arms cover i can be treated similarly. Any palindrome in T' whose right arm covers i is an extension of some maximal palindrome in T ending at $i - 1$. It is known that the number of possible characters immediately preceding such maximal palindromes is $O(\log n)$ [12]. Therefore, $\sigma_i \in O(\log n)$ holds, and thus, the query time is $O(\delta(n, \sigma) + (\log \log n)^2 + \beta_{rem})$.

Type R3. Let $w = T[b..e]$ be a MUPS of T and let $v = T[b + 1..e - 1]$. Further let $T[b_{l1}..e_{l1}]$ and $T[b_{r1}..e_{r1}]$ be the leftmost and the rightmost occurrence of v in T except for $T[b + 1..e - 1]$. We define interval $\rho_w = \{k \mid k \notin [b + 1, e - 1] \text{ and } k \in [b_{r1}, e_{l1}]\}$. Note that ρ_w can be empty. If the editing position i is in ρ_w, then the only occurrence of v in T' is $T'[b + 1..e - 1]$, i.e., v is unique in T'. Thus, w is a removed MUPS of Type R3. Contrary, if $i \notin [b, e]$ and $i \notin \rho_w$, there are at least two occurrences of v in T', i.e., w cannot be a MUPS of Type R3.

In the preprocessing phase, we first compute the set of intervals $\mathcal{R} = \{\rho_w \mid$ w is a MUPS of $T\}$. $\mathcal{R}$ can be computed by traversing over the suffix tree of T enhanced with additional explicit nodes, each of which represents a substring $T[b + 1..e - 1]$ for each MUPS $T[b..e]$ of T. Also, we apply the preprocessing for stabbing queries to $\mathcal{R}$. The total time for preprocessing is $O(n)$.

Given a query $sub(i, s)$, compute all intervals in $\mathcal{R}$ stabbed by position i by answering a stabbing query. They correspond to MUPSs of Type R3. The query time is $O(1 + \gamma_{rem})$, where γ_{rem} is the number of MUPSs of Type R3.

To summarize, we can compute all MUPSs to be removed after a single-character substitution in $O(\delta(n, \sigma) + (\log \log n)^2 + \alpha_{rem} + \beta_{rem} + \gamma_{rem})$ time.

4.2   Computing MUPSs to Be Added

Next, we propose an algorithm to detect MUPSs to be added after a substitution. As in Sect. 4.1, we categorize MUPSs to be added into three types:

A1) A MUPS of T' that covers i.
A2) A MUPS of T' that does not cover i and is repeating in T.
A3) A MUPS of T' that does not cover i and is unique but not minimal in T.

Furthermore, we categorize MUPSs of Type A1 into two sub-types:

A1-1) A MUPS of T' that covers position i in its arm.
A1-2) A MUPS of T' centered at editing position i.

Type A1-1. A MUPS of Type A1-1 is a shrink of some maximal palindrome in T' covering editing position i in its arm. Further, such a maximal palindrome in T' corresponds to some 1-mismatch maximal palindrome in T, which covers i as a mismatch position. Thus, we preprocess for arms of each 1-mismatch maximal palindrome in T. For MUPSs of Type A1, we utilize the following observation:

Observation 1. *For any palindrome v covering position i in T', v is unique in T' if and only if $|inbeg_{T',i}(v)| = 1$ and $|xbeg_{T',i}(v)| = 0$.*

In the preprocessing phase, we first consider sorting extended arms of 1-mismatch maximal palindromes in T. Let EA be the multiset of strings that consists of the extended right arms and the reverse of the extended left arms of all 1-mismatch maximal palindromes in T. Note that each string in EA can be represented in constant space since it is a substring of T or T^R. Let MA' be a lexicographically sorted array of all elements in EA. Here, the order between the same strings can be arbitrary. Also, for each string in EA, we consider a quadruple of the form (par, pos, chr, rnk) where $par \in \{odd, even\}$ represents the parity of the length of the corresponded 1-mismatch maximal palindrome, pos is the mismatched position on the *opposite* arm, chr is the mismatched character on the extended arm, and rnk is the rank of the extended arm in MA'. Let MA be a radix sorted array of these quadruples. It can be seen that for each triple (p, i, s) of parity p, mismatched position i, and mismatched character s, all elements corresponding to the triple are stored continuously in MA. We denote by $MA_{p,i,s}$ the subarray of MA consists of such elements. In other words, $MA_{p,i,s}$ is a sorted array of extended arms of maximal palindromes of parity p covering position i in T' when the i-th character of T is substituted by s.

Now let us focus on odd-palindromes. Even-palindromes can be treated similarly. We construct the suffix tree of T and make the loci of strings in EA explicit. We also make the loci of the extended right arm of every odd-palindrome in T explicit. Simultaneously, we *mark* the nodes corresponding to the extended right arms and apply the preprocessing for the nearest marked ancestor (NMA) queries to the marked tree. We denote the tree by ST_{odd}. Next, we initialize the set $\mathcal{A}_{A1,1} = \emptyset$. The set $\mathcal{A}_{A1,1}$ will become an index of MUPSs of Type A1-1 when the preprocessing is finished. For each non-empty $MA_{odd,i,s}$ and for each string w in $MA_{odd,i,s}$, we do the followings: Let x_w be the odd-palindrome whose extended right arm is w when $T[i]$ is substituted by s. Let u and v are the preceding and the succeeding string of w in $MA_{odd,i,s}$ (if such palindromes do not exist, they are empty). Further let $\ell_w = \max\{lcp(u,w), lcp(w,v)\}$. When $T[i]$ is substituted by s, any shrink y of x_w such that y covers position i and the arm-length of y is at least ℓ_w has only one occurrence which covers position i in T', i.e., $|inbeg_{T',i}(y)| = 1$. Next, we query the NMA for the node corresponding to w on ST_{odd}. Let ℓ'_w be the length of the extended right arm obtained by the NMA query. When $T[i]$ is substituted by s, any shrink y' of x_w such that the arm-length of y' is at least ℓ'_w, has no occurrences which do not cover position i in T', i.e., $|xbeg_{T',i}(y')| = 0$. Thus, by Observation 1, the shrink $y^\star$ of x_w of arm-length $\max\{\ell_w, \ell'_w\}$ is a MUPS of Type A1-1 for the query $sub(i,s)$, if such $y^\star$ exists. In such a case, we store the information about $y^\star$ (i.e., its center and radius) into $\mathcal{A}_{A1,1}$ using (odd, i, s) as the key. After finishing the above preprocessing for all strings in MA, we sort all elements in $\mathcal{A}_{A1,1}$ by their keys.

Since each element in EA is a substring of $T\$T^R\#$, they can be sorted in $O(n + |EA|) = O(n)$ time by Corollary 1. Namely, MA' can be computed in linear time, and thus MA too. By Lemma 2, tree ST_{odd} can be constructed in $O(n)$ time. Also, we can answer each NMA query and LCP query in constant time after $O(n)$ time preprocessing. Hence, the total preprocessing time is $O(n)$.

Given a query $sub(i,s)$, we compute all MUPSs of Type A1-1 by searching for elements in $\mathcal{A}_{A1,1}$ with keys (odd, i, s) and $(even, i, s)$. An element with each of the keys can be found in $O(\log \min\{\sigma, \log n\})$ time. Thus, all MUPSs of Type A1-1 can be computed in $O(\log \min\{\sigma, \log n\} + \alpha'_{add})$ time where α'_{add} is the number of MUPSs of Type A1-1.

Type A1-2. The MUPS of Type A1-2 is a shrink of the maximal palindrome in T' centered at i. By definition, there is at most one MUPS of Type A1-2.

In the preprocessing phase, we again construct MA and related data structures as in Type A1-1. Further, we apply the $O(n)$-time preprocessing of Lemma 6 for Problem 1. The total preprocessing time is $O(n)$.

Given substitution query $sub(i,s)$, we compute the MUPS centered at i in T' as follows (if it exists): It is clear that $T'[i..i] = s$ is the MUPS of Type

A1-2 if s is a unique character in T'. In what follows, we consider the other case. Let w be the maximal palindrome centered at i in T'. First, we compute the maximum lcp value ℓ_w between Rarm_w and extended arms in $\mathsf{MA}_{\mathsf{odd},i,s}$. Then, any shrink y of w, such that the arm-length of y is at least ℓ_w, has no occurrences which cover position i in T', i.e., $|inbeg_{T',i}(y)| = 1$. We can compute ℓ_w in $O(\log\min\{\sigma, \log n\})$ time, combining LCE queries and binary search. Note that rarm_w occurs at $i + 1$ in both T and T' while Rarm_w might be absent from T. Next we compute the arm-length ℓ'_w of the shortest palindrome v such that $center(v) = i$ and $|xbeg_{T',i}(v)| = 0$, i.e., v is absent from T. Since the shrink $\tilde{v}$ of w of arm-length $\ell'_w - 1$ is the longest palindrome such that $center(\tilde{v}) = i$ and $\tilde{v}$ occurs in T, we can reduce the problem of computing ℓ'_w to Problem 1. Thus, we can compute ℓ'_w in $O(\delta(n, \sigma) + (\log\log n)^2)$ time by Lemma 6. Similar to the case of Type A1-1, by Observation 1, the shrink y^* of x_w of arm-length $\max\{\ell_w, \ell'_w\}$ is a MUPS of Type A1-2, if such y^* exists. Therefore, the MUPS of Type A1-2 can be computed in $O(\delta(n, \sigma) + (\log\log n)^2)$ time.

Type A2. A MUPS of Type A2 occurs at least twice in T, and there is only one occurrence not covering editing position i. For a palindrome w repeating in T, let $T[b_{l1}..e_{l1}]$ and $T[b_{l2}..e_{l2}]$ be the leftmost and the second leftmost occurrence of w in T. Further, let $T[b_{r1}..e_{r1}]$ and $T[b_{r2}..e_{r2}]$ be the rightmost and the second rightmost occurrence of w in T. We define interval ρ_w as the intersection of all occurrences of w except for the leftmost one, i.e., $\rho_w = \{k \mid k \notin [b_{l1}, e_{l1}] \text{ and } k \in [b_{r1}, e_{l2}]\}$. Similarly, we define interval $\tilde{\rho}_w$ as the intersection of all occurrences of w except for the rightmost one. Note that ρ_w and $\tilde{\rho}_w$ can be empty. Then, w is unique after the i-th character is edited if and only if $i \in \rho_w \cup \tilde{\rho}_w$. Thus, any MUPS of Type A2 is a palindrome corresponding to some interval in $\rho_w \cup \tilde{\rho}_w$ stabbed by i. To avoid accessing intervals that do not correspond to the MUPSs to be added, we decompose each ρ_w. It is easy to see that for any shrink v of w, $\rho_v \subset \rho_w$ holds. Also, if $T[i]$, with $i \in \rho_v$, is edited, then both w and v become unique in T', i.e., w cannot be a MUPS of T'. For each unique palindrome w in T, we decompose ρ_w into at most three intervals $\rho_w = \rho_w^1 \rho_{w'} \rho_w^2$ where $w' = w[2..|w| - 1]$. Similarly, we decompose $\tilde{\rho}_w$ into $\tilde{\rho}_w = \tilde{\rho}_w^1 \tilde{\rho}_{w'} \tilde{\rho}_w^2$. Then, w is a MUPS of Type A2 if and only if $i \in \rho_w^1 \cup \rho_w^2 \cup \tilde{\rho}_w^1 \cup \tilde{\rho}_w^2$.

In the preprocessing phase, we first construct the eertree of T and the suffix tree of T enhanced with additional explicit nodes for all distinct palindromes in T. Next, we compute at most four (leftmost, second leftmost, rightmost, second rightmost) occurrences of each palindrome in T by traversing the enhanced suffix tree. At the same time, we compute ρ_w and $\tilde{\rho}_w$ for each palindrome in w. Next, we sequentially access distinct palindromes by traversing $\mathsf{EERTREE}(T)$ in a pre-order manner. Then, for each palindrome w, we decompose ρ_w and $\tilde{\rho}_w$ based on the rules as mentioned above. Finally, we apply the preprocessing for stabbing queries to the $O(n)$ intervals obtained. The total preprocessing time is $O(n)$.

Given a query $sub(i, s)$, we compute all intervals stabbed by position i. The palindromes corresponding to the intervals are MUPSs of Type A2. Hence, the query time is $O(1 + \beta_{add})$, where β_{add} is the number of MUPSs of Type A2.

Type A3. A MUPS of Type A3 is unique but not minimal in T. Such a unique palindrome u in T contains a MUPS $w \neq u$ of T as a shrink. Since u is a MUPS of T', w is repeating in T', i.e., w is a removed MUPS of Type R2. Contrary, consider a MUPS w of Type R2, which is repeating in T'. Then, the shortest unique extension of w in T' is an added MUPS of Type A3, if it exists. The preprocessing for Type A3 is almost the same as for Type R2. We store a bit more information for Type A3 in addition to the information in $\mathcal{A}_{R2}$.

In the preprocessing phase, we first apply the $O(n)$-time preprocessing of Lemma 6 for Problem 1. Next, we initialize the set $\mathcal{A}_{A3} = \emptyset$. This set $\mathcal{A}_{A3}$ will become an index of MUPSs of Type A3 when the preprocessing is finished. For each MUPS $w = T[b..e]$ of T, we process the followings: For the beginning position $j \neq b$ of each occurrence of Larm_w in T, we compute the lcp value ℓ_j between $T[j + |\mathsf{Larm}_w|..|T|]\$$ and $T[[c]..|T|]\$$ with allowing one mismatch where c is the center of w in T. If ℓ_j is smaller than rarm_w, then we do nothing for this occurrence since w cannot occur at j after any single-character substitution. Otherwise, let $q = j + |\mathsf{Larm}_w| - 1 + d$ be the first mismatched position in T. When the q-th character of T is substituted by the character $\mathsf{rarm}_w[d]$, $w = T[b..e]$ occurs at $j \neq b$, i.e., it is a MUPS of Type R2. Unlike for Type R2, we add the *pair* of MUPS and (1-mismatched) lcp value $(T[b..e], \ell_j)$ into $\mathcal{A}_{A3}$ with the pair of index and character $(q, \mathsf{rarm}_w[d])$ as the key. In addition, symmetrically, we update $\mathcal{A}_{A3}$ for each occurrence of Rarm_w in T. After finishing the above processes for every MUPS of T, we then sort the elements of $\mathcal{A}_{A3}$ by radix sort on the keys. If there are multiple identical elements with the same key, we unify them into a single element. Also, if there are multiple elements with the same key, we store them in a linear list. By Lemma 1, the total number of occurrences of arms of MUPSs is $O(n)$, and hence, the total preprocessing time is $O(n)$.

Given a query $sub(i, s)$, we query Problem 1 with the same pair (i, s) as the input. Then, we complete checking whether there exists a MUPS of Type A3 centered at i. For the remaining MUPSs of Type 3, we retrieve the MUPSs of Type A3 using the index $\mathcal{A}_{A3}$ as in the query algorithm for Type R2. This can be done in $O(\log \min\{\sigma, \log n\} + \gamma_{add})$ time where γ_{add} is the number of MUPSs of Type A3. Therefore, the total query time of Type A3 is $O(\delta(n, \sigma) + (\log \log n)^2 + \gamma_{add})$.

To summarize, we obtain our main theorem:

Theorem 3. *After $O(n)$-time preprocessing, we can compute the set of MUPSs after a single-character substitution in $O(\delta(n, \sigma) + (\log \log n)^2 + d) \subset O(\log n)$ time where d is the number of changes of MUPSs.*

With a little modification, we obtain the following:

Corollary 2. *If $\sigma \in O(1)$, after $O(n)$-time preprocessing, we can compute the set of MUPSs after a single-character substitution in $O(1 + d)$ time.*

Acknowledgements. We would like to thank Associate Professor Shunsuke Inenaga (Kyushu University) for the valuable discussions on simplifying our algorithms.

References

1. Abedin, P., Hooshmand, S., Ganguly, A., Thankachan, S.V.: The heaviest induced ancestors problem revisited. In: Navarro, G., Sankoff, D., Zhu, B. (eds.) Annual Symposium on Combinatorial Pattern Matching, CPM 2018, 2–4 July 2018, Qingdao, China. LIPIcs, vol. 105, pp. 20:1–20:13. Schloss Dagstuhl - Leibniz-Zentrum für Informatik (2018). https://doi.org/10.4230/LIPIcs.CPM.2018.20
2. Amir, A., Boneh, I.: Dynamic palindrome detection. CoRR abs/1906.09732 (2019). arXiv:1906.09732
3. Amir, A., Boneh, I., Charalampopoulos, P., Kondratovsky, E.: Repetition detection in a dynamic string. In: Bender, M.A., Svensson, O., Herman, G. (eds.) 27th Annual European Symposium on Algorithms, ESA 2019, 9–11 September 2019, Munich/Garching, Germany. LIPIcs, vol. 144, pp. 5:1–5:18. Schloss Dagstuhl - Leibniz-Zentrum für Informatik (2019). https://doi.org/10.4230/LIPIcs.ESA.2019.5
4. Amir, A., Charalampopoulos, P., Iliopoulos, C.S., Pissis, S.P., Radoszewski, J.: Longest common factor after one edit operation. In: Fici, G., Sciortino, M., Venturini, R. (eds.) SPIRE 2017. LNCS, vol. 10508, pp. 14–26. Springer, Cham (2017). https://doi.org/10.1007/978-3-319-67428-5_2
5. Amir, A., Charalampopoulos, P., Pissis, S.P., Radoszewski, J.: Longest common substring made fully dynamic. In: Bender, M.A., Svensson, O., Herman, G. (eds.) 27th Annual European Symposium on Algorithms, ESA 2019, 9–11 September 2019, Munich/Garching, Germany. LIPIcs, vol. 144, pp. 6:1–6:17. Schloss Dagstuhl - Leibniz-Zentrum für Informatik (2019). https://doi.org/10.4230/LIPIcs.ESA.2019.6
6. Apostolico, A., Breslauer, D., Galil, Z.: Parallel detection of all palindromes in a string. Theor. Comput. Sci. **141**(1 & 2), 163–173 (1995). https://doi.org/10.1016/0304-3975(94)00083-U
7. Bille, P., Gawrychowski, P., Gørtz, I.L., Landau, G.M., Weimann, O.: Longest common extensions in trees. Theor. Comput. Sci. **638**, 98–107 (2016). https://doi.org/10.1016/j.tcs.2015.08.009
8. Charalampopoulos, P., Gawrychowski, P., Pokorski, K.: Dynamic longest common substring in polylogarithmic time. In: Czumaj, A., Dawar, A., Merelli, E. (eds.) 47th International Colloquium on Automata, Languages, and Programming, ICALP 2020, 8–11 July 2020, Saarbrücken, Germany (Virtual Conference). LIPIcs, vol. 168, pp. 27:1–27:19. Schloss Dagstuhl - Leibniz-Zentrum für Informatik (2020). https://doi.org/10.4230/LIPIcs.ICALP.2020.27
9. Droubay, X., Justin, J., Pirillo, G.: Episturmian words and some constructions of de Luca and Rauzy. Theor. Comput. Sci. **255**(1–2), 539–553 (2001). https://doi.org/10.1016/S0304-3975(99)00320-5
10. Farach-Colton, M., Ferragina, P., Muthukrishnan, S.: On the sorting-complexity of suffix tree construction. J. ACM **47**(6), 987–1011 (2000). https://doi.org/10.1145/355541.355547
11. Funakoshi, M., Mieno, T.: Minimal unique palindromic substrings after single-character substitution. CoRR abs/2105.11693 (2021). https://arxiv.org/abs/2105.11693
12. Funakoshi, M., Nakashima, Y., Inenaga, S., Bannai, H., Takeda, M.: Computing longest palindromic substring after single-character or block-wise edits. Theor. Comput. Sci. **859**, 116–133 (2021). https://doi.org/10.1016/j.tcs.2021.01.014

13. Gawrychowski, P., Tomohiro, I., Inenaga, S., Köppl, D., Manea, F.: Tighter bounds and optimal algorithms for all maximal α-gapped repeats and palindromes. Theory of Computing Systems **62**(1), 162–191 (2017). https://doi.org/10.1007/s00224-017-9794-5

14. Inoue, H., Nakashima, Y., Mieno, T., Inenaga, S., Bannai, H., Takeda, M.: Algorithms and combinatorial properties on shortest unique palindromic substrings. J. Discrete Algorithms **52–53**, 122–132 (2018). https://doi.org/10.1016/j.jda.2018.11.009

15. Kociumaka, T., Kubica, M., Radoszewski, J., Rytter, W., Walen, T.: A linear-time algorithm for seeds computation. ACM Trans. Algorithms 16(2), 27:1–27:23 (2020). https://doi.org/10.1145/3386369

16. Manacher, G.K.: A new linear-time "on-line" algorithm for finding the smallest initial palindrome of a string. J. ACM **22**(3), 346–351 (1975). https://doi.org/10.1145/321892.321896

17. Matsubara, W., Inenaga, S., Ishino, A., Shinohara, A., Nakamura, T., Hashimoto, K.: Efficient algorithms to compute compressed longest common substrings and compressed palindromes. Theor. Comput. Sci. **410**(8–10), 900–913 (2009). https://doi.org/10.1016/j.tcs.2008.12.016

18. Mieno, T., Watanabe, K., Nakashima, Y., Inenaga, S., Bannai, H., Takeda, M.: Palindromic trees for a sliding window and its applications. Inf. Process. Lett., 106174 (2021). https://doi.org/10.1016/j.ipl.2021.106174. ISSN 0020-0190

19. Rubinchik, M., Shur, A.M.: EERTREE: an efficient data structure for processing palindromes in strings. Eur. J. Comb. **68**, 249–265 (2018). https://doi.org/10.1016/j.ejc.2017.07.021

20. Rubinchik, M., Shur, A.M.: Palindromic k-factorization in pure linear time. In: Esparza, J., Král', D. (eds.) 45th International Symposium on Mathematical Foundations of Computer Science, MFCS 2020, 24–28 August 2020, Prague, Czech Republic. LIPIcs, vol. 170, pp. 81:1–81:14. Schloss Dagstuhl - Leibniz-Zentrum für Informatik (2020). https://doi.org/10.4230/LIPIcs.MFCS.2020.81

21. Schmidt, J.M.: Interval stabbing problems in small integer ranges. In: Dong, Y., Du, D.-Z., Ibarra, O. (eds.) ISAAC 2009. LNCS, vol. 5878, pp. 163–172. Springer, Heidelberg (2009). https://doi.org/10.1007/978-3-642-10631-6_18

22. Urabe, Y., Nakashima, Y., Inenaga, S., Bannai, H., Takeda, M.: Longest Lyndon substring after edit. In: Navarro, G., Sankoff, D., Zhu, B. (eds.) Annual Symposium on Combinatorial Pattern Matching, CPM 2018, 2–4 July 2018, Qingdao, China. LIPIcs, vol. 105, pp. 19:1–19:10. Schloss Dagstuhl - Leibniz-Zentrum für Informatik (2018). https://doi.org/10.4230/LIPIcs.CPM.2018.19

23. Watanabe, K., Nakashima, Y., Inenaga, S., Bannai, H., Takeda, M.: Fast algorithms for the shortest unique palindromic substring problem on run-length encoded strings. Theory Comput. Syst. **64**(7), 1273–1291 (2020). https://doi.org/10.1007/s00224-020-09980-x

24. Weiner, P.: Linear pattern matching algorithms. In: 14th Annual Symposium on Switching and Automata Theory, Iowa City, Iowa, USA, 15–17 October 1973, pp. 1–11. IEEE Computer Society (1973). https://doi.org/10.1109/SWAT.1973.13

Permutation-Constrained Common String Partitions with Applications

Manuel Lafond[1] and Binhai Zhu[2(✉)]

[1] Department of Computer Science, Université de Sherbrooke,
Sherbrooke, QC J1K 2R1, Canada
manuel.lafond@usherbrooke.ca

[2] Gianforte School of Computing, Montana State University,
Bozeman, MT 59717, USA
bhz@montana.edu

Abstract. We introduce and study a new combinatorial problem based on the famous Minimum Common String Partition (MCSP) problem, which we call Permutation-constrained Common String Partition (PCSP for short). In PCSP, we are given two sequences/genomes s and t with the same length and a permutation π on $[k]$, the question is to decide whether it is possible to decompose s and t into k blocks that conform with the permutation π. The main result of this paper is that if s and t are both d-occurrence (i.e., each letter/gene appears at most d times in s and t), then PCSP is FPT. We also study a variant where the input specifies whether each matched pair of block needs to be preserved as is, or reversed. With this result on PCSP, we show that a series of genome rearrangement problems are FPT as long as the input genomes are d-occurrence.

Keywords: Genome rearrangement · Permutation-constrained common string partition · Minimum common string partition · FPT

1 Introduction

Computing the distance between genomes is a fundamental problem in biology; in fact, the reversal and the corresponding breakpoint concept on sections of some chromosome (when a whole genome was not obtainable) were studied as early as in 1926 by Sturtevant and Dobzhansky [20,21]. The signed reversals were confirmed in radish genome in 1988 [18]. Since then, a lot of rearrangement operations were studied, for instance, reversals, transpositions and block interchanges, etc. (Here we focus more on operations on unichromosomal genomes— singleton genomes, each being a sequence, we skip the details for genomic operations applied on multichromosomal genomes, like translocation and the more general double-cut-and-join operations).

Let us first describe some popular string operations. A *transposition* takes two adjacent substrings and swaps them. A *block-interchange* takes two non-overlapping substrings and swaps them. The recent *p-cut* partitions a string

T. Lecroq and H. Touzet (Eds.): SPIRE 2021, LNCS 12944, pp. 47–60, 2021.
https://doi.org/10.1007/978-3-030-86692-1_5

into p substrings and permutes them in any way [2]. A *flip* switches the sign of each symbol of a chosen substring. A *reversal* first applies a flip to a substring, and then reverses the order of the characters. An *unsigned reversal* only reverses the order of the characters of a chosen substring. If $\mathcal{P}$ is a set of string operations, $d_\mathcal{P}(s,t)$ is the minimum number of operations from $\mathcal{P}$ to apply to s and transform it into t. We also call $d_\mathcal{P}(s,t)$ the corresponding *distance* under operations in $\mathcal{P}$.

Most of these problems are NP-hard under different circumstances. The transposition distance problem [4] and the unsigned reversal distance problem [6] are NP-hard even if s and t are permutations (i.e., 1-occurrence strings). The block-interchange distance problem is polynomially solvable for permutations [8]; however, when s and t are sequences over a fixed-size alphabet the problem becomes NP-hard [9]. Other types of distances that remain to be incorporated into our framework have recently been studied, including the NP-hard tandem duplication distance [16] and the copy-number distance [15,19].

In many practical applications, the value $\ell = d_\mathcal{P}(s,t)$ is small, and it makes sense to consider the FPT (fixed-parameter tractable) complexity of these problems—using parameter ℓ. When s and t are permutations, both the transposition distance problem and the unsigned reversal distance problem are FPT since each such operation reduces at most three breakpoints [17]. When s and t are d-occurrence sequences (i.e., each letter appears at most d times in s and t), whether these problems are FPT is unknown, to the best of our knowledge. The d-occurrence variant is of practical interest in bioinformatics, since in genome comparison, repeated occurrences represent duplicated genes. Duplication is a rare event and the number of duplicates is usually less than a dozen. For instance, plants have undergone up to three rounds of whole genome duplications, resulting in a number of duplicates bounded by eight (see e.g. [22]).

Using a general framework first stated by Mahajan et al. [17], if ℓ operations can be applied on s to obtain t, then we can partition s and t into $O(\ell)$ substrings, hereafter called *blocks*, such that the blocks of t are a permutation of the blocks of s. It turns out that the problem of decomposing s and t (possibly with letter duplications) into a minimum set of common blocks was studied in another context called ortholog assignment [7]. The problem was formally coined as Minimum Common String Partition (MCSP), which received quite some algorithmic attention due to its NP-hardness status [10,13]. The best known approximation (to minimize the number of blocks) has a factor of $O(\log n)$ [12]. For these reasons, the fixed-parameter tractability of MCSP was considered in 2008 [11]. For d-occurrence strings, the problem was first shown to be FPT in [3,14]. In 2014, it was shown that MCSP parameterized by the solution size ℓ is FPT [5].

Although MCSP has been used to approximate various string distances, the applications of MCSP to compute *exact* distances is not always clear. When trying to solve, for instance, the transposition distance problem using MCSP, different string partitions with the same number of blocks lead to different transposition distances. It would seem necessary to enumerate all MCSP solutions to compute our string distances exactly. In fact, we show that restricting an MCSP instance to find blocks ordered according to a permutation π is sufficient.

Our Results. In this paper, we thus first establish a precise relationship between MCSP and string distance problems in terms of fixed-parameter tractability. We propose a string operation that generalizes all the ones mentioned above, and then show that if MCSP is FPT when the matched substrings in the partitions must satisfy a given block permutation, then computing the string distance under our general operation is FPT. Our new variant of MCSP is called the Permutation-constrained Common String Partition problem (PCSP for short) and we show that when s and t are both d-occurrence strings, PCSP is FPT in $\ell + d$, where ℓ is the number of desired blocks in the partition. This result holds if some matched pairs of blocks should be the reverse of one another or not, making the result applicable to several combinations of allowed operations.

The recent results by Bulteau, et al. imply that PCSP, parameterized by the number of blocks ℓ only, is W[1]-hard [1], thereby motivating the need for two parameters ℓ and d. Our algorithmic framework is based on the work of Bulteau et al. [3], who showed that MCSP is FPT in $d + \ell$. The algorithm of [3] uses a *match graph* that depicts which characters could be corresponding in a block partition. A branching strategy finds exactly one match per block in a solution. Our approach relies on a similar strategy, but requires non-trivial extensions to the framework. First, we want to find a block partition that uses a specific permutation π, and so our branching algorithm must maintain a set of matched substrings that always agrees with π. Moreover, we cannot assume that we want to find a solution that minimizes the number of blocks, since none may agree with π. This prevents us from using structural results on optimal solutions (see e.g. Lemma 2 in [3]), which in turn requires a deeper case analysis. Finally, in our general framework, some matched substrings in a solution might be the reverse of one another. We develop novel machinery in the match graph to ensure that characters are matched consistently, i.e. that our branching matches all characters in a common substring in a forward or reverse manner, and not both.

As a consequence of our results, we show that several string distances are FPT since they reduce to PCSP. The main results that can be derived from this PCSP reduction is that for strings s and t in which each symbol occurs at most d times, and with a pair of strings at distance at most k,

- the transposition distance can be computed in time $k^{O(k)}d^{6k+2}n$.
- the block interchange distance can be computed in time $k^{O(k)}d^{8k+2}n$.
- the flip, reversal and unsigned reversal distances can be computed in time $k^{O(k)}d^{4k+2}n$
- the p-cut distance can be computed in time $(pk)^{O(pk)}d^{2k(p-1)+2}n$.
- for any subset $\mathcal{P}$ of operations among transpositions, block interchanges, reversals, unsigned reversals or flips, computing $d_{\mathcal{P}}(s,t)$ can be done in time $k^{O(k)}d^{8k+2}n$,

where n is the length of the input strings. Due to space constraints, all proofs can be found in the full version of this paper.

2   Preliminary Notions

For $\ell \in \mathbb{N}$, we denote $[\ell] = \{1, \ldots, \ell\}$. If $X = (x_1, \ldots, x_l)$ is an ordered sequence of arbitrary elements, we denote $set(X) = \{x_i : i \in [l]\}$. The empty string is denoted λ. For a string s, we write $s[i]$ for its i-th character. We denote by $s[a, b]$ the substring of s containing positions a to b, inclusively. If $b < a$, then $s[a, b] = \lambda$. The concatenation of strings s_1 and s_2 is denoted $s_1 s_2$. For clarity, a string s may be denoted $s = (a_1, a_2, \ldots, a_n)$, where each a_i is an individual symbol. We denote a partition of a string s by $s_1|s_2|\ldots|s_\ell$, where s is equal to the concatenation $s_1 s_2 \ldots s_\ell$. We allow any of the s_i's to be the empty string, which will have implications later on. Each s_i substring is called a *block* of the partition. Moreover, for each non-empty s_i, the position of the last character of s_i in s is called a *breakpoint*. Observe that position n is always a breakpoint.

A *marker* is a specific character of a string s. Each character $s[i]$ is a distinct marker, even if other positions contain the same symbol. For concreteness, one may think of a marker as a pair (s, i) where s is the string containing the marker, and i is its position—although we shall not use this notation. Two markers are *adjacent* if they are at consecutive positions in s, in either order. If a marker x of s is positioned before another marker y of s, we write $x \prec y$. For a marker x and an integer i, we write $x + i$ to denote the marker located i positions to the right of x, and $x - i$ for the marker i positions to the left of x. For markers $x \prec y$, we write $s[x, y]$ for the substring that starts at x and ends at y, inclusively.

A *string function* is a function $h : \Sigma^* \to \Sigma^*$ that maps strings to strings. We require that $h(\lambda) = \lambda$. We say that h is *length-preserving* if $|h(s)| = |s|$ for any string s. Unless stated otherwise, we shall assume that $h(s)$ can be computed in time $O(|s|)$ if h is length-preserving. The identity function is denoted id and satisfies $id(s) = s$ for all s. Other functions of interest include those that apply on an alphabet in which each symbol x has a negation $-x$, and $-(-x) = x$:

- the *flip* function negates every symbol of s; $flip((a_1, \ldots, a_n)) = (-a_1, \ldots, -a_n)$;
- the *rev* function flips and reverses each symbol; $rev((a_1, \ldots, a_n)) = (-a_n, \ldots, -a_1)$;
- the *urev* function reverses the symbols; $urev((a_1, \ldots, a_n)) = (a_n, \ldots, a_1)$.

In the general Permutation-Constrained Common String Partition (PCSP), we receive strings s and t, an integer ℓ, a permutation π of $[\ell]$ and string functions $f_1, \ldots, f_\ell$. We must partition s and t in a way that the i-th block of t is equal to the $\pi(i)$-th block of s after applying the f_i function to that block.

The Permutation-Constrained Common String Partition (PCSP) problem

Input. Strings s and t of equal length, integer ℓ, permutation π of $[\ell]$, and a sequence of string functions $F = (f_1, f_2, \ldots, f_\ell)$.

Question. Does there exist partitions $s_1|\ldots|s_\ell$ and $t_1|\ldots|t_\ell$ of s and t, respectively, such that $t_i = f_i(s_{\pi(i)})$ for each $i \in [\ell]$?

The ℓ parameter therefore refers to the desired number of substrings in our partitions. Recall that in the introduction, we used parameter k as the distance between two genomes. The exact relationship between ℓ and k depends on the specific distance of interest, as we shall detail later in the paper. Notice that in the classical MCSP problem, π and the f_i functions are not present (i.e. each f_i is the identity). The incorporation of string functions allow blocks to be affected by operations such as flips or reversals. As we mentioned, PCSP is W[1]-hard in parameter ℓ even if each f_i is the identity [1].

Generic String Operations

One of our goals is to establish a relationship between PCSP and string distances based on certain types of operations. We focus on operations that, given a string s, can (1) partition the string; (2) apply a string function on each block; and (3) apply a permutation on the blocks from a given list. Note that a similar string operation was proposed in [2], but without string functions and permutation list.

To define this more precisely, let s be a string, let p be an integer, let $H = (h_1, \ldots, h_p)$ be a sequence of p string functions, and let Π be a list of permutations of $[p]$. We call the triple (p, H, Π) an *operation*. Given a string s, *applying* (p, H, Π) on s modifies the string by applying the following steps:

- partition s into a set of p substrings $s_1|s_2|\ldots|s_p$ (some of which may be λ);
- for each $i \in [p]$, let $s_i' = h_i(s_i)$;
- choose $\pi \in \Pi$ and return the string $s' = s'_{\pi(1)} s'_{\pi(2)} \cdots s'_{\pi(p)}$.

Let $\mathcal{P}$ be a set of operations, which can be thought of as a set of moves that are permitted. If, for each $(p', H, \Pi) \in \mathcal{P}$, we have $p' \leq p$, then we call $\mathcal{P}$ a set of p-operations. Let us write $s \xrightarrow{\mathcal{P}} t$ if there exists $(p, H, \Pi) \in \mathcal{P}$ that can be applied on s to obtain t. Also write $s \xrightarrow{\mathcal{P}, k} t$ if there exist strings $s_2, \ldots, s_{k-1}$ such that $s \xrightarrow{\mathcal{P}} s_2 \xrightarrow{\mathcal{P}} \ldots \xrightarrow{\mathcal{P}} s_{k-1} \xrightarrow{\mathcal{P}} t$. Given strings s and t, we denote by $d_{\mathcal{P}}(s, t)$ the minimum value of k that satisfies $s \xrightarrow{\mathcal{P}, k} t$.

Relationship with Known String Operations

The (p, H, Π)-operations generalize several well-known string operations. For instance, a *transposition* takes two consecutive substrings and exchanges them. This takes the form $s_1|s_2|s_3|s_4 \to s_1|s_3|s_2|s_4$, which can be modeled by putting $p = 4, H = \{id\}^4$, and Π only allowing to exchange blocks 2 and 3. Similarly, a *block interchange* swaps two arbitrary substrings. This takes the form $s_1|s_2|s_3|s_4|s_5 \to s_1|s_4|s_3|s_2|s_5$, which takes $p = 5, H = \{id\}^5$ and Π allowing to swap blocks 2 and 4. The p-cut defined in [2], which allows cutting into p pieces and permuting them in any way, can also be modeled similarly.

For examples that do not use the identity function, the *reversal* operation, takes the form $s_1|s_2|s_3 \to s_1|rev(s_2)|s_3$, hence $p = 3$, $H = (id, rev, id)$ and Π only allowing the identity permutation. A *tandem duplication* replaces a substring $s[a, b]$ by $s[a, b]s[a, b]$. We can apply $s_1|s_2|s_3 \to s_1|s_2 s_2|s_3$, with $p = 3$ and $h_2(s_2) = s_2 s_2$ (and $h_1 = h_3 = id$).

3   Reduction to PCSP for Simple String Functions

We will show that if the H functions allowed in the operations are "simple", computing $d_{\mathcal{P}}(s,t)$ reduces to PCSP. We will need the following.

Lemma 1. *Let $\mathcal{P}$ be a set of p-operations that only use length-preserving string functions. Then $d_{\mathcal{P}}(s,t)$ can be computed in time $O(|\mathcal{P}|^k n^{pk} p^{pk})$ where $n = |s|$.*

This can be proved easily using an algorithm that brute-forces every possible sequence of k operations. We can now define our notion of simplicity.

Definition 1. *A string function f is* simple *if f is length-preserving and either:*

- $f(s) = f(s_1)f(s_2)\dots f(s_\ell)$ *for any string s and partition $s_1|\dots|s_\ell$ of s; or*
- $f(s) = f(s_\ell)f(s_{\ell-1})\dots f(s_1)$ *for any string s and partition $s_1|\dots|s_\ell$ of s;*

It is obvious that $id, flip, rev$ and $urev$ are simple, for instance. Also notice that if $s = \langle x_1, x_2, \dots, x_n \rangle$, then $f(s) \in \{\langle f(x_1),\dots,f(x_n)\rangle, \langle f(x_n),\dots,f(x_1)\rangle\}$, i.e. the function is applied to each character individually, and we decide to reverse or not. For that reason, a simple function can be represented as a lookup table that maps each symbol of Σ to another symbol, plus one bit that distinguishes whether to reverse or not. In what follows, we shall assume that each simple string function is given as a lookup table of size $O(|\Sigma|) = O(n)$.

Definition 2. *Let $\mathbb{H}$ be a set of simple string functions. We say that an operation (p, H, Π) is* $\mathbb{H}$-restricted *if $set(H) \subseteq \mathbb{H}$.*

Let $\mathbb{H} = \{h_1, \dots, h_r\}$ be a set of simple string functions. By $\mathbb{H}\langle k \rangle$, we mean the set of functions obtained by composing any k functions or less from $\mathbb{H}$:

$$\mathbb{H}\langle k \rangle = \bigcup_{k'=1}^{k} \{h_{a_{k'}} \circ h_{a_{k'-1}} \circ \dots \circ h_{a_1} : a_1, \dots, a_{k'} \in [r]\}$$

By convention, put $\mathbb{H}\langle 0 \rangle = \{id\}$. By $\mathbb{H}\langle k \rangle^\ell$, we mean the set of vectors of ℓ functions from $\mathbb{H}\langle k \rangle$. Before proceeding, we must show that $\mathbb{H}\langle k \rangle$ function are simple, and that their lookup tables can be computed easily.

Lemma 2. *Let $\mathbb{H}$ be a set of simple functions on alphabet Σ. Assume that each function $f \in \mathbb{H}$ is represented as a lookup table of size $|\Sigma|$, plus one bit for whether f reverses its input or not. Then $\mathbb{H}\langle k \rangle$ can be computed in time $O(k|\Sigma||\mathbb{H}\langle k \rangle|)$. Moreover, any $h \in \mathbb{H}\langle k \rangle$ is simple and, for any string s, $h(s)$ can be computed in time $O(|s|)$.*

We can now show that distances that use operations based on simple string functions reduce to PCSP. Beforehand, we need tools to represent permuted matches between a common partition in a succinct manner. We define a special alphabet for this purpose. Let F be the set of all length-preserving string functions (which includes id). We will treat the set $\Sigma_F = \mathbb{N} \times F$ as a special alphabet, each symbol consisting of a pair with an integer and a function. For any $f \in F$ and any $(i, h) \in \Sigma_F$, we put $f((i, h)) = (i, f \circ h)$. For simplicity, we will write i instead of (i, id), with the understanding that no integer i is part of the alphabet of our strings of interest s and t (this is without loss of generality). We thus write $f(i)$ instead of $f((i, id)) = (i, f)$. It is therefore understood that for any $i, j \in \mathbb{N}$ and $f, f' \in F$, $f(i)$ and $f'(j)$ are both symbols of Σ_F, and they are the same symbol if and only if $i = j$ and $f = f'$.

Lemma 3. *Let $\mathbb{H}$ be a set of simple string functions and let $\mathcal{P}$ be a set of $\mathbb{H}$-restricted p-operations.*

Then $d_{\mathcal{P}}(s, t) \leq k$ if and only if there exists a partition $s_1 | \ldots | s_\ell$ of s, a partition $t_1 | \ldots | t_\ell$ of t, a permutation π of $[\ell]$, and $f_1, \ldots, f_\ell \in \mathbb{H}\langle k \rangle$ such that all the following conditions hold:

1. *$\ell \leq 1 + k(p-1)$;*
2. *$t_i = f_i(s_{\pi(i)})$ for each $i \in [\ell]$;*
3. *$d_{\mathcal{P}}(\langle 1, 2, \ldots, \ell \rangle, \langle f_1(\pi(1)), f_2(\pi(2)), \ldots, f_\ell(\pi(\ell)) \rangle) \leq k$.*

There is a clear intuition behind this lemma. If $d_{\mathcal{P}}(s, t) \leq k$, then each of the k operations creates at most p breakpoints, and there are at most $\ell \leq 1 + k(p-1)$ of them. These breakpoints define a partition of s and t into corresponding blocks, and if this partition into ℓ substrings is known, we could replace each i-th block by the symbol i in s, and by $f_i(\pi(i))$ in t.

Lemma 3 shows that the exact content of the matched blocks does not truly matter, as we may reduce to computing a distance between strings of length ℓ.

Theorem 1. *Let $\mathbb{H}$ be a set of simple string functions, and let $\mathcal{P}$ be a set of $\mathbb{H}$-restricted p-operations.*

Assume that any PCSP instance (s, t, ℓ, π, F) satisfying $F \in \mathbb{H}\langle k \rangle^\ell$ can be solved in time $g(\ell, n)$. Then deciding whether $d_{\mathcal{P}}(s, t) \leq k$ can be done in time $O((pk)^{3pk+1} \cdot |\mathbb{H}\langle k \rangle|^{pk} \cdot |\mathcal{P}|^k \cdot g(1 + k(p-1), n))$.

The correctness of Theorem 1 and the complexity can be derived from Algorithm 1, which uses Lemma 3 directly to compute $d_{\mathcal{P}}$.

```
1  function computeDistance(s, t, P, F, k)
2      Compute ℍ⟨k⟩
3      foreach ℓ ∈ [1 + k(p − 1)] do
4          foreach permutation π of [ℓ] do
5              foreach F = (f₁, . . . , fℓ) ∈ ℍ⟨k⟩ℓ do
6                  Compute PCSP on input (s, t, ℓ, π, F)
7                  if the answer is yes then
8                      Compute dₚ(⟨1, . . . , ℓ⟩, ⟨f₁(π(1)), . . . , fℓ(π(ℓ))⟩)
9                      if the distance is at most k then
10                         return yes
11     return no
```

Algorithm 1: Main $d_{\mathcal{P}}$ algorithm

4   FPT Algorithms for d-occurrence PCSP with Reversals

In this section, we show that if F only contains $id, flip, rev$ and $urev$, and if each symbol occurs at most d times in the input strings, then PCSP is FPT in parameter $d + \ell$. Since empty blocks are allowed, we will show that in PCSP, we do not have the obligation of satisfying the given permutation exactly—we can have missing blocks, provided that the order of the blocks found agrees with π. For the rest of this section, we assume that we are given strings s and t, an integer ℓ, a permutation π, and functions $(f_1, \ldots, f_\ell)$.

For a string w, a *numbered partition* of w is a sequence of ordered pairs $(w_{a_1}, a_1), \ldots, (w_{a_r}, a_r)$ where $w_{a_1} | \ldots | w_{a_r}$ is a partition of w and $a_1, \ldots, a_r \in [\ell]$. We denote a numbered partition by $P_w = (w_{a_1}, a_1) | \ldots | (w_{a_r}, a_r)$. If $a_1 < a_2 < \ldots < a_r \leq \ell$, then P_w is called an *ordered partition*. Let $P_s = (s_{a_1}, a_1) | \ldots | (s_{a_r}, a_r)$ and $P_t = (t_{b_1}, b_1) | \ldots | (t_{b_r}, b_r)$ be ordered partitions of s and t, respectively. with $r \leq \ell$. We say that (P_s, P_t) *agrees* with π and F if, for each $b_i \in \{b_1, \ldots, b_r\}$, we have $\pi(b_i) \in \{a_1, \ldots, a_r\}$ and $t_{b_i} = f_{b_i}(s_{\pi(b_i)})$.

Roughly speaking, an ordered partition allows us to "tag" each block with its position in a solution to PCSP. There may be holes between some a_i and a_{i+1}, which correspond to empty blocks in a solution.

Lemma 4. *The instance (s, t, π, ℓ, F) is a Yes-instance of PCSP if and only if there exist ordered partitions $(s_{a_1}, a_1) | \ldots | (s_{a_r}, a_r)$ and $(t_{b_1}, b_1) | \ldots | (t_{b_r}, b_r)$ that agree with π and F.*

4.1   Bounded Occurrences and Reversal Functions

From now on, we assume that each symbol occurs at most d times in s, and at most d times in t. We assume that s and t are on an alphabet Σ where each symbol x has a unique reverse $-x$, where $-(-x) = x$. We also assume that each function of F is in $\{id, flip, rev, urev\}$. For a marker x of s or t, denote by $S(x) \in \{s, t\}$ the string that contains x. Our goal is to find an ordered partition that agrees with π and F.

Our algorithm is based on the approach of Bulteau, et al. [3] for the classical common string partition problem (noting that this approach also allows deletions of "extra" symbols). This algorithm guesses a pair of matching markers in each block of a solution. We apply the same idea here, but develop new tools to account for the permutation π and the F functions.

A *fixed-match* is a tuple (x, y, a_i, b_j) where x is a marker of s, y is a marker of t, $a_i, b_j \in [\ell]$, $a_i = \pi(b_j)$, and $y = f_{b_i}(x)$ (this is slightly abusing notation, since x and y are markers, not strings—it is understood here that x and y are treated as their symbols in this equality). The interpretation of (x, y, a_i, b_j) is that in the desired ordered partition, x belongs to block (s_{a_i}, a_i) of s, y belongs to block (t_{b_j}, b_j) of t, and $t_{b_j} = f_{b_j}(s_{\pi(b_j)})$. Of course, the blocks are unknown, but x and y serve as a witness of their existence. Also note that a_i is redundant in the definition of a fixed-match, since it is given by $\pi(b_j)$. However, it will help making some notions clearer later on.

A *fixed-match set* is a set M of fixed-matches such that, for any distinct $(x, y, a_i, b_j), (x', y', a_p, b_q) \in M$, we have $x \neq x', y \neq y'$, and $b_j \neq b_q$ (which implies $a_i \neq a_p$). The idea is that a fixed-match set provides at most one match per block, and our goal is to eventually find *exactly* one fixed-match per block of a solution, if any. With the following, we ensure that fixed-match sets can satisfy the requirements of an ordered partition.

Definition 3. *A fixed-match set M is order-consistent if, for any (x, y, a_i, b_j), $(x', y', a_p, b_q) \in M$, $x \prec x'$ implies $a_i < a_p$ and $y \prec y'$ implies $b_j < b_q$.*

Assuming that a solution is known, we know which substring t_j corresponds to which $f_j(s_{\pi(j)})$ substring. However, we also need a way to determine which markers are in correspondence, which depends on whether f_j reverses the string or not. Assume that $t_j = f_j(s_{\pi(j)})$ for some substrings t_j of t and $s_{\pi(j)}$ of s. For each $i \in [|s_{\pi(j)}|]$, if $f_{b_j} \in \{id, flip\}$, match the i-th marker of $s_{\pi(j)}$ with the i-th marker of t_j, and otherwise match the i-th marker of $s_{\pi(j)}$ with the $(|t_j| - i + 1)$-th marker of t_j. Two markers x of $s_{\pi(j)}$ and y of t_j matched as described above are call *matched with respect to $s_{\pi(j)}$ and t_j*. In essence, this matching simply indicates the new location of the x marker after applying f_j. We next relate fixed-match sets with numbered partitions (which are not necessarily ordered).

Definition 4. *Let M be a fixed-match set with r elements. We say that M is complete if there exist numbered partitions $P_s = (s_{a_1}, a_1) | \ldots | (s_{a_r}, a_r)$ of s and $P_t = (t_{b_1}, b_1) | \ldots | (t_{b_r}, b_r)$ of t such that all the following conditions hold:*

1. *for each $b_j \in \{b_1, \ldots, b_r\}$, $t_{b_j} = f_{b_j}(s_{\pi(b_j)})$;*
2. *for each $(x, y, a_i, b_j) \in M$, x is in the s_{a_i} substring, y is in the t_{b_j} substring, and x, y are matched with respect to s_{a_i} and t_{b_j}.*

We call (P_s, P_t) a completion of M. Furthermore, we say that M is completable if there exists a fixed-match set $M' \supseteq M$ such that M' is complete.

Note that the definition of complete does not require $a_1 < \ldots < a_r$ nor $b_1 < \ldots < b_r$. In fact, not all complete fixed-match sets correspond to a solution, as order-consistency is required.

Lemma 5. *There exist ordered partitions of s and t that agree with π and F if and only if there exists a complete and order-consistent fixed-match set M.*

4.2 Constructing a Complete and Order-Consistent Fixed-Match Set

In the rest of the section, we aim to construct a complete and order-consistent fixed-match set. We adopt a branching strategy that adds one fixed-match at a time. If, at any point in the search tree, we reach a fixed-match set M that is not order-consistent, it cannot lead to the desired numbered partitions and we can stop trying to complete M (this is because we can only add elements to a fixed-match set, and adding elements cannot repair the lack of order-consistency).

Termination Rule. If M is not order-consistent, then M is not part of a solution.

This is where we extend the framework of [3]. Let M be a fixed-match set, and let u be a marker of s or t that is not in any fixed-match of M. Let $l_{s,t,M}(u)$ (resp. $r_{s,t,M}(u)$) be the closest marker to the left (resp. right) of u that belongs to some fixed-match of M. We may write $l_M(u)$ and $r_M(u)$ if s and t are clear from the context. Note that $l_M(u)$ or $r_M(u)$ may not exist, in which case they are left undefined. Let $(x, y, a_i, b_j) \in M$. We are interested in pairs of markers that could belong to the same block as x and y. For this purpose, let $\{u, v\}$ be a pair of markers that are not in the same string. Assume that u is in s and v is in t. There are four types of matches that are represented in Fig. 1.

- (LL match) $\{u, v\}$ is a *left-left match* of (x, y, a_i, b_j) if $f_{b_j} \in \{id, flip\}$, $r_M(u) = x, r_M(v) = y$ and $t[v, y] = f_{b_j}(s[u, x])$;
- (RR match) $\{u, v\}$ is a *right-right match* of (x, y, a_i, b_j) if $f_{b_j} \in \{id, flip\}$, $l_M(u) = x, l_M(v) = y$ and $t[y, v] = f_{b_j}(s[x, u])$;
- (LR match) $\{u, v\}$ is a *left-right match* of (x, y, a_i, b_j) if $f_{b_j} \in \{rev, urev\}$, $r_M(u) = x, l_M(v) = y$ and $t[y, v] = f_{b_j}(s[u, x])$;
- (RL match) $\{u, v\}$ is a *right-left match* of (x, y, a_i, b_j) if $f_{b_j} \in \{rev, urev\}$, $l_M(u) = x, r_M(v) = y$ and $t[v, y] = f_{b_j}(s[x, u])$.

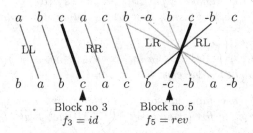

Fig. 1. A simple example depicting the four types of matches. The black edges represent fixed-matches, red edges are LL matches, green are RR, pink are LR and blue are RL. Hypothetically, the leftmost black edge belongs to some fixed-match (x, y, a_i, b_j) satisfying $b_j = 3$ (its block number), and $f_3 = id$. Similarly, the rightmost black edge belongs to a fixed-match (x', y', a_p, b_q) with $b_q = 5$ and $f_5 = rev$.

Note that $\{u, v\}$ can be multiple types of matches simultaneously. Also note that we defined $\{u, v\}$ as an *unordered* pair, and assumed that u was in s and v was in t. This means that if $\{u, v\}$ is a left-right match, for instance, then $\{v, u\}$ is also a left-right match. This is an important notational detail: when a pair $\{x, y\}$ of markers from s and t is given, x could be in s and y in t, or vice-versa. In most cases, this will not matter, but since this may impact the LR and RL types, we will need to specify the strings of origin of x and y when required.

We will view the markers and matches as a bipartite edge-colored multigraph. Given a fixed-match set M, the graph $G_{s,t}(M)$ contains one vertex for each marker in s or t. We will not make the distinction between markers and vertices from now on, and will omit the s, t subscript when they are clear from the context. We add one edge for each match of any type, and label the edge by the type. That is, for each $(x, y, a_i, b_j) \in M$, add an edge $\{u, v\}$ labeled LL (respectively RR, LR, RL) if $\{u, v\}$ is a left-left (resp. right-right, left-right, right-left) match of (x, y, a_i, b_j). Furthermore, for each $(x, y, a_i, b_j) \in M$, add the edge $\{x, y\}$ and label it $fixed$.

Note that it is possible that an edge $\{u, v\}$ is present more than once, but each has a distinct label. The basic properties of $G(M)$ are listed in the following.

Lemma 6. *Let u be a vertex of $G(M)$. Then u is incident to at most two edges, and u is not incident to two edges with the same label. Moreover, if u is in a fixed-match in M, then u is incident to exactly one edge, which is labeled $fixed$.*

It follows from the above that $G(M)$ has maximum degree 2 and consists of a collection of disjoint cycles and paths. As in [3], our goal is to identify a subset of vertices that prevents us from having a complete fixed-match set.

Definition 5. *Let M be a fixed-match set, let $M' \supseteq M$ be a complete fixed-match set and let (P_s, P_t) be a completion of M'. Let u be a marker of s or t. Then u is unseen by M in (P_s, P_t) if, for any $(x, y, a_i, b_j) \in M$, u is not in the same substring as x in P_s and u is not in the same substring as y in P_t.*

Lemma 7. *Let M be a fixed-match set and suppose that, for any complete $M' \supseteq M$ and any completion (P_s, P_t) of M', there exists a marker unseen by M in (P_s, P_t). Then M is not complete.*

An *odd M-path* is a connected component of $G(M)$ with an odd number of vertices. Note that because $G(M)$ is bipartite and because each vertex has degree at most 2, such a connected component must be a path. Also note that a vertex of degree 0 is an odd path. By Lemma 6, such a path cannot contain $fixed$ edges, and does not contain two consecutive edges with the same label.

Lemma 8. *Suppose that M is completable, and suppose that there is an odd M-path $(u_1, \dots, u_h)$ in $G(M)$. Then for any complete $M' \supseteq M$ and any completion (P_s, P_t) of M', at least one of $u_1, \dots, u_h$ is unseen by M in (P_s, P_t).*

The idea of the proof is that if we assume that each u_i is not unseen by some completion (P_s, P_t), u_1 must be matched with u_2, then u_3 with u_4, and so on until u_h cannot be matched with any marker.

Note that if, in an odd path $(u_1, \dots, u_h)$, the u_1 marker corresponds to symbol $w \in \Sigma$, then each of $u_1, \dots, u_h$ is a marker that corresponds to symbol w or $-w$. Since each has at most d occurrences in s and in t, the maximum number of vertices in an odd path is $4d$. Therefore, the number of possible vertices on such a path to add to a fixed-match of M is bounded by a function of d, as well as the number of "partners" of these vertices in a fixed-match. In fact, this is the only reason why our branching algorithm does not work for any given sequence of simple string functions F, since if F has arbitrary simple functions, we cannot guarantee that the number of vertices on an odd path is a function of d only. Moreover in general, the number of possible partners may depend on $|F|$.

Branching Rule. If there is an odd M-path $(u_1, \ldots, u_h)$, branch into all ways of adding a fixed-match that contains one of $u_1, \ldots, u_h$ to M.

Our next step is to show that once all odd paths have been handled, we have reached a complete fixed-match set. This can be shown by induction: if we find a *fixed* edge $\{x, y\}$ that has an LL, RR, LR or RL edge next to it, we can assume that this latter edge can belong to the same substring as x and y in a solution. We can remove it and apply induction. The main difficulty is that we must ensure that the graph resulting from this removal has no odd path.

Lemma 9. *A fixed-match set M is complete if and only if $G(M)$ has no odd path.*

The main consequence is that branching over odd paths is all that is required. We can now describe a concrete algorithm (with $M = \emptyset$ initially).

Theorem 2. *If F is a subset of $\{id, flip, rev, urev\}$, then PCSP can be solved in time $O(d^{2\ell}(8\ell)^{\ell}n)$.*

```
1  function PCSP(s, t, ℓ, π, F, M)
2     if M is not order-consistent then
3        |  return "No solution"
4     if G(M) has an odd path (u₁, ..., uₕ) then
5        |  if |M| = ℓ then
6        |     |  return "No solution"
7        |  foreach uᵢ on the path and each marker v with the same symbol
       |        or negated symbol, such that v is not in a fixed-match of M do
8        |     |  foreach bᵢ that is not already in a fixed-match of M do
9        |     |     |  Call PCSP(s, t, ℓ, π, F, M ∪ {(u, v, π(bᵢ), bᵢ)})
10       |     |     |  if a positive answer was returned then
11       |     |     |     |  return "Yes"
12       |  return "No solution"
13    else
14       |  return "Yes"
```

Algorithm 2: Main *PCSP* algorithm

Consequences. Theorem 1 and Theorem 2 can be combined to deduce that several distances can be computed in FPT time, provided that each symbol occurs at most d times. It suffices to express these distances as operations and plug in the complexity values. See full version for details.

5   Conclusion

We studied a new version of the famous Minimum Common String Partition problem, called the Permutation-constrained Common String Partition (PCSP), where we are given two sequences s and t with the same content and a permutation π on $[k]$, and the problem is to decide whether the two sequence can be decomposed into k blocks fitting π. Although PCSP is W[1]-hard when parameterized by k, we show that it is FPT (with parameters k and d) if s and t are both d-occurrence sequences. This results in a series of FPT results for genome rearrangements when the input sequences are d-occurrence.

References

1. Bulteau, L., Fellows, M., Komusiewicz, C., Rosamond, F.: Parameterized string equations. arXiv arXiv:2104.14171 (2021)
2. Bulteau, L., Fertin, G., Jean, G., Komusiewicz, C.: Sorting by multi-cut rearrangements. In: Bureš, T., et al. (eds.) SOFSEM 2021. LNCS, vol. 12607, pp. 593–607. Springer, Cham (2021). https://doi.org/10.1007/978-3-030-67731-2_43
3. Bulteau, L., Fertin, G., Komusiewicz, C., Rusu, I.: A fixed-parameter algorithm for minimum common string partition with few duplications. In: Darling, A., Stoye, J. (eds.) WABI 2013. LNCS, vol. 8126, pp. 244–258. Springer, Heidelberg (2013). https://doi.org/10.1007/978-3-642-40453-5_19
4. Bulteau, L., Fertin, G., Rusu, I.: Sorting by transpositions is difficult. SIAM J. Discret. Math. **26**(3), 1148–1180 (2012). https://doi.org/10.1137/110851390
5. Bulteau, L., Komusiewicz, C.: Minimum common string partition parameterized by partition size is fixed-parameter tractable. In: Proceedings of 25th ACM-SIAM Symposium on Discrete Algorithms, SODA 2014, pp. 102–121. SIAM (2014)
6. Caprara, A.: Sorting permutations by reversals and Eulerian cycle decompositions. SIAM J. Discret. Math. **12**(1), 91–110 (1999). https://doi.org/10.1137/S089548019731994X
7. Chen, X., et al.: Assignment of orthologous genes via genome rearrangement. IEEE/ACM Trans. Comput. Biol. Bioinform. **2**(4), 302–315 (2005). https://doi.org/10.1109/TCBB.2005.48
8. Christie, D.A.: Sorting permutations by block-interchanges. Inf. Process. Lett. **60**(4), 165–169 (1996). https://doi.org/10.1016/S0020-0190(96)00155-X
9. Christie, D.A., Irving, R.W.: Sorting strings by reversals and by transpositions. SIAM J. Discret. Math. **14**(2), 193–206 (2001). https://doi.org/10.1137/S0895480197331995
10. Chrobak, M., Kolman, P., Sgall, J.: The greedy algorithm for the minimum common string partition problem. ACM Trans. Algorithms **1**(2), 350–366 (2005). https://doi.org/10.1145/1103963.1103971
11. Damaschke, P.: Minimum common string partition parameterized. In: Crandall, K.A., Lagergren, J. (eds.) WABI 2008. LNCS, vol. 5251, pp. 87–98. Springer, Heidelberg (2008). https://doi.org/10.1007/978-3-540-87361-7_8
12. Ganczorz, M., Gawrychowski, P., Jez, A., Kociumaka, T.: Edit distance with block operations. In: Proceedings of ESA 2018, LIPIcs, vol. 112, pp. 33:1–33:14. Schloss Dagstuhl-Leibniz-Zentrum fuer Informatik (2018)
13. Goldstein, A., Kolman, P., Zheng, J.: Minimum common string partition problem: hardness and approximations. Eur. J. Combin. **12**, R50 (2005)
14. Jiang, H., Zhu, B., Zhu, D., Zhu, H.: Minimum common string partition revisited. J. Comb. Optim. **23**(4), 519–527 (2012). https://doi.org/10.1007/s10878-010-9370-2
15. Lafond, M., Zhu, B., Zou, P.: Genomic problems involving copy number profiles: Complexity and algorithms. In: Proceedings of CPM 2020, LIPIcs, vol. 161, pp. 22:1–22:15. Schloss Dagstuhl-Leibniz-Zentrum fuer Informatik (2020)

16. Lafond, M., Zhu, B., Zou, P.: The tandem duplication distance is NP-hard. In: Proceedings of STACS 2020, LIPIcs, vol. 154, pp. 15:1–15:15. Schloss Dagstuhl-Leibniz-Zentrum fuer Informatik (2020)
17. Mahajan, M., Rama, R., Raman, V., Vijaykumar, S.: Approximate block sorting. Int. J. Found. Comput. Sci. **17**(2), 337–356 (2006). https://doi.org/10.1142/S0129054106003863
18. Makaroff, C., Palmer, J.: Mitochondrial DNA rearrangements and transcriptional alternatives in the male sterile cytoplasm of Ogura radish. Mol. Cell. Biol. **8**, 1474–1480 (1988)
19. Qingge, L., He, X., Liu, Z., Zhu, B.: On the minimum copy number generation problem in cancer genomics. In: Proceedings of the 10th ACM Conference on Bioinformatics, Computational Biology, and Health Informatics, BCB 2018, pp. 260–269. ACM Press, New York (2018). https://doi.org/10.1145/3233547.3233586
20. Sturtevant, A.: A crossover reducer in drosophila melanogaster due to inversion of a section of the third chromosome. Biol. Zent. Bl. **46**, 697–702 (1926)
21. Sturtevant, A., Dobzhansky, T.: Inversions in the third chromosome of wild races of Drosophila pseudoobscura, and their use in the study of the history of the species. Proc. Nat. Acad. Sci. U.S.A. **22**, 448–450 (1936)
22. Zheng, C., Kerr Wall, P., Leebens-Mack, J., de Pamphilis, C., Albert, V.A., Sankoff, D.: Gene loss under neighborhood selection following whole genome duplication and the reconstruction of the ancestral Populus genome. J. Bioinform. Comput. Biol. **7**(03), 499–520 (2009)

All Instantiations of the Greedy Algorithm for the Shortest Common Superstring Problem are Equivalent

Maksim S. Nikolaev$^{(\boxtimes)}$ (iD)

Steklov Institute of Mathematics at St. Petersburg, Russian Academy of Sciences,
Saint Petersburg, Russia

Abstract. In the Shortest Common Superstring problem (SCS), one needs to find the shortest superstring for a set of strings. While SCS is NP-hard and MAX-SNP-hard, the Greedy Algorithm "choose two strings with the largest overlap; merge them; repeat" achieves a constant factor approximation that is known to be at most 3.5 and conjectured to be equal to 2. The Greedy Algorithm is not deterministic, so its instantiations with different tie-breaking rules may have different approximation factors. In this paper, we show that it is not the case: all factors are equal. To prove this, we show how to transform a set of strings so that all overlaps are different whereas their ratios stay roughly the same.

Keywords: Superstring · Shortest common superstring · Approximation · Greedy algorithms · Greedy conjecture

1 Introduction

In the Shortest Common Superstring problem (SCS), one is given a set of strings and needs to find the shortest string that contains each of them as a substring. Applications of this problem include genome assembly [12,19] and data compression [3,4,15]. We refer the reader to the survey [5] for an overview of SCS as well as its applications and algorithms.

SCS is known to be NP-hard [4] and even MAX-SNP-hard [1], but it admits constant-factor approximation in polynomial time. The best known approximation ratios are $2\frac{11}{23}$ due to Mucha [11] (see [7, Section 2.1] for an overview of the previous approximation algorithms and inapproximability results). While these approximation algorithms use many sophisticated techniques, the 30 years old *Greedy Conjecture* [1,15–17] claims that the trivial *Greedy Algorithm* (GA) "choose two strings with the largest overlap; merge them; repeat" is a factor 2 approximation (in fact, this is the best possible approximation factor: consider a dataset $\mathcal{S} = \{c(ab)^n, (ab)^n c, (ba)^n\}$). Ukkonen [18] shows that for a fixed alphabet, GA can be implemented in linear time.

The research is supported by Russian Science Foundation (18-71-10042).

T. Lecroq and H. Touzet (Eds.): SPIRE 2021, LNCS 12944, pp. 61–67, 2021.
https://doi.org/10.1007/978-3-030-86692-1_6

Blum et al. [1] prove that GA returns a 4-approximation of SCS, and Kaplan and Shafrir [8] improve this bound to 3.5. A slight modification of GA gives a 3-approximation of SCS [1], and other greedy algorithms are studied from theoretical [1,13] and practical perspectives [2,14].

It is known that the Greedy Conjecture holds for the case when all input strings have length at most 4 [9]. Also, the Greedy Conjecture holds if GA happens to merge strings in a particular order [10,20]. GA gives a 2-approximation of a different metric called compression [16]. The compression is defined as the sum of the lengths of all input strings minus the length of a superstring (hence, it is the number of symbols saved with respect to a naive superstring resulting from concatenating the input strings).

GA is not deterministic as we do not specify how to break ties in case when there are many pairs of strings with maximum overlap. For this reason, different instantiations of GA (that is, GA with a tie-breaking rule) may produce different superstrings for the same input and hence they may have different approximation factors. In fact, if S contains only strings of length 2 or less or if S is a set of k-substrings of an unknown string, then there are instantiations of GA [6], that find *the exact* solution, whereas in general GA fails to do so.

The original Greedy Conjecture states that *any* instantiation of GA is a factor 2 approximation. As this is still widely open, it is natural to try to prove the conjecture at least for *some* instantiations. This could potentially be easier not just because this is a weaker statement, but also because a particular instantiation of GA may decide how to break ties by asking an omniscient oracle. In this paper, we show that this weak form of Greedy Conjecture is in fact equivalent to the original one. More precisely, we show, that if *some* instantiation of GA is a factor λ approximation, then *all* instantiations are factor λ approximation.

To prove this, we introduce the so-called *Perturbing Procedure*, that, for a given dataset $S = \{s_1, \ldots, s_n\}$, a parameter $m \gg n$, and a sequence of greedy non-trivial merges (merges of strings with a non-empty overlap), constructs a new dataset $S' = \{s'_1, \ldots, s'_n\}$, such that, for all $i \neq j$, s'_i is roughly m times longer than s_i, the overlap of s'_i and s'_j is roughly m times longer than the overlap of s_i and s_j, and the mentioned greedy sequence of non-trivial merges for S is *the only* such sequence for S'.

2   Preliminaries

Let $|s|$ be the length of a string s and $\mathrm{ov}(s,t)$ be the *overlap* of strings s and t, that is, the longest string y, such that $s = xy$ and $t = yz$. In this notation, a string xyz is *a merge* of strings s and t. By ε we denote the empty string. By $\mathrm{OPT}(S)$ we denote an optimal superstring for the dataset S.

Without loss of generality we may assume that the set of input strings S contains no string that is a substring of another. This assumption implies that in any superstring all strings occur in some order: if one string begins before another, then it also ends before. Hence, we can consider only superstrings that can be obtained from some permutation $(s_{\sigma(1)}, \ldots, s_{\sigma(n)})$ of $\mathcal{S}$ after merging adjacent strings. The length of such superstring $s(\sigma)$ is simply

$$|s(\sigma)| = \sum_{i=1}^{n} |s_i| - \sum_{i=1}^{n-1} |\operatorname{ov}(s_{\sigma(i)}, s_{\sigma(i+1)})|. \tag{1}$$

Let A be an instantiation of GA (we denote this by $A \in$ GA). By σ_A we denote the permutation corresponding to a superstring $A(\mathcal{S})$ constructed by A, and by $(l_A(1), r_A(1)), \ldots, (l_A(n-1), r_A(n-1))$, we denote the order of merges: strings $s_{l_A(i)}$ and $s_{r_A(i)}$ are merged at step i. By the definition of GA we have

$$|\operatorname{ov}(s_{l_A(i)}, s_{r_A(i)})| \geq |\operatorname{ov}(s_{l_A(j)}, s_{r_A(j)})|, \quad \forall i < j < n,$$

and if, for some i, $|\operatorname{ov}(s_{l_A(i)}, s_{r_A(i)})| = 0$, then the same holds for any $i' > i$. We denote the first such i by T_A and this is the first trivial merge (that is, one with the empty overlap), after which all the merges are trivial. Note that just before step T_A, all the remaining strings have empty overlaps, so the resulting superstring is just a concatenation of them in some order and this order does not affect the length of the result. If there were no trivial merges, we set $T_A = n$.

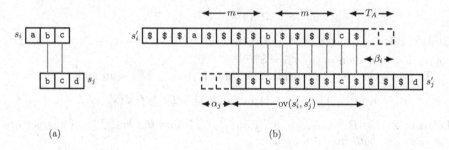

(a)   (b)

Fig. 1. (a) strings s_i and s_j from $\mathcal{S}$. (b) the resulting strings s_i' and s_j' after perturbing; here, $m = 4$, $T_A = 3$, $\alpha_i = 1$, $\beta_i = 2$, $\alpha_j = 2$ and $\beta_j = T_A$; since $\alpha_j = \beta_i = 2$, we may conclude that s_i and s_j were merged by A at step 2.

3 Perturbing Procedure

Here, we describe the mentioned procedure that eliminates ties. Consider a dataset $\mathcal{S}$, an instantiation $A \in$ GA and *a sentinel* \$—a symbol that does not occur in $\mathcal{S}$, and a parameter m whose value will be determined later. For every string $s_i = c_1 c_2 \ldots c_{n_i} \in \mathcal{S}$ define a string

$$s_i' = \$^{m-\alpha_i} c_1 \$^m c_2 \$^m c_3 \$^m \ldots \$^m c_{n_i} \$^{T_A - \beta_i}, \tag{2}$$

where

1. α_i is the number of step such that $r_A(\alpha_i) = i$, if such step exists and is less than T_A, and $\alpha_i = T_A$ otherwise; note that if $\alpha_i < T_A$ then s_i is *the right part* of a non-trivial merge at step α_i;
2. β_i is the number of step such that $l_A(\beta_i) = i$, if such step exists and is less than T_A, and $\beta_i = T_A$ otherwise; note that if $\beta_i < T_A$ then s_i is *the left* part of a non-trivial merge at step β_i.

Basically, we insert the string $\$^m$ before every character of s_i and then remove some $\$$'s from the beginning of the string and add some $\$$'s to its end (see Fig. 1). The purpose of this removal and addition is to perturb slightly overlaps of equal length, so there are no longer any ties in non-trivial merges.

We denote the resulting set of perturbed strings $\{s'_1, \ldots, s'_n\}$ by $\mathcal{S}'$, and all entities related to this dataset we denote by adding a prime (for example, σ'_A). Let us derive some properties of $\mathcal{S}'$.

Lemma 1. *For all $i \neq j$, $k \neq l$, $m > 2n$*

1. *if $|\operatorname{ov}(s_i, s_j)| = d > 0$, then $|\operatorname{ov}(s'_i, s'_j)| = (m+1)d - \alpha_j + T_A - \beta_i$;*
2. *if $|\operatorname{ov}(s_i, s_j)| = 0$, then $|\operatorname{ov}(s'_i, s'_j)| = \min\{T_A - \beta_i, m - \alpha_j\}$;*
3. *perturbing procedure preserves order on overlaps of different lengths, that is, if $|\operatorname{ov}(s_i, s_j)| > |\operatorname{ov}(s_k, s_l)|$, then $|\operatorname{ov}(s'_i, s'_j)| > |\operatorname{ov}(s'_k, s'_l)|$.*

Proof. Let $\operatorname{ov}(s_i, s_j)$ be $c_1 c_2 \ldots c_d$. Consider the string

$$u = \$^{m-\alpha_j} c_1 \$^m \ldots \$^m c_d \$^{T_A - \beta_i}.$$

Clearly, u is the overlap of s'_i and s'_j and $|u| = (m+1)d - \alpha_j + T_A - \beta_i$. Also, if $|\operatorname{ov}(s_i, s_j)| = 0$ then $\operatorname{ov}(s'_i, s'_j) = \$^{\min\{T_A - \beta_i, m - \alpha_j\}}$.

To prove the last statement, note that $\alpha_j + \beta_i \leq 2T_A < m$ and

$$|\operatorname{ov}(s'_i, s'_j)| > (m+1)|\operatorname{ov}(s_k, s_l)| + T_A \geq |\operatorname{ov}(s'_k, s'_l)|.$$

Lemma 2. *Let $B \in \mathrm{GA}$. Then $T_A = T'_A = T'_B$ and the first $T_A - 1$ merges are the same for both instantiations.*

Proof. We prove by induction that $l_A(t) = l'_A(t) = l'_B(t)$ and $r_A(t) = r'_A(t) = r'_B(t)$ for all $t < T_A$.

Case $t = 1$. As A is greedy, then $k_1 := |\operatorname{ov}(s_{l_A(1)}, s_{r_A(1)})| \geq |\operatorname{ov}(s_i, s_j)|$, for all $i \neq j$, $(i,j) \neq (l_A(1), r_A(1))$. Hence

$$|\operatorname{ov}(s'_i, s'_j)| \leq (m+1)k_1 - \alpha_j + T_A - \beta_i$$
$$< (m+1)k_1 - 1 + T_A - 1 = |\operatorname{ov}(s'_{l_A(1)}, s'_{r_A(1)})|,$$

and $l'_A(1) = l'_B(1) = l_A(1)$ as well as $r'_A(1) = r'_B(1) = r_A(1)$.

Suppose that the statement holds for all $t \leq t' < T_A - 1$. Note that at moment $t = t' + 1$ the sum $\alpha_j + \beta_i$ is strictly greater than $2t$ unless $(i,j) = (l_A(t), r_A(t))$. Similarly to the base case, we have

$$|\operatorname{ov}(s'_i, s'_j)| \leq (m+1)k_t - \alpha_j + T_A - \beta_i$$
$$< (m+1)k_t - t + T_A - t = |\operatorname{ov}(s'_{l_A(t)}, s'_{r_A(t)})|,$$

where $k_t = |\operatorname{ov}(s_{l_A(t)}, s_{r_A(t)})|$, and the induction step is proven.

Now note that starting from step T_A all the remaining strings in S have empty overlaps and hence so do the remaining strings in S', as for all of them $\beta_i = T_A$ and the minimum in paragraph 2 of Lemma 1 is equal to zero. Thus, $T_A = T'_A = T'_B$ and the lemma is proven.

Corollary 1. *As all non-trivial merges coincide, $|A(S')| = |B(S')|$.*

4 Equivalence of Instantiations

Theorem 1. *If some instantiation A of GA achieves a λ-approximation, then so does any other instantiation.*

Proof. Assume the opposite and consider $B \in \mathrm{GA}$ as well as a dataset S such that $|B(S)| > \lambda|\mathrm{OPT}(S)|$. Let $S' = S'(B, m)$ be the corresponding perturbed dataset, where $m > 2n$ will be specified later.

Note that $|s'_i|/m \to |s_i|$ and $|\operatorname{ov}(s'_i, s'_j)|/m \to |\operatorname{ov}(s_i, s_j)|$ as m approaches infinity, thanks to Lemma 1.1–2. Then $|\mathrm{OPT}(S')|/m \to |\mathrm{OPT}(S)|$, since

$$\frac{1}{m}|\mathrm{OPT}(S')| = \frac{1}{m}\min_\sigma \left\{ \sum_{i=1}^{n} |s'_i| - \sum_{i=1}^{n-1} |\operatorname{ov}\left(s'_{\sigma(i)}, s'_{\sigma(i+1)}\right)| \right\}$$

$$\to \min_\sigma \left\{ \sum_{i=1}^{n} |s_i| - \sum_{i=1}^{n-1} |\operatorname{ov}\left(s_{\sigma(i)}, s_{\sigma(i+1)}\right)| \right\} = |\mathrm{OPT}(S)|,$$

$|B(S')|/m \to |B(S)|$ and hence $|A(S')|/m \to |B(S)|$, by Corollary 1.

As $|B(S)| - \lambda|\mathrm{OPT}(S)| > 0$, we can choose m so that $|B(S')| - \lambda|\mathrm{OPT}(S')|$ as well as $|A(S')| - \lambda|\mathrm{OPT}(S')|$ are positive. Hence A is not a factor λ approximation.

Corollary 2. *To prove (or disprove) the Greedy Conjecture, it is sufficient to consider datasets satisfying some of the following three properties:*

1. *there are no ties between non-empty overlaps, that is, datasets where all the instantiations of the greedy algorithm work the same;*
2. *there are no empty overlaps: $\operatorname{ov}(s_i, s_j) \neq \varepsilon, \forall i \neq j$;*
3. *all non-empty overlaps are (pairwise) different: $|\operatorname{ov}(s_i, s_j)| \neq |\operatorname{ov}(s_k, s_l)|$, for all $i \neq j$, $k \neq l$, $(i, j) \neq (k, l)$.*

Proof. 1. Follows directly from the proof of Theorem 1, as we always can use the dataset S' instead of S.

2. Append \$ to each string of S'. Then, every two strings have non-empty overlap that at least contains \$, and in general $T_A = T'_A = T'_B$ from Lemma 2 does not hold (T'_A and T'_B are always n). However, the first T_A merges are still the same and after them all the remaining strings have overlaps of length 1 and then the lengths of the final solutions are the same as well.

3. Append $\$^{n(T_A - \beta_i)}$ to each string of $\mathcal{S}'$ instead of $\$^{T_A - \beta_i}$. Then

$$|\operatorname{ov}\left(s_i', s_j'\right)| = (m+1)|\operatorname{ov}\left(s_i, s_j\right)| - \alpha_j + nT_A - n\beta_i,$$

provided m is large enough, and $\alpha_j + n\beta_i \neq \alpha_k + n\beta_l$ if $(i,j) \neq (k,l)$. Repeating the proofs of Lemmas 1 and 2 with this version of $\mathcal{S}'$, we obtain this statement of the corollary.

To combine several of this properties (for example second and third), it is sufficient to sequentially apply the corresponding transformations on the original dataset: at first we get a dataset $\mathcal{S}'$ from $\mathcal{S}$ as in paragraph 2, then we treat $\mathcal{S}'$ (already without empty overlaps) as original and transform it to a dataset $\mathcal{S}''$ according to paragraph 3 using a different sentinel instead of $\$$.

5 Conclusion

In this paper we revealed the equivalence of greedy algorithms for the shortest common superstring problem. This means, in particular, that proving or disproving the Greedy Conjecture is difficult not due to the non-deterministic nature of the Greedy Algorithm, but due to the complexity of the overlaps structure.

Acknowledgments. Many thanks to Alexander Kulikov for valuable discussions and proofreading the text, and the anonymous reviewers for their useful comments.

References

1. Blum, A., Jiang, T., Li, M., Tromp, J., Yannakakis, M.: Linear approximation of shortest superstrings. In: STOC 1991, pp. 328–336. ACM (1991). https://doi.org/10.1145/179812.179818
2. Cazaux, B., Juhel, S., Rivals, E.: Practical lower and upper bounds for the shortest linear superstring. In: SEA 2018, vol. 103, pp. 18:1–18:14. LIPIcs (2018). https://doi.org/10.4230/LIPIcs.SEA.2018.18
3. Gallant, J.K.: String compression algorithms. Ph.D. thesis, Princeton (1982)
4. Gallant, J., Maier, D., Storer, J.A.: On finding minimal length superstrings. J. Comput. Syst. Sci. **20**(1), 50–58 (1980). https://doi.org/10.1016/0022-0000(80)90004-5
5. Gevezes, T.P., Pitsoulis, L.S.: The shortest superstring problem. In: Rassias, T.M., Floudas, C.A., Butenko, S. (eds.) Optimization in Science and Engineering, pp. 189–227. Springer, New York (2014). https://doi.org/10.1007/978-1-4939-0808-0_10
6. Golovnev, A., Kulikov, A.S., Logunov, A., Mihajlin, I., Nikolaev, M.: Collapsing superstring conjecture. In: Approximation, Randomization, and Combinatorial Optimization. Algorithms and Techniques (APPROX/RANDOM 2019). Schloss Dagstuhl-Leibniz-Zentrum fuer Informatik (2019). https://doi.org/10.4230/LIPIcs.APPROX-RANDOM.2019.26
7. Golovnev, A., Kulikov, A.S., Mihajlin, I.: Approximating shortest superstring problem using de bruijn graphs. In: Fischer, J., Sanders, P. (eds.) CPM 2013. LNCS, vol. 7922, pp. 120–129. Springer, Heidelberg (2013). https://doi.org/10.1007/978-3-642-38905-4_13

8. Kaplan, H., Shafrir, N.: The greedy algorithm for shortest superstrings. Inf. Process. Lett. **93**(1), 13–17 (2005). https://doi.org/10.1016/j.ipl.2004.09.012
9. Kulikov, A.S., Savinov, S., Sluzhaev, E.: Greedy conjecture for strings of length 4. In: Cicalese, F., Porat, E., Vaccaro, U. (eds.) CPM 2015. LNCS, vol. 9133, pp. 307–315. Springer, Cham (2015). https://doi.org/10.1007/978-3-319-19929-0_26
10. Laube, U., Weinard, M.: Conditional inequalities and the shortest common superstring problem. Int. J. Found. Comput. Sci. **16**(06), 1219–1230 (2005). https://doi.org/10.1142/S0129054105003777
11. Mucha, M.: Lyndon words and short superstrings. In: SODA 2013, pp. 958–972. SIAM (2013). https://doi.org/10.1137/1.9781611973105.69
12. Pevzner, P.A., Tang, H., Waterman, M.S.: An Eulerian path approach to DNA fragment assembly. Proc. Natl. Acad. Sci. U.S.A. **98**(17), 9748–9753 (2001). https://doi.org/10.1073/pnas.171285098
13. Rivals, E., Cazaux, B.: Superstrings with multiplicities. In: CPM 2018, vol. 105, pp. 21:1–21:16 (2018). https://doi.org/10.4230/LIPIcs.CPM.2018.21
14. Romero, H.J., Brizuela, C.A., Tchernykh, A.: An experimental comparison of two approximation algorithms for the common superstring problem. In: ENC 2004, pp. 27–34. IEEE (2004). https://doi.org/10.1109/ENC.2004.1342585
15. Storer, J.A.: Data compression: methods and theory. Computer Science Press Inc., (1987)
16. Tarhio, J., Ukkonen, E.: A greedy approximation algorithm for constructing shortest common superstrings. Theor. Comput. Sci. **57**(1), 131–145 (1988). https://doi.org/10.1016/0304-3975(88)90167-3
17. Turner, J.S.: Approximation algorithms for the shortest common superstring problem. Inf. Comput. **83**(1), 1–20 (1989). https://doi.org/10.1016/0890-5401(89)90044-8
18. Ukkonen, E.: A linear-time algorithm for finding approximate shortest common superstrings. Algorithmica **5**(1–4), 313–323 (1990). https://doi.org/10.1007/BF01840391
19. Waterman, M.S.: Introduction to Computational Biology: Maps, Sequences and Genomes. CRC Press, Boca Raton (1995). https://doi.org/10.1201/9780203750131
20. Weinard, M., Schnitger, G.: On the greedy superstring conjecture. SIAM J. Discret. Math. **20**(2), 502–522 (2006). https://doi.org/10.1137/040619144

String Covers of a Tree

Jakub Radoszewski$^{(\boxtimes)}$ ⓘ, Wojciech Rytter ⓘ, Juliusz Straszyński ⓘ,
Tomasz Waleń ⓘ, and Wiktor Zuba ⓘ

University of Warsaw, Warsaw, Poland
{jrad,rytter,jks,walen,w.zuba}@mimuw.edu.pl

Abstract. We consider covering labeled trees by a collection of paths
with the same string label, called a (string) cover of a tree. We
show how to compute all covers of a directed (rooted) labeled tree in
$\mathcal{O}(n \log n / \log \log n)$ time and all covers of an undirected labeled tree in
$\mathcal{O}(n^2)$ time and space or in $\mathcal{O}(n^2 \log n)$ time and $\mathcal{O}(n)$ space. We also
show several essential differences between covers in standard strings and
covers in trees.

1 Introduction

We consider undirected and directed trees with at least 2 nodes and edges labeled
by single symbols. The label of a simple path $v_1 \overset{a_1}{-} v_2 \overset{a_2}{-} v_3 \overset{a_3}{-} \ldots \overset{a_{k-1}}{-} v_k$ is the
string $W = a_1 a_2 \ldots a_{k-1}$. Let us note that the label of the reverse path from v_k
to v_1 is W^R, the reverse of W. We say that a non-empty string W **covers** a tree
if each edge is on a simple path labeled by W. In case of rooted trees we consider
only edges directed bottom-up towards the root (the symmetric ordering top-
down is equivalent with respect to coverings). Figure 1 shows the covers of an
example undirected tree.

A standard string can be considered as a directed path, hence covers of a
directed path can be found using one of the known algorithms for computing
covers of strings [4,7,24,25]. However, covering an undirected simple path (see
e.g. Fig. 2) is a very different problem and is much harder than covering a string.
It is equivalent to covering with two strings W and W^R. A nontrivial almost
linear time algorithm for covering an undirected path is implied by the algorithm
in [27, Section 3] about covering a string with two equal-length strings.

Covers in directed trees are also quite different from covers in strings. A
directed tree can be seen as a collection of strings corresponding to leaf-to-root
paths, but a cover of a tree does not necessarily cover each of these strings; see
Fig. 3, where the string *ababaaba* is not a cover of a string corresponding to the
leftmost leaf-to-root path, though it is a cover of the whole tree. However, the
following fact can be shown with the aid of induction.

Work supported by the Polish National Science Center, grant no. 2018/31/D/ST6/
03991.

T. Lecroq and H. Touzet (Eds.): SPIRE 2021, LNCS 12944, pp. 68–82, 2021.
https://doi.org/10.1007/978-3-030-86692-1_7

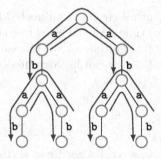

Fig. 1. The string *aab* is a cover of this undirected tree T; all its occurrences are shown in green. The tree has 14 distinct covers: *aab, aabab, ab, abaabab, abab, ba, baa, baab, baabab, baba, babaa, babaab, babaaba, babaabab.* Observe that the bottom-up directed version of T rooted in the top node has only two covers: *ba* and *baba*.

$$1 \overset{a}{-} 2 \overset{a}{-} 3 \overset{a}{-} 4 \overset{a}{-} 5 \overset{a}{-} 6 \overset{b}{-} 7 \overset{a}{-} 8 \overset{a}{-} 9 \overset{a}{-} 10 \overset{a}{-} 11 \overset{a}{-} 12$$

Fig. 2. An undirected path $a^m b a^m$ of length $n = 2m + 1$ has $\Omega(n)$ different aperiodic covers, of the form $a^m b a^i$ or $a^i b a^m$, for $i = 0, \ldots, m$. A standard string of length n can only have $\mathcal{O}(\log n)$ aperiodic covers; see [4].

Observation 1. *If S is a cover of a directed tree T, then it is a cover of at least one of the strings corresponding to leaf-to-root branches.*

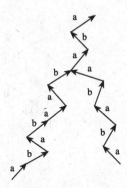

Fig. 3. The displayed directed tree has 3 covers *aba, ababa, ababaaba*. In case of standard strings a shorter cover is a suffix of the longer one. This does not work for covers of directed trees; *ababa* is not a suffix of *ababaaba*

This work extends the rich study of covers in non-standard settings—e.g., 2-dimensional [9,10,26], Abelian [20,21,23], parameterized and order-preserving [17], and on indeterminate [1,3,11,16] and weighted strings [5,16]—to labeled trees. Moreover, we continue the line of work on algorithmic and combinatorial properties of palindromes, powers and runs in labeled trees, which have different properties than in strings [8,10,12–14,18,19,29].

Our Results: We show that all covers of a directed labeled tree can be computed in $\mathcal{O}(n \log n / \log \log n)$ time if the labels of tree edges belong to an integer alphabet, i.e., are integers of magnitude $n^{\mathcal{O}(1)}$. In case of undirected trees with labels over any alphabet, all covers can be computed in $\mathcal{O}(n^2)$ time and space or in $\mathcal{O}(n^2 \log n)$ time and $\mathcal{O}(n)$ space.

2   Preliminaries

In this section we introduce several algorithms and combinatorial properties related to labeled and unlabeled trees. If $u_1 \rightsquigarrow u_k = (u_1, u_2, \ldots, u_{k-1}, u_k)$ is a (simple) path in a labeled tree T, then its *length dist*(u_1, u_k) is defined as the number of edges (i.e., $k-1$) and its *string label* is the concatenation of labels of edges $(u_1, u_2), \ldots, (u_{k-1}, u_k)$.

Let T be a rooted labeled tree. We assume that all edges are directed towards the root. For a node v of T, by $label_d(v)$ we denote the string label of a path from v to its ancestor at distance d and by $label(v)$ we denote the string label of a path from v to the root.

2.1   The Table of Prefixes

For a string S, we denote $\mathsf{TreePREF}_S[v] = \max\{d \geq 0 \;:\; label_d(v) = S[1..d]\}$, where $S[1..d]$ is a length-d prefix of string S; see Fig. 4.

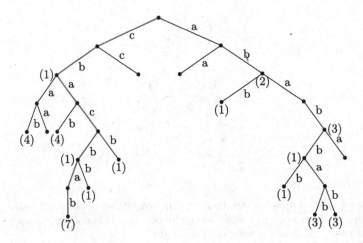

Fig. 4. An example of $\mathsf{TreePREF}_S$ array, where $S = babcabc$ (red path, bottom-up). The nonzero values of the array are shown in brackets. (Color figure online)

To compute this array we use the concept of a suffix tree of a labeled tree T that was introduced by Kosaraju [22]. A *compacted trie* is a trie in which maximal paths whose inner nodes have degree 2 are represented as single edges with string labels. Usually such labels are not stored explicitly, but as pointers to a base string (or base strings); only the first letters are stored. The remaining nodes are called explicit, whereas the nodes that are removed due to compactification are called implicit. The *suffix tree of a rooted labeled tree* T is a compacted trie of the strings $label(v)$ for all nodes v in T. An efficient construction of the suffix tree of a tree was given by Shibuya [28].

Fact 1 (Shibuya [28]) The suffix tree of a rooted tree with n nodes over an integer alphabet has size $\mathcal{O}(n)$ and can be constructed in $\mathcal{O}(n)$ time.

Lemma 1. *For a directed labeled tree T with n nodes, the array* TreePREF$_S$ *can be computed in $\mathcal{O}(n)$ time.*

Proof. We create a path from the root of T leading to a new leaf s with $label(s) = S$; let T' be the resulting tree. Then we compute the suffix tree $\mathcal{T}$ of the tree T'. For each node u of T', by $where(u)$ let us denote the node of $\mathcal{T}$ with path label $label(u)$. For each node u of T' that originates from T, TreePREF$_S[u]$ is the depth of the lowest common ancestor (LCA) of $where(u)$ and $where(s)$. We use the fact that LCA queries can be answered in $\mathcal{O}(1)$ time after linear time preprocessing [6]. $\square$

2.2 Summing Second Heights of All Nodes

We are interested in computing sums of heights of nodes. Let us define $height(v)$ to be the number of nodes on the longest path from v to a leaf.

Remark 1. The sum of heights of all nodes can be quadratic, for example for a simple directed path.

The situation changes if we consider for each node its second highest child. Denote by $sec\text{-}height(v)$ the second largest height of a child of v (possibly $sec\text{-}height(v) = height(v) - 1$ if there are two children with the same largest height). If v has only one child then we define $sec\text{-}height(v) = 0$.

Proposition 1. $\sum_{v \in T} sec\text{-}height(v) \leq n$.

Proof. For a node v we define $MaxPath(v)$ as a longest path from v to its leaf. Initially we choose (one of possibly many) $MaxPath(root)$, then we remove this path (both nodes and edges) and choose longest paths for roots of resulting subtrees. We continue in this way and obtain a decomposition of the tree into node-disjoint longest paths; see Fig. 5.

Let $FirstChild(v)$ denote a child of v which belongs to the same path in the decomposition and $SecondChild(v)$ denote a child $w \neq FirstChild(v)$ of v of largest height. Let V' be the set of nodes with at least two children each. Then

$$\sum_v sec\text{-}height(v) = \sum_{v \in V'} |MaxPath(SecondChild(v))| \leq n,$$

since all selected longest paths are node-disjoint ($|p|$ denotes the number of nodes in a path p). $\square$

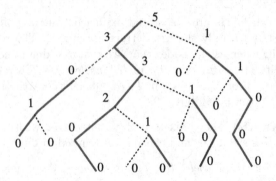

Fig. 5. A tree T with the values *sec-height*(v). The longest paths are distinguished. Observe that $\sum_v$ *sec-height*$(v) < |T|$.

3   Covers of Directed Trees

Let T be a rooted labeled tree with $n \geq 2$ nodes. Let *Subtree*(v) denote the set of descendant nodes of v, including v. For a set M of marked nodes and nodes u, v in T we define

$$\mathsf{Bottom}(v) = \text{a node } u \in M \cap \textit{Subtree}(v) \text{ with minimal } dist(u,v),$$
$$\mathsf{chain}(u) = \{v : \mathsf{Bottom}(v) = u\}.$$

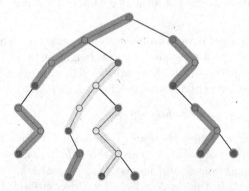

Fig. 6. Chain decomposition; each chain is the set of nodes v such that $\mathsf{Bottom}(v) = u$, where u is its bottom node (marked node, black in the figure). Red nodes are top nodes of nontrivial chains; nodes constituting single element chains are black and red at the same time. We have $\mathsf{maxgap} = 4$; the two chains of size 4 are drawn in orange. (Note that the chain containing the root has size 3, since the root node is not counted.) (Color figure online)

Let $u \in M$. Our algorithm keeps an invariant that $\mathsf{chain}(u)$ is a path. We denote by $\|\mathsf{chain}(u)\|$ the number of non-root nodes on $\mathsf{chain}(u)$, called the *size* of the chain. If $v \in$ chain(u), then by Top(v) we denote the topmost node of chain(u). We define:

$$\text{maxgap}(M) = \max\{\|\text{chain}(v)\| : v \in T\}.$$

In other words, maxgap(M) is the maximal number of edges from a bottom node to the root, if the root is in the same chain, or to the lowest node in a different chain; see Fig. 6. Here $\|\text{chain}(v)\| = \infty$ if v does not belong to any chain.

Observation 2. *Let S be the label of a chosen leaf-to-root path, d be a positive integer, and let $M = \{w : \text{TreePREF}_S(w) \geq d\}$. The tree T has a cover of length d if and only if maxgap$(M) \leq d$.*

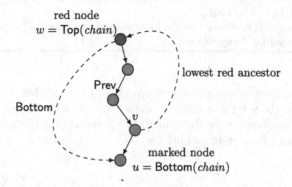

Fig. 7. Structure of a chain $u \rightsquigarrow w$. The top node w keeps the value Bottom$[w] = u$. (Color figure online)

Algorithm 1: Covers-in-directed-tree

$S := label(leaf, root)$ for some *leaf*;
Compute TreePREF$_S[v]$ for all $v \in T$;
$m := \min\{\text{TreePREF}_S[v] : v \text{ is a leaf}\}$;
foreach v in T **do** TreePREF$_S[v] := \min(\text{TreePREF}_S[v], m)$;
Initialize;
for $d := m$ **down to** *1* **do**

    $NEW := \{v : \text{TreePREF}_S[v] = d\} \setminus leaves(T)$;
    foreach v in NEW **do** MarkAndUpdate(v);

    if maxgap$(M) \leq d$ **then** Report a cover of length d;

For each $u \in M$, the values len$[u] := \|\text{chain}(u)\|$ and Top$[u] := $ Top(u) are stored. Moreover, we assume that the node $w = $ Top$[u]$ is colored in red and stores u, as Bottom$[w] := $ Bottom(w). If $v \notin M$, we store, as Prev$[v]$, the child of v on the path from v to Bottom(v). If $v \notin M$, we will compute Top(v) as the lowest red ancestor of v; this will be the bottleneck of the algorithm. See Fig. 7.

We use the algorithm Covers-in-directed-tree shown as Algorithm 1 with auxiliary functions. The algorithm considers all possible lengths d of a cover in descending order and stores the chain decomposition implied by the set M of Observation 2. Non-leaf nodes are sorted by their TreePREF$_S$ values using buckets in $\mathcal{O}(n)$ time. To some extent this algorithm resembles Moore and Smyth's algorithm [24,25] for computing covers of strings (working in a reversed order).

Function Initialize

Comment: Creates chains for $M = leaves(T)$
foreach v in T **do**
    Compute $Leaf[v]$ as the leftmost nearest leaf in the subtree of v;
foreach $leaf$ in $leaves(T)$ **do**
    $NewChain(leaf, v)$ where v is the topmost node with $Leaf[v] = leaf$;
Compute the initial Prev$[\cdot]$ values;

The function $NewChain(u, w)$ creates a chain $u \rightsquigarrow w$, whereas the function $FindChain(v) = (u \rightsquigarrow w)$ first computes $w = $ Top(v) as the lowest red ancestor of v and then computes $u = $ Bottom$[w]$. We assume that the nodes of T are pre-ordered from left to right in Initialize.

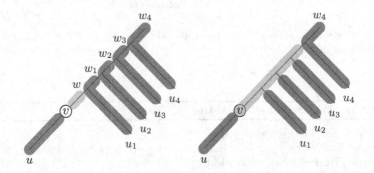

Fig. 8. Illustration of execution of one call to MarkAndUpdate(v). The successive values of Top$(chain_1)$ are $w, w_1, \ldots, w_3$. Finally Top$(chain_1) = w_3$. The nodes $p = parent(w)$, $p_1 = parent(w_1)$, $p_2 = parent(w_2)$ are changing their chains and Prev values. For each of them *chain-height* (for a definition, see below) is decreasing, also their chain goes through a different child. The newly marked node v is *winning* with u_1, u_2, u_3.

In addition to the pseudocodes, the len$[\cdot]$ array is updated whenever operations on chains are performed. Moreover, the maximum (finite value) in this array, i.e., maxgap(M), never increases because a newly created chain only extends if it "wins" with an existing one (see Fig. 8). Hence, it is sufficient to store an array of n buckets containing all len$[\cdot]$ values and retain as maxgap(M)

Function MarkAndUpdate(v)

Comment: Inserts v to M and updates the chains decomposition; see Fig. 8
$chain := FindChain(v)$;
$chain_1 := NewChain(v, \mathsf{Top}(chain))$;
$\mathsf{Top}(chain) := \mathsf{Prev}[v]$; color $\mathsf{Top}(chain)$ in red;
while $\mathsf{Top}(chain_1) \neq root$ **do**
    $chain_2 := FindChain(parent(\mathsf{Top}(chain_1)))$;
    if not $Crossover(chain_1, chain_2)$ **then**
        break;
    // Otherwise we say that v *wins* with w
update $\mathsf{maxgap}(M)$;

---

the maximum index of a non-empty bucket, which never increases. Overall this works in time linear in the number of operations on chains plus $\mathcal{O}(n)$.

To analyze the complexity of the algorithm, let us define *chain-height*$(v) = dist(v, \mathsf{Bottom}(v))$.

Observation 3. *In the algorithm Covers-in-directed-tree, after a node p changes its chain in the function Crossover, we have chain-height$(p) \leq$ sec-height$(p) + 1$. Afterwards, each time p changes its chain in the function Crossover, its chain-height decreases. Consequently, p changes its chain $\mathcal{O}(sec\text{-}height(p))$ times.*

Lemma 2. *The algorithm Covers-in-directed-tree works in $\mathcal{O}(\beta(n))$ time, where $\beta(n)$ is the cost of answering on-line n lowest red ancestor queries.*

Proof. It is enough to show that the total number of "wins" performed in the algorithm is $\mathcal{O}(n)$. Then the total number of iterations of the while-loop in the function MarkAndUpdate is linear.

Due to Observation 3 the total number of wins in the function MarkAndUpdate is amortized by the sum of all numbers *sec-height*(p). Now the thesis follows from Proposition 1.   □

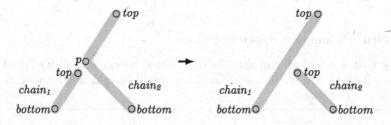

Fig. 9. Graphical illustration of one call to Crossover($chain_1, chain_2$).

To answer Top queries, we can use the following result from [2], where in this case the marked nodes are the red nodes.

Function Crossover($chain_1$, $chain_2$)

Comment: Works as shown in Fig. 9. We have $parent(\mathsf{Top}(chain_1)) \in chain_2$.

$p := parent(\mathsf{Top}(chain_1))$;

if $dist(p, \mathsf{Bottom}(chain_1)) \geq dist(p, \mathsf{Bottom}(chain_2))$ **then**

    return false;

else

    uncolor $\mathsf{Top}(chain_1)$;

    $(\mathsf{Top}(chain_2), \mathsf{Prev}[p], \mathsf{Top}(chain_1)) :=$
    $(\mathsf{Prev}[p], \mathsf{Top}(chain_1), \mathsf{Top}(chain_2))$;

    color $\mathsf{Top}(chain_2)$ in red; **return true**;

Fact 2. Lowest red ancestor queries for a dynamic set of marked nodes in a tree of size n can be answered in $\mathcal{O}(\log n/\log\log n)$ time.

Now, the main result of this section follows from Lemma 2 and Fact 2.

Theorem 1. *All covers in a directed tree with n nodes can be computed in $\mathcal{O}(n \log n/\log\log n)$ time.*

4   Covers of Undirected Trees

Covering an undirected tree is much harder than that of a directed tree. Even the case when the tree is a simple undirected path is nontrivial. For example, the shortest standard cover of a Fibonacci string *abaababaabaab* (and the corresponding directed path) is *abaab*, however it is not true in the undirected case, where the shortest cover is *ab*. A serious difference between covers in directed and covers in undirected trees is shown in the generic example below.

Example 1. Take a full binary tree T_k of height k, subdivide each edge, and then label the higher edge obtained in this division with a, and the lower edge with b (Fig. 10 shows the tree T_4). T_k has $2^{k+2} - 3$ nodes. This tree has $\Omega(k) = \Omega(\log n)$ different covers of the same length (this cannot happen for covers of strings).

4.1 $\mathcal{O}(n^2)$ Time and Space Solution

We say that a set M of simple paths in a tree T *covers* T if each edge of T is on some path in M. We define an auxiliary problem that will be used to test candidates for a cover.

Strips Covering Problem

Input: A set M of simple paths in an undirected tree T (given by their endpoints).

Output: YES if M covers T, NO otherwise.

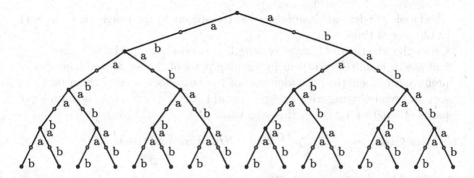

Fig. 10. The undirected tree T_4. It has 44 distinct covers and there are many distinct aperiodic covers of the same length. Generally there are $k(2k+3)$ covers in T_k. The label of every path that starts in a leaf, has length at least two and if it ends with an edge with a letter b, then it has a subpath with label aa, is a cover of T_k. The reverses of such labels are also covers. There are 4 distinct covers of length 6: $bababa, ababab, babaab, baabab$.

Lemma 3. *The Strips Covering Problem can be solved in $\mathcal{O}(|T| + |M|)$ time.*

Proof. Let us root T in an arbitrarily chosen node. We can reduce the problem to the case when each path $\pi \in M$ is a bottom-up path. For a path $\pi = u \rightsquigarrow v$ we compute $w = LCA(u, v)$ and replace π by two paths $u \rightsquigarrow w, v \rightsquigarrow w$. This reduction works in linear time [6].

We use counters for nodes. For each node w initially $count[w] = 0$. Next for each non-empty bottom-up path (u, w) we set $count[u]+=1$, $count[w]-=1$. An edge $(w, parent(w))$ is covered by M if and only if the sum of counters of all descendants of w (including w) is positive. These sums can be easily computed in a bottom-up manner within the required complexity. In the end M covers T if and only if all edges are covered.   □

Lemma 4. *An undirected labeled tree with n nodes has at most $2n - 2$ covers.*

Proof. Take any leaf; the only edge connected to this leaf must be covered by a first or last letter of the cover. An occurrence of the cover must appear on some path in the tree, and each such path is determined by the two nodes on its ends. As one end is fixed (the chosen leaf), there can only be $n - 1$ such paths. This gives at most $2(n - 1)$ possible cover candidates.   □

Theorem 2. *All covers of an undirected tree with n nodes can be computed in $\mathcal{O}(n^2)$ time and space.*

Proof. We group all $\mathcal{O}(n^2)$ paths in the tree by their string labels, and for each of the $k = \mathcal{O}(n)$ candidates from Lemma 4 check if all paths in the group with the same string label as the candidate cover the whole tree. If $M_1, \ldots, M_k$ are these sets of paths, then the solution using Lemma 3 works in $\mathcal{O}(kn + \sum_{i=1}^{k} |M_i|) = \mathcal{O}(n^2)$ time since the sets M_i are pairwise disjoint.

For grouping paths by their string labels we assign identifiers in $\{1, \ldots, n^2\}$ to all paths using the following algorithm:

- The labels of edges are renumbered with integers in the range $\{1, \ldots, n-1\}$ in $\mathcal{O}(n \log n)$ time.
- Given the identifiers of paths of length i, we compute paths of length $i + 1$ and assign identifiers to them by forming pairs of the form: (the identifier of prefix path of length i, the identifier of the last edge), sorting them by radix sort and renumbering with integers from $\{1, \ldots, n^2\}$. If p_{i+1} is the number of paths of length $i + 1$, then this step takes $\mathcal{O}(p_{i+1} + n)$ time.

This gives $\mathcal{O}(n \log n + \sum_{i=2}^{n}(p_i + n)) = \mathcal{O}(n^2)$ time and space. $\qquad \square$

4.2 $\mathcal{O}(n^2 \log n)$ Time and $\mathcal{O}(n)$ Space

Let us recall that there are $\mathcal{O}(n)$ candidates for a cover (see Lemma 4). In this section we show how to check if each of them is a cover in $\mathcal{O}(n \log n)$ time. We use the following auxiliary problem together with a centroid decomposition.

Anchored Covering Problem
Input: Labeled tree T, its node r and a string C; $|T| = n$, $|C| = m$.
Output: All edges of T that are covered by occurrences of C in T that pass through the node r.

Lemma 5. *Anchored Covering Problem for a tree with n nodes can be solved in $\mathcal{O}(n)$ time.*

Proof. Let us root T in r. We denote by $T_1, T_2, \ldots, T_t$ the subtrees of r. For each $i \in \{1, \ldots, t\}$, we define P_i, S_i as the set of nodes $v \in T_i$ such that the label of the path $v \rightsquigarrow r$ is a prefix and a reverse of a suffix of C, respectively. For $v \in T_i$ we have $v \in P_i$ ($v \in S_i$) if $\mathsf{TreePREF}_C[v] = depth(v)$ ($\mathsf{TreePREF}_{C^R}[v] = depth(v)$), respectively, where $depth(v)$ denotes the distance of v from the root r), so these sets can be computed in $\mathcal{O}(n)$ time (cf. Lemma 1).

A node v is called **good** if it is an endpoint of an undirected path labeled C and passing through r.

Claim 1. $v \in T_i$ is good if there is a node $u \in T_j, j \neq i$ such that

$$(depth(u) + depth(v) = m) \text{ and } (u \in S_j, v \in P_i \text{ or } u \in P_j, v \in S_i).$$

We use the function MarkNodes to mark good nodes. In the pseudocode we denote by $depths(W)$ a list of depths of nodes in W. We represent each of the sets $PrefDepths$, $SufDepths$ by its characteristic vector so that adding a set $depths(W)$ to them can be done in time proportional to the number of inserted elements.

In the algorithm AnchoredCovering we are implementing the claim and process i's in the ascending and descending orders; in the first pass $PrefDepths$, $SufDepths$ consist of depths of P_j, S_j of $u \in T_j$ for all $j < i$ and in the second pass for $j > i$. Each pass works in $\mathcal{O}(n)$ time. In the end all edges that are

---

Function MarkNodes(i)

Comment: (some) good nodes in the subtree T_i are marked

foreach v in P_i **do**
    if $m - depth(v) \in SufDepths$ **then**
        mark v as a good node;
foreach v in S_i **do**
    if $m - depth(v) \in PrefDepths$ **then**
        mark v as a good node;
$PrefDepths := PrefDepths \cup depths(P_i)$;
$SufDepths := SufDepths \cup depths(S_i)$;

Algorithm 2: AnchoredCovering

$PrefDepths := SufDepths := \emptyset$;
for $i := 1$ **to** t **do** $MarkNodes(i)$;
$PrefDepths := SufDepths := \emptyset$;
for $i := t$ **down to** 1 **do** $MarkNodes(i)$;

return *all edges on paths from marked nodes to* r;

---

located on paths $v \rightsquigarrow root$ for good nodes v are computed in $\mathcal{O}(n)$ time with a simple bottom-up traversal. □

Let $T_1, T_2, \ldots, T_t$ be the connected components obtained after removing a node r from T. The node r is called a *centroid* of T if $|T_i| \leq n/2$ for all T_i.

The *centroid decomposition* of T, $CDecomp(T)$, is defined recursively as:

$$CDecomp(T) = \{(T, r)\} \cup \bigcup_{i=1}^{t} CDecomp(T_i).$$

Every tree has a centroid, see [15], and a centroid of a tree can be computed in $\mathcal{O}(n)$ time.

The recursive definition of $CDecomp(T)$ implies the following bounds.

Fact 3. For a tree T with n nodes, the total size of all subtrees in $CDecomp(T)$ is $\mathcal{O}(n \log n)$. The decomposition $CDecomp(T)$ can be computed in $\mathcal{O}(n \log n)$ time and $\mathcal{O}(n)$ additional space (excluding the space used for storing the output).

Theorem 3. *One can check if a given string covers an undirected tree with n nodes in $\mathcal{O}(n \log n)$ time and $\mathcal{O}(n)$ space.*

Proof. A known property of a centroid decomposition is that for every path π in T there exists an element $(T', r') \in CDecomp(T)$ such that π is a path in T' that passes through r'. We generate the pairs (T', r') forming $CDecomp(T)$

one by one, for each of them solve an instance of the Anchored Covering Problem, and mark all edges returned by each instance. The complexity follows by Fact 3. □

Corollary 1. *All covers of an undirected tree can be computed in $\mathcal{O}(n^2 \log n)$ time and $\mathcal{O}(n)$ space.*

References

1. Alatabbi, A., Rahman, M.S., Smyth, W.F.: Computing covers using prefix tables. Discret. Appl. Math. **212**, 2–9 (2016). https://doi.org/10.1016/j.dam.2015.05.019
2. Alstrup, S., Husfeldt, T., Rauhe, T.: Marked ancestor problems. In: 39th Annual Symposium on Foundations of Computer Science, FOCS '98, pp. 534–544. IEEE Computer Society, Palo Alto, California, USA (1998). https://doi.org/10.1109/SFCS.1998.743504
3. Antoniou, P., Crochemore, M., Iliopoulos, C.S., Jayasekera, I., Landau, G.M.: Conservative string covering of indeterminate strings. In: Holub, J., Zdárek, J. (eds.) Proceedings of the Prague Stringology Conference 2008, Prague, Czech Republic, 1–3 September 2008, pp. 108–115. Prague Stringology Club, Department of Computer Science and Engineering, Faculty of Electrical Engineering, Czech Technical University in Prague (2008). http://www.stringology.org/event/2008/p10.html
4. Apostolico, A., Farach, M., Iliopoulos, C.S.: Optimal superprimitivity testing for strings. Inf. Process. Lett. **39**(1), 17–20 (1991). https://doi.org/10.1016/0020-0190(91)90056-N
5. Barton, C., Kociumaka, T., Liu, C., Pissis, S.P., Radoszewski, J.: Indexing weighted sequences: neat and efficient. Inf. Comput. **270**, 104462 (2020). https://doi.org/10.1016/j.ic.2019.104462
6. Bender, M.A., Farach-Colton, M.: The level ancestor problem simplified. In: Rajsbaum, S. (ed.) LATIN 2002. LNCS, vol. 2286, pp. 508–515. Springer, Heidelberg (2002). https://doi.org/10.1007/3-540-45995-2_44
7. Breslauer, D.: An on-line string superprimitivity test. Inf. Process. Lett. **44**(6), 345–347 (1992). https://doi.org/10.1016/0020-0190(92)90111-8
8. Brlek, S., Lafrenière, N., Provençal, X.: Palindromic complexity of trees. In: Potapov, I. (ed.) DLT 2015. LNCS, vol. 9168, pp. 155–166. Springer, Cham (2015). https://doi.org/10.1007/978-3-319-21500-6_12
9. Charalampopoulos, P., Radoszewski, J., Rytter, W., Waleń, T., Zuba, W.: Computing covers of 2D-strings. In: Gawrychowski, P., Starikovskaya, T. (eds.) 32nd Annual Symposium on Combinatorial Pattern Matching, CPM 2021, 5–7 July 2021, Wrocław, Poland. LIPIcs, vol. 191, pp. 12:1–12:20. Schloss Dagstuhl - Leibniz-Zentrum für Informatik (2021). https://doi.org/10.4230/LIPIcs.CPM.2021.12
10. Crochemore, M., et al.: The maximum number of squares in a tree. In: Kärkkäinen, J., Stoye, J. (eds.) CPM 2012. LNCS, vol. 7354, pp. 27–40. Springer, Heidelberg (2012). https://doi.org/10.1007/978-3-642-31265-6_3
11. Crochemore, M., Iliopoulos, C.S., Kociumaka, T., Radoszewski, J., Rytter, W., Waleń, T.: Covering problems for partial words and for indeterminate strings. Theor. Comput. Sci. **698**, 25–39 (2017). https://doi.org/10.1016/j.tcs.2017.05.026
12. Funakoshi, M., Nakashima, Y., Inenaga, S., Bannai, H., Takeda, M.: Computing maximal palindromes and distinct palindromes in a trie. In: Holub, J., Zdárek, J. (eds.) Prague Stringology Conference 2019, Prague, Czech Republic, 26–28

August 2019, pp. 3–15. Czech Technical University in Prague, Faculty of Information Technology, Department of Theoretical Computer Science (2019). http://www.stringology.org/event/2019/p02.html

13. Gawrychowski, P., Kociumaka, T., Rytter, W., Waleń, T.: Tight bound for the number of distinct palindromes in a tree. In: Iliopoulos, C.S, Puglisi, S., Yilmaz, E. (eds.) SPIRE 2015. LNCS, vol. 9309, pp. 270–276. Springer, Cham (2015). https://doi.org/10.1007/978-3-319-23826-5_26

14. Gawrychowski, P., Kociumaka, T., Rytter, W., Waleń, T.: Tight bound for the number of distinct palindromes in a tree. CoRR **abs/2008.13209** (2020). arXiv:2008.13209

15. Harary, F.: Graph Theory. Reading, Addison-Wesley, Boston, MA (1994)

16. Iliopoulos, C.S., Mohamed, M., Mouchard, L., Perdikuri, K., Smyth, W.F., Tsakalidis, A.K.: String regularities with don't cares. Nordic J. Comput. **10**(1), 40–51 (2003)

17. Kikuchi, N., Hendrian, D., Yoshinaka, R., Shinohara, A.: Computing covers under substring consistent equivalence relations. In: Boucher, C., Thankachan, S.V. (eds.) SPIRE 2020. LNCS, vol. 12303, pp. 131–146. Springer, Cham (2020). https://doi.org/10.1007/978-3-030-59212-7_10

18. Kociumaka, T., Pachocki, J., Radoszewski, J., Rytter, W., Waleń, T.: Efficient counting of square substrings in a tree. Theor. Comput. Sci. **544**, 60–73 (2014). https://doi.org/10.1016/j.tcs.2014.04.015

19. Kociumaka, T., Radoszewski, J., Rytter, W., Waleń, T.: String powers in trees. Algorithmica **79**(3), 814–834 (2017). https://doi.org/10.1007/s00453-016-0271-3

20. Kociumaka, T., Radoszewski, J., Wiśniewski, B.: Subquadratic-time algorithms for abelian stringology problems. In: Kotsireas, I.S., Rump, S.M., Yap, C.K. (eds.) MACIS 2015. LNCS, vol. 9582, pp. 320–334. Springer, Cham (2016). https://doi.org/10.1007/978-3-319-32859-1_27

21. Kociumaka, T., Radoszewski, J., Wiśniewski, B.: Subquadratic-time algorithms for abelian stringology problems. AIMS Med. Sci. **4**(3), 332–351 (2017). https://doi.org/10.3934/ms.2017.3.332

22. Kosaraju, S.R.: Efficient tree pattern matching (preliminary version). In: 30th Annual Symposium on Foundations of Computer Science, Research Triangle Park, North Carolina, USA, 30 October–1 November 1989, pp. 178–183. IEEE Computer Society (1989). https://doi.org/10.1109/SFCS.1989.63475

23. Matsuda, S., Inenaga, S., Bannai, H., Takeda, M.: Computing abelian covers and abelian runs. In: Holub, J., Zdárek, J. (eds.) Proceedings of the Prague Stringology Conference 2014, Prague, Czech Republic, 1–3 September 2014, pp. 43–51. Department of Theoretical Computer Science, Faculty of Information Technology, Czech Technical University in Prague (2014). http://www.stringology.org/event/2014/p05.html

24. Moore, D.W.G., Smyth, W.F.: An optimal algorithm to compute all the covers of a string. Inf. Process. Lett. **50**(5), 239–246 (1994). https://doi.org/10.1016/0020-0190(94)00045-X

25. Moore, D.W.G., Smyth, W.F.: A correction to "An optimal algorithm to compute all the covers of a string". Inf. Process. Lett. **54**(2), 101–103 (1995). https://doi.org/10.1016/0020-0190(94)00235-Q

26. Popa, A., Tanasescu, A.: An output-sensitive algorithm for the minimization of 2-dimensional string covers. In: Gopal, T.V., Watada, J. (eds.) TAMC 2019. LNCS, vol. 11436, pp. 536–549. Springer, Cham (2019). https://doi.org/10.1007/978-3-030-14812-6_33

27. Radoszewski, J., Straszyński, J.: Efficient computation of 2-covers of a string. In: Grandoni, F., Herman, G., Sanders, P. (eds.) 28th Annual European Symposium on Algorithms, ESA 2020, 7–9 September 2020, Pisa, Italy (Virtual Conference). LIPIcs, vol. 173, pp. 77:1–77:17. Schloss Dagstuhl - Leibniz-Zentrum für Informatik (2020). https://doi.org/10.4230/LIPIcs.ESA.2020.77
28. Shibuya, T.: Constructing the suffix tree of a tree with a large alphabet. IEICE Trans. Fundam. Electron. Commun. Comput. Sci. **E86–A(5)**, 1061–1066 (2003)
29. Sugahara, R., Nakashima, Y., Inenaga, S., Bannai, H., Takeda, M.: Computing runs on a trie. In: Pisanti, N., Pissis, S.P. (eds.) 30th Annual Symposium on Combinatorial Pattern Matching, CPM 2019, June 18–20, 2019, Pisa, Italy. LIPIcs, vol. 128, pp. 23:1–23:11. Schloss Dagstuhl - Leibniz-Zentrum für Informatik (2019). https://doi.org/10.4230/LIPIcs.CPM.2019.23

Compression

Grammar Index by Induced Suffix Sorting

Tooru Akagi[1], Dominik Köppl[2(✉)], Yuto Nakashima[1], Shunsuke Inenaga[1,3],
Hideo Bannai[2], and Masayuki Takeda[1]

[1] Department of Informatics, Kyushu University, Fukuoka, Japan
{toru.akagi,yuto.nakashima,inenaga,takeda}@inf.kyushu-u.ac.jp
[2] M&D Data Science Center, Tokyo Medical and Dental University, Tokyo, Japan
{koeppl.dsc,hdbn.dsc}@tmd.ac.jp
[3] PRESTO, Japan Science and Technology Agency, Kawaguchi, Japan

Abstract. We propose a new compressed text index built upon a grammar compression based on induced suffix sorting [Nunes et al., DCC'18]. We show that this grammar exhibits a locality sensitive parsing property, which allows us to specify, given a pattern P, certain substrings of P, called *cores*, that are similarly parsed in the text grammar whenever these occurrences are extensible to occurrences of P. Supported by the cores, given a pattern of length m, we can locate all its occ occurrences in a text T of length n within $\mathcal{O}(m \lg |\mathcal{S}| + \mathrm{occ}_C \lg |\mathcal{S}| \lg n + \mathrm{occ})$ time, where $\mathcal{S}$ is the set of all characters and non-terminals, occ is the number of occurrences, and occ_C is the number of occurrences of a chosen core C of P in the right hand side of all production rules of the grammar of T. Our grammar index requires $\mathcal{O}(g)$ words of space and can be built in $\mathcal{O}(n)$ time using $\mathcal{O}(g)$ working space, where g is the sum of the lengths of the right hand sides of all production rules. We practically evaluate that our proposed index excels at locating long patterns in highly-repetitive texts. Our implementation is available at https://github.com/TooruAkagi/GCIS_Index.

Keywords: Grammar compression · Locality sensitive parsing · Induced suffix sorting · Text indexing data structure

1 Introduction

Compressed text indexes have become the standard tool for maintaining highly-repetitive texts when full-text search queries like locating all occurrences of a pattern are of importance. When working on indexes on highly-repetitive data, a desired property is to have a *self-index*, i.e., a data structure that supports queries on the underlying text without storing the text in its plain form. One type of such self-indexes are *grammar indexes*, which are an augmentation of the admissible grammar [20] produced by a grammar compressor. Grammar indexes

© Springer Nature Switzerland AG 2021
T. Lecroq and H. Touzet (Eds.): SPIRE 2021, LNCS 12944, pp. 85–99, 2021.
https://doi.org/10.1007/978-3-030-86692-1_8

exhibit strong compression ratios for semi-automatically generated or highly-repetitive texts. Unlike other indexes that perform pattern matching stepwise character-by-character, some grammar indexes have locality sensitive parsing properties, which allow them to match certain non-terminals of the admissible grammar built upon the pattern with the non-terminals of the text. Such a property helps us to perform fewer comparisons, and thus speeds up pattern matching for particularly long patterns, which could be large gene sequences in a genomic database or source code files in a database maintaining source code. Here, our focus is set on indexes that support locate(P) queries retrieving the starting positions of all occurrences of a given pattern P in a given text.

1.1 Our Contribution

Our main contribution is the discovery of a locality sensitive parsing property in the grammar produced by the grammar compression by induced sorting (GCIS) [29], which helps us to answer locate with an index built upon GCIS with the following bounds:

Theorem 1. *Given a text T of length n, we can compute an indexing data structure on T in $\mathcal{O}(n)$ time, which can locate all occ occurrences of a given pattern of length m in $\mathcal{O}(m \lg |\mathcal{S}| + occ_C \lg n \lg |\mathcal{S}| + occ)$ time, where $\mathcal{S}$ is the set of characters and non-terminals of the GCIS grammar and occ_C is the number of occurrences in the right side of the production rules of the GCIS grammar of a selected core of the pattern, where a core is a string of symbols of the grammar of P defined in Sect. 4.1. Our index uses $\mathcal{O}(g)$ words of working space, where g is the sum of the lengths of the right hand sides of all production rules.*

Similar properties hold for other grammars such as the signature encoding [25], ESP [7], HSP [16], the Rsync parse [17], or the grammar of [3, Sect. 4.2]. A brief review of these and other self-indexes follows.

1.2 Related Work

With respect to indexing a grammar for answering locate, the first work we are aware of is due to [5] who studied indices built upon so-called *straight-line programs (SLPs)*. An SLP is a context-free grammar representing a single string in the Chomsky normal form.

Other research focused on particular types of grammar, such as the ESP-index [24,31,32], an index [4] combining Re-Pair [22] with the Lempel–Ziv-77 parsing [34], a dynamic index [26] based on signature encoding [25], the Lyndon SLP [33], or the grammar index of [3]. For the experiments in Sect. 5, we will additionally have a look at other self-indexes capable of locate-queries. There, we analyze Burrows–Wheeler-transform (BWT) [2]-based approaches, namely the FM-index [15] and the r-index [18].

Finally, the grammar GCIS has other interesting properties besides being locality sensitive. [28] showed how to compute the suffix array and the longest-common-prefix array from GCIS during a decompression step restoring the original text. Recently, [8] showed how to compute the BWT directly from the GCIS grammar.

2 Preliminaries

With lg we denote the logarithm to base two (i.e., $\lg = \log_2$). Given two integers i, j, we denote the interval $[i..j] = \{i, i + 1, \ldots, j - 1, j\}$, with $[i..j] = \{\}$ if $i > j$. Our computational model is the standard word RAM with machine word size $\Omega(\lg n)$, where n denotes the length of a given input string $T[1..n]$, which we call *the text*, whose characters are drawn from an integer alphabet Σ of size $n^{\mathcal{O}(1)}$. We call the elements of Σ *characters*. For a string $S \in \Sigma^*$, we denote with $S[i..]$ its i-th suffix, and with $|S|$ its length. The order $<$ on the alphabet Σ induces a lexicographic order on Σ^*, which we denote by $\prec$.

2.1 Induced Suffix Sorting

SAIS [27] is a linear-time algorithm for computing the suffix array [23]. We briefly review the parts of SAIS important for constructing the GCIS grammar. SAIS assigns each suffix a type, which is either L or S: $T[i..]$ is an L suffix if $T[i..] \succ T[i + 1..]$, or $T[i..]$ is an S suffix otherwise, i.e., $T[i..] \prec T[i + 1..]$, where we stipulate that $T[|T|]$ is always type S. Since it is not possible that $T[i..] = T[i+1..]$, SAIS assigns each suffix a type. An S suffix $T[i..]$ is additionally an S^* suffix (also called LMS suffix in [27]) if $T[i-1..]$ is an L suffix. The substring between two succeeding S^* suffixes is called an *LMS substring*. In other words, a substring $T[i..j]$ with $i < j$ is an LMS substring if and only if $T[i..]$ and $T[j..]$ are S^* suffixes and there is no $k \in [i + 1..j - 1]$ such that $T[k..]$ is an S^* suffix. Regarding the defined types, we make no distinction between suffixes and their starting positions (e.g., the statements that (a) $T[i]$ is type L and (b) $T[i..]$ is an L suffix are equivalent). In fact, we can determine L and S positions solely based on their succeeding positions with the equivalent definition: if $T[i] > T[i + 1]$, then $T[i]$ is L; if $T[i] < T[i + 1]$, then $T[i]$ is S; finally, if $T[i] = T[i + 1]$, then $T[i]$ has the same type as $T[i + 1]$.

The LMS substrings of $\#T$ for $\#$ being a special character smaller than all characters appearing in T such that $\#T$ starts with an S^* position, induce a factorization of $T = F_1 \cdots F_z$, where each factor starts with an LMS substring. We call this factorization *LMS-factorization*. By replacing each factor F_i by the lexicographic rank of its respective LMS substring[1], we obtain a string $T^{(1)}$ of these ranks. We recurse on $T^{(1)}$ until we obtain a string $T^{(\tau_T - 1)}$ whose rank-characters are all unique or whose LMS-factorization consists of at most two factors.

---

[1] For SAIS to work, it uses a slightly different order on the LMS substrings, called LMS-order. It differs from the lexicographic order when comparing two LMS substrings, where one of them is a prefix of the other. In such a case, the LMS-order would give the longer string a smaller rank.

2.2   Constructing the Grammar

We assign each computed factor $F_j^{(h)}$ a non-terminal $X_j^{(h)}$ such that $X_j^{(h)} \to F_j^{(h)}$, but omit the delimiter #. The order of the non-terminals $X_j^{(h)}$ is induced by the lexicographic order of their respective LMS-substrings. We now use the non-terminals instead of the lexicographic ranks in the recursive steps. If we set $X^{(\tau_T)} \to T^{(\tau_T - 1)}$ as the start symbol, we obtain a context-free grammar $\mathcal{G}_T := (\Sigma, \Gamma, \pi, X^{(\tau_T)})$, where Γ is the set of non-terminals and a function $\pi : \Gamma \to (\Sigma \cup \Gamma)^+$ that applies (production) rules. For simplicity, we stipulate that $\pi(c) = c$ for $c \in \Sigma$. Let g denote the sum of the lengths of the right hand sides of all grammar rules. We say that a non-terminal ($\in \Gamma$) or a character ($\in \Sigma$) is a *symbol*, and denote the set of characters and non-terminals with $\mathcal{S} := \Sigma \cup \Gamma$. We understand π also as a string morphism $\pi : \mathcal{S}^* \to \mathcal{S}^*$ by applying π on each symbol of the input string. This allows us to define the *expansion* $\pi^*(X)$ of a symbol X, which is the iterative application of π until obtaining a string of characters, i.e., $\pi^*(X) \subset \Sigma^*$ and $\pi^*(X^{(\tau_T)}) = T$. Since $\pi(X)$ is deterministically defined, we use to say *the right hand side of X* for $\pi(X)$.

Lemma 1 ([29]). *The GCIS grammar $\mathcal{G}_T$ can be constructed in $\mathcal{O}(n)$ time. $\mathcal{G}_T$ is reduced, meaning that we can reach all non-terminals of Γ from $X^{(\tau_T)}$.*

$\mathcal{G}_T$ can be visualized by its derivation tree $\mathcal{T}_T$, which has $X^{(\tau_T)}$ as its root. Each rule $X_k^{(h)} \to X_i^{(h-1)} \cdots X_j^{(h-1)}$ defines a node $X_k^{(h)}$ having $X_i^{(h-1)}, \ldots, X_j^{(h-1)}$ as its children. The height of $\mathcal{T}_T$ is $\tau_T = \mathcal{O}(\lg n)$ because the number of LMS substrings of $T^{(h)}$ is at most half of the length of $T^{(h)}$ for each recursion level h. The leaves of $\mathcal{T}_T$ are the terminals at height 0 that constitute the characters of the text T. Reading the nodes on height $h \in [0..\tau_T - 1]$ from left to right gives $T^{(h)}$ with $T^{(0)} = T$. Note that we use $\mathcal{T}_T$ only as a conceptional construct since it would take $\mathcal{O}(n)$ words of space. Instead, we merge (identical) subtrees of the same non-terminal together to form a directed acyclic graph DAG, which is implicitly represented by π as follows:

By construction, each non-terminal appears exactly in one height of $\mathcal{T}_T$. We can therefore separate the non-terminals into the sets $\Gamma^{(1)}, \ldots, \Gamma^{(\tau_T)}$ such that a non-terminal of height h belongs to $\Gamma^{(h)}$. More precisely, π maps a non-terminal on height $h > 1$ to a string of symbols on height $h - 1$. Hence, the grammar is acyclic.

3   GCIS Index

In what follows, we want to show that we can augment $\mathcal{G}_T$ with auxiliary data structures for answering locate. Our idea stems from the classic pattern matching algorithm with the suffix tree [19, APL1]. The key difference is that we search the core of a pattern in the right hand sides of the rules. For that, we make use of the generalized suffix tree GST built upon the right hand sides of all rules separated by a special delimiter symbol $ being smaller than all symbols.

Specifically, we rank the rules such that $\{X_1, \ldots, X_{|\Gamma|}\} = \Gamma$ (this ranking will be fixed later), and set $R := \pi(X_1)\$\pi(X_2)\$ \cdots \pi(X_{|\Gamma|})\$$. Since we have a budget of $\mathcal{O}(g)$ words, we can afford to use a plain pointer-based tree topology. Each leaf λ stores a pointer to the non-terminal $X^{(h)}$ and an offset o such that $\pi(X^{(h)})[o..]$ is a prefix of λ's string label. Next, we need the following operations on GST: First, $\text{lca}(u, v)$ gives the lowest common ancestor (LCA) of two nodes u and v. We can augment GST with the data structure of [1] in linear time and space in the number of nodes of GST. This data structure answers lca in constant time. Next, $\text{child}(u, c)$ gives the child of the node u connected to u with an edge having a label starting with $c \in \Gamma$. Our GST implementation answers child in $\mathcal{O}(\lg |\mathcal{S}|)$ time. For that, each node stores the pointers to its children in a binary search tree with the first symbol of each connecting edge as key. Finally, $\text{string\_depth}(v)$ returns the string depth of a node v, i.e., the length of its string label, which is the string read from the edge labels on the path from the root to v. We can compute and store the string depth of each node during its construction. The operation child allows us to compute the *locus* of a string S, i.e., the highest GST node u whose string label has S as a prefix, in $\mathcal{O}(|S| \lg |\mathcal{S}|)$ time. For each $\pi(X)$, we augment the locus u of $\pi(X)\$$ with a pointer to X such that we can perform $\text{lookup}(S)$ returning the non-terminal X with $\pi(X) = S$ or an invalid symbol $\perp$ if such an X does not exist. The time is dominated by the time for computing the locus of S. Finally, all leaves in suffix order are stored in a linked list such that we can traverse the leaves in lexicographic order with respect to their corresponding suffixes.

Linkage to the Grammar. Each rule $X \in \Gamma$ stores an array $X.P$ of $|\pi(X)|$ pointers to the leaves in GST such that the $X.P[i]$ points to the leaf that points back to X and has offset i (its string label has $\pi(X)[i..]$ as a prefix). Additionally, each rule X stores the length of $\pi(X)$, an array $X.L$ of all expansion lengths of all its prefixes, i.e., $X.L[i] := \sum_{j=1}^{i} |\pi^*(\pi(X)[j])|$, and an array $X.R$ of the lengths of the right hand sides of all its prefixes, i.e., $X.R[i] := \sum_{j=1}^{i} |\pi(\pi(X)[j])|$.

LCE Queries. Each internal node v stores a pointer to the leftmost leaf in the subtree rooted at v. With that we can use the function $\text{lce}(X, Y, i, j)$ returning the *longest common extension* (LCE) of $\pi(X)[i..]$ and $\pi(Y)[j..]$ for $X, Y \in \Gamma$ and $i \in [1..|\pi(X)|], j \in [1..|\pi(Y)|]$. We can answer $\text{lce}(X, Y, i, j)$ by selecting the leaves $X.P[i]$ and $Y.P[j]$, retrieve the LCA $\text{lca}(X.P[i], Y.P[j])$ of both leaves, and take its string depth, all in constant time. More strictly speaking, we return $\min(|\pi(X)[i..]|, |\pi(Y)[j..]|, \text{string\_depth}(\text{lca}(X.P[i], Y.P[j])))$, since the delimiter $\$$ is not a unique character, but appears at each end of each right hand side in the underlying string R of GST.

Complexity Bounds. GST can be computed in $\mathcal{O}(g)$ time [13]. The grammar index consists of the GCIS grammar, GST built upon $|R| = g + |\Gamma|$ symbols, and augmented with a data structure for lca [1]. This all takes $\mathcal{O}(g)$ space. Each non-terminal is augmented with an array $X.P$ of pointers to leaves, $X.L$ and $X.R$ storing the expansion lengths of all prefixes of $\pi(X)$, which take again $\mathcal{O}(g)$ space when summing over all non-terminals.

4   Pattern Matching Algorithm

Like [30, Sect. 2], our idea is to first fix a core C of a given pattern P, find the occurrences of C in the text, and then try to extend all these occurrences to occurrences of P.

4.1   Cores

A core C is a string of symbols of the GCIS grammar $\mathcal{G}_P$ built on the pattern P with the following property: given C consists of consecutive nodes on height $h \geq 0$ in $\mathcal{T}_T$, if there is an occurrence of C in $\mathcal{T}_T$ being a set of nodes on height h that have not the same parent node on height $h+1$, then the expansion of this occurrence of C does *not* lead to an occurrence of P. So for each occurrence of C in $\mathcal{T}_T$ whose expansion is contained in an occurrence of P, this occurrence is a (not necessarily proper) substring of the right hand side of a rule of $\mathcal{G}_T$.

We qualify a core by the difference in the number of occurrences of P and C in $\mathcal{T}_T$. On the one hand, although a character $P[i]$ always qualifies as a core, the appearance of $P[i]$ in T is unlikely to be an evidence of an occurrence of P.

On the other hand, the non-terminal covering most of the characters of P might not be a core. Hence, we aim for the highest possible non-terminal, for which we are sure that it exhibits the core property.

Finding a Core. We determine a core C of P during the computation of the GCIS grammar $\mathcal{G}_P$ of P. During this computation, we want to assure that we only create a new non-terminal for a factor F whenever $\text{lookup}(F) = \bot$; if $\text{lookup}(F) = X$, we borrow the non-terminal X from $\mathcal{G}_T$. By doing so, we ensure that non-terminals of $\mathcal{G}_P$ and $\mathcal{G}_T$ are identical whenever their right hand sides of their productions are equal. In detail, if we create the factors $P^{(h)} = F_1^{(1)} \cdots F_{z_h}^{(h)}$, we first retrieve $Y_i^{(h)} := \text{lookup}(P^{(h)})$ for each $i \in [2..z_h - 1]$. If one of the lookup-queries returns $\bot$, we abort since we can be sure that the pattern does not occur in T. That is because all non-terminals $Y_2^{(h)}, \ldots Y_{z_h-2}^{(h)}$ classify as cores. To see this, we observe that prepending or appending symbols to $P^{(h)}$ does not change the factors $F_2^{(h)}, \ldots, F_{z-1}^{(h)} =: C^{(h)}$.

Correctness. We show that prepending or appending characters to $F_1^{(h)} \; C^{(h)}$ $F_{z_h}^{(h)}$ does not modify the computed factorization of $C^{(h)} = F_2^{(h)}, \ldots, F_{z-1}^{(h)}$. What we show is that we cannot change the type of any position $C^{(h)}[i]$ to S*: Firstly, the type of a position (S or L) depends only on its succeeding position, and hence prepending cannot change the type of a position in $C^{(h)}$. Secondly, appending characters can either prolong $F_{z_h}^{(h)}$ or create a new factor $F_{z_h+1}^{(h)}$ since $F_{z_h}^{(h)}$ starts with S*, and therefore appending cannot change $C^{(h)}$. An additional insight is that on the one side, prepending character can only introduce a new factor or extend $F_1^{(h)}$. On the other side, appending characters can introduce at most one new S* position in $F_{z_h}^{(h)}$ that can make it split into two factors. We will need this observation later for extending the core to the pattern.

The construction of $\mathcal{G}_P$ iterates the LMS factorization until we are left with a string of symbols $P^{(\tau_P)}$ whose LMS factorization consists of at most two factors. In that case, we partition $P^{(\tau_P)}$ into three substrings C_pCC_s with C_p and C_s possibly empty, and defined by one of the following mutually exclusive conditions: (1) If the LMS factorization consists of two non-empty factors $F_1 \cdot F_2$, then C_p is F_1. (2) Given $P^{(\tau_P)} = (P^{(\tau_P)}[j_1])^{c_1} \cdots (P^{(\tau_P)}[j_k])^{c_k}$ is the run-length-encoded representation of P with $1 = j_1 < \ldots < j_k = |P^{(\tau_P)}|$, $c_{j_i} \geq 1$ for $i \in [1..k]$, and $P[j_i] \neq P[j_{i+1}]$ for $i \in [1..k-1]$, we set $C_s \leftarrow (P^{(\tau_P)}[j_k])^{c_k}$ if $P^{(\tau_P)}[j_k] < P^{(\tau_P)}[j_{k-1}]$. (3) In the other cases, C_p and/or C_s are empty.

To see why C is a core, we only have to check the case when C_s is empty. The other cases have already been covered by the aforementioned analysis of the cores on the lower heights. If C_s is empty, then C ends with P, and as a border case, the last position of C is S*. In that case, appending a symbol smaller than $P[m]$ to $F_1^{(h)}C$ changes the type of the last position of C to L. If we append a symbol larger than $P[m]$, then the last position of C becomes S, but does not become S* since $P^{(\tau_P)}[j_k] > P^{(\tau_P)}[j_{k-1}]$ due to construction (otherwise C_s would not be empty).

In total, there are symbols $A^{(1)}, \ldots, A^{(\tau_P-1)}$ and $S^{(1)}, \ldots, S^{(\tau_P-1)}$ such that

$$P = \pi(F_1^{(1)} \cdots F_1^{(\tau_P-2)} A^{(1)} \cdots A^{(\tau_P-2)} CS^{(\tau_P-2)} \cdots S^{(1)} F_{z_{\tau_P-2}}^{(\tau_P-2)} \cdots F_{z_1}^{(1)}), \quad (1)$$

and $A^{(h)}, S^{(h)} \in \Gamma^{(h)}$ are cores of P, while $F_1^{(h)}, F_{z_h}^{(h)} \in (\Gamma^{(h-1)})^*$ are factors.

4.2 Matching with GST

Having C, we now switch to GST and use it to find all DAG parents of C, whose number we denote by $occ_C \in \mathcal{O}(g)$. This number is also the number of occurrences of C in the right hand sides of all rules of $\mathcal{G}_T$. Having these parents, we want to find all lowest DAG ancestors of C whose expansions are large enough to not only cover C but also P by extending C to its left and right side—see Fig. 1 for a sketch. We proceed as follows: We first compute the locus v of C in GST in $\mathcal{O}(|C| \lg |\mathcal{S}|)$ time via child. Subsequently, we take the pointer to the leftmost leaf in the subtree rooted at v, and then process all leaves in this subtree by using the linked list of leaves. For each such leaf λ, we compute a path in form of a list λ_L from the non-terminal containing C on its right hand side up to an ancestor of it that has an expansion large enough to cover P if we would expand the contained occurrence of C to P. We do so as follows: Each of these leaves stores a pointer to a non-terminal X and a starting position i such that we know that $\pi^*(X)[i..]$ starts with $\pi^*(C)$. By knowing the expansion lengths $X.L[|\pi(X)|]$, $X.L[i-1]$, and $|\pi^*(C)|$, we can judge whether the expansion of X has enough characters to be able to extend its occurrence of C to P. If it has enough characters, we put (X,i) onto λ_L such that we know that $\pi^*(X)[X.L[i-1]+1..]$ has C as a prefix. If X does not have enough characters, we exchange C with X and recurse on finding a non-terminal with a larger expansion. By doing so, we visit at most $\tau_T = \mathcal{O}(\lg n)$ non-terminals per occurrence of C in the right hand sides of $\mathcal{G}_T$.

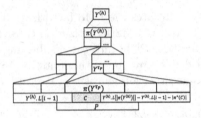

Fig. 1. Deriving a non-terminal $Y^{(h)}$ with $\pi^*(Y^{(h)})$ containing P from a non-terminal $Y^{(\tau_P)}$ with $\pi^*(Y^{(\tau_P)})$ containing $\pi^*(C)$. The expansion of none of the descendants of $Y^{(h)}$ towards $Y^{(\tau_P)}$ is large enough for extending its contained occurrence of $\pi^*(C)$ to an occurrence of P. We can check the expansion lengths of the substrings in $\pi(Y^{(h)})$ via the array $Y^{(h)}.L$.

We perform all operations in $\mathcal{O}(\mathrm{occ}_C \tau_T \lg |\mathcal{S}|)$ time because we query child in every recursion step.

The previous step computes, for each accessed leaf λ, a list λ_L containing a DAG path $(Y^{(h)}, \ldots, Y^{(\tau_P)})$ of length $\mathcal{O}(\tau_T)$ and an offset $o^{(\tau_P)}$ such that $Y^{(\tau_P)}[o^{(\tau_P)}..]$ starts with C. By construction, these paths cover all occurrences of C in $\mathcal{T}_T$. Note that we process the DAG node $Y^{(\tau_P)}$ (but for different offsets $o^{(\tau_P)}$) as many times as C occurs in $\pi(Y^{(\tau_P)})$). In what follows, we try to expand the occurrence of C captured by $Y^{(\tau_P)}$ and $o^{(\tau_P)}$ to an occurrence of P.

Naively, we would walk down from $Y^{(\tau_P)}[o^{(\tau_P)}]$ to the character level and extend the substring $\pi^*(C)$ in both directions by character-wise comparison with P. However, this would take $\mathcal{O}(\mathrm{occ}_C m \tau_T)$ time since such a non-terminal $Y^{(h)}$ is of height $\mathcal{O}(\tau_T)$. Our claim is that we can perform the computation in $\mathcal{O}(m + \mathrm{occ}_C \tau_T)$ time with the aid of lce and an amortization argument.

For that, we use Eq. (1), which allows us to use LCE queries in the sense that we can try to extend an occurrence of C with an already extended occurrence (that maybe does not match P completely). For the explanation, we only focus on extending all occurrences of C to the right to CC_s (the left side side is done symmetrically). We maintain an array D of length τ_P storing pairs $(X^{(h)}, \ell_h)$ for each height $h \in [1..\tau_P - 1]$ such that $\pi(X^{(h)})$ has the currently longest extension of length ℓ_h with the core $S^{(h-1)}$ of P in common (cf. Eq. (1)). By maintaining D, we can first query lce with the specific non-terminal in D, and then resort to plain symbol comparison. We descend to the child where the mismatch happens and recurse until reaching the character level of $\mathcal{T}_T$. This all works since by the core property the mismatch of a child means that there is a mismatch in the expansion of this child. Since a plain symbol comparison with matching symbols lets us exchange the currently used non-terminal in D with a longer one, we can bound (a) the total number of naive symbol matches to $\mathcal{O}(m)$ and (b) the total number of naive symbol mismatches and LCE queries to $\mathcal{O}(\mathrm{occ}_C \tau_T)$.

Finding the Starting Positions. It is left to compute the starting position in T of each occurrence captured by an element in W. We can do this similarly to computing the pre-order ranks in a tree: For each pair $(X, \ell) \in W$, climb up DAG from X to the root while accumulating the expansion lengths of all left siblings of the nodes we visit (we can make use of $X.L$ for that). If this accumulated length is s, then $\ell + s$ is the starting position of the occurrence captured by (X, ℓ). However, this approach would cost $\mathcal{O}(\tau_T)$ time per element of W. Here, we use the amortization argument of [6, Sect. 5.2], which works if we augment, in a pre-computation step, each non-terminal X in Γ with (a) a pointer to the lowest ancestor Y_X on every path from X to the DAG root that has X at least twice as a descendant, and (b) the lengths of the expansions of the left siblings of the child of Y_X being a parent of X or X itself. By doing so, when taking a pointer of a non-terminal X to its ancestor Y_X, we know that X has another occurrence in DAG (and thus there is another occurrence of P). Therefore, we can charge the cost of climbing up the tree with the amount of occurrences occ of the pattern.

Total Time. To sum up, we spent $\mathcal{O}(m \lg |\mathcal{S}|)$ time for finding C, $\mathcal{O}(\mathrm{occ}_C \tau_T \lg |\mathcal{S}|)$ time for computing the non-terminals covering C, $\mathcal{O}(m + \mathrm{occ}_C \tau_T)$ time for reducing these non-terminals to W, and $\mathcal{O}(\mathrm{occ})$ time for retrieving the starting positions of the occurrences of P in T from W. To be within our $\mathcal{O}(g)$ space bounds, we can process each DAG parent of C individually, and keep only D globally stored during the whole process. The total additional space is therefore $\mathcal{O}(\tau_T) \subset \mathcal{O}(g)$ for maintaining D and a path for each occurrence of C.

5   Implementation and Experiments

The implementation deviates from theory with respect to the rather large hidden constant factor in the $\mathcal{O}(g)$ words of space. We drop GST, and represent DAG with multiple arrays. For that, we first enumerate the non-terminals as follows: The height and the lexicographic order induce a natural order on the non-terminals in Γ, which are ranked by first their height and secondly by the lexicographic order of their right hand sides, such that we can represent $\Gamma = \{X_1, \ldots, X_{|\Gamma|}\}$. By stipulating that all characters are lexicographically smaller than all non-terminals, we obtain the property that $\pi(X_i) \prec \pi(X_{i+1})$ for all $i \in [1..|\Gamma| - 1]$. In the following, we first present a plain representation of DAG, called GCIS-nep, then give our modified locate algorithm, and subsequently present a compressed version of DAG using universal coding, called GCIS-uni. Finally, we evaluate both implementations in Sect. 5.

   Our first implementation, called GCIS-nep[2], represents each symbol with a 32-bit integer. We use $R := \prod_{i=1}^{|\Gamma|} \pi(X_i)$ again, but omit the delimiters $\$$ separating the right hand sides. To find the right hand side of a non-terminal X_i, we create an array of positions $Q[1..|\Gamma|]$ such that $Q[i]$ points to the starting position of $\pi(X_i)$ in R. Finally, we create an array $L[1..|\Gamma|]$ storing the length of the expansion $|\pi^*(X_i)|$ in $L[i]$, for each non-terminal X_i. Due to the stipulated order of the symbols, the strings $R[Q[i]..Q[i+1] - 1]$ are sorted in ascending

---

[2] GCIS-nep stands for GCIS with non-terminals encoded plainly.

Table 1. Sizes of the used datasets and the indexes stored on disk. Sizes are in megabytes [MB].

Dataset	Input size	GCIS-nep	GCIS-uni	ESP-index	FM-index	r-index
COMMONCRAWL	221.180	220.119	138.856	156.006	122.575	454.124
DNA	403.927	527.553	327.852	297.001	216.153	2123.817
EINSTEIN.DE	92.758	1.139	0.428	0.697	40.291	1.1458
ENGLISH.001.2	104.857	14.784	7.489	10.464	46.981	14.389
FIB41	267.914	0.001	0.001	0.001	71.305	0.007
INFLUENZA	154.808	23.373	13.871	15.729	53.066	28.775
KERNEL	257.961	21.298	10.469	12.545	125.087	28.947
RS.13	216.747	0.002	0.001	0.002	57.653	0.009
TM29	268.435	0.002	0.001	0.002	69.347	0.009
WORLD LEADERS	46.968	5.415	2.573	3.611	21.097	5.627

order. Hence, we can evaluate lookup(S) for a string S in $\mathcal{O}(|S|\lg|\mathcal{S}|)$ time by a binary search on Q with $i \mapsto R[Q[i]..Q[i+1]-1]$ as keys.

Locate. Our implementation follows theory for computing $\mathcal{G}_P$ and C (cf. Sect. 4.1) in the same time bounds, but deviates after computing the core C: To find all non-terminals whose right hand sides contain C, we linearly scan the right hand sides of all non-terminals on height τ_P, which we can do cache-friendly since the right-hands of R are sorted by the height of their respective non-terminals. This takes $\mathcal{O}(g+|C|)$ time in total with a pattern matching algorithm [21].

Finally, for extending a found occurrence of the core C to an occurrence of P, we follow the naive approach to descend DAG to the character level and compare the expansion with P character-wise, which results in $\mathcal{O}(\mathrm{occ}_C|P|\tau_T)$ time. The total time cost is $\mathcal{O}(g+|P|(\mathrm{occ}_C\tau_T+\lg|\mathcal{S}|))$.

GCIS-uni. To save space, we can leverage universal code to compress the right hand sides of the productions. First, we observe that Q and the first symbols $F := \pi(X_1)[1],\ldots,\pi(X_{|\Gamma|})[1]$ form an ascending sequence, such that we represent both Q and F in Elias–Fano coding [11]. Next, we observe that each right hand side $\pi(X_i)$ form a bitonic sequence: the ranks of the first ℓ_i symbols are non-decreasing, while rest of the ranks are non-increasing. Our idea is to store ℓ_i and the rest of $\pi(X_i)[2..]$ in delta-coding, i.e., $\Delta[i][k] := |\pi(X_i)[k]-\pi(X_i)[k-1]|$ for $k \in [2..|\pi(X_i)|]$, which is stored in Elias-γ code [12]. Although $\pi(X_i)[k]-\pi(X_i)[k-1] < 0$ for $k > \ell_i$, we can decode $\pi(X_i)[k]$ by subtracting instead of adding the difference to $\pi(X_i)[k-1]$ as usual in delta-coding. Hence, we can replace R with Δ, but need to adjust Q such that $Q[i]$ points to the first bit of $\Delta[i]$. Finally, like in the first variant, we store the expansion lengths of all non-terminals in L. Here, we separate L in a first part using 8 bits per entry, then 16 bits per entry, and finally 32 bits per entry. To this end, we represent L by three arrays, start with filling the first array, and continue with filling

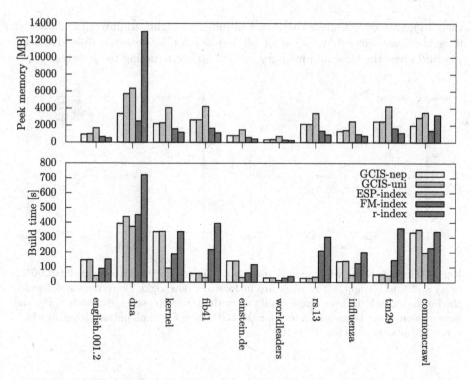

Fig. 2. Maximum memory consumption *(top)* and time *(bottom)* during the construction of the indexes.

the next array whenever we process a value whose bit representation cannot be stored in a single entry of the current array. Since Elias–Fano code supports constant-time random access and Elias-γ supports constant-time linear access, we can decode $\pi(X_i)$ by accessing $F[i]$ and then sequentially decode $\Delta[i]$. Hence, we can simulate GCIS-nep with this compressed version without sacrificing the theoretical bounds. We call the resulting index GCIS-uni.

Experiments. In the following we present an evaluation of our C++ implementation and different self-indexes for comparison, which are the FM-index [14], the ESP-Index [32], and the r-index [18][3]. All code has been compiled with gcc-10.2.0 in the highest optimization mode -O3. We ran all our experiments with an Intel Xeon CPU X5670 clocked at 2.93 GHz running Arch Linux.

Our datasets shown in Table 1 are from the Pizza&Chili and the tudocomp [9] corpus.[4] With respect to the index sizes, we have the empirically ranking GCIS-uni < ESP-index < GCIS-nep, followed by one of the BWT-based indexes. While

---

[3] See   https://github.com/mpetri/FM-Index,   https://github.com/tkbtkysms/esp-index-I, and https://github.com/nicolaprezza/r-index, respectively.

[4] To save space, we renamed the datasets COMMONCRAWL.ASCII.TXT and EINSTEIN.DE.TXT to COMMONCRAWL and EINSTEIN.DE, respectively.

the r-index needs less space than the FM-index on highly-compressible datasets, it is the least favorable option of all indexes for less-compressible datasets. Figure 2 gives the time and memory needed for constructing the indexes.

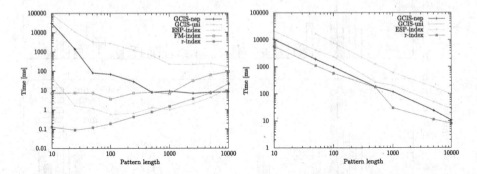

Fig. 3. Time for locate while scaling the pattern length on the datasets ENGLISH.001.2 (left) and FIB41 (right). The plots are in logscale. The right figure does not feature the FM-index, which takes considerably more time than the other approaches. For the same reason, there is no data shown for the ESP-index for small pattern lengths, which needs 170 s on average for $|P| = 10$.

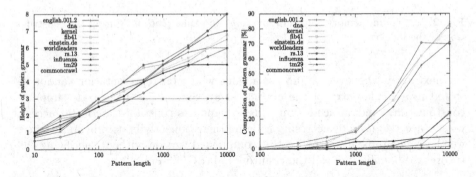

Fig. 4. *Left:* The average height τ_P of $\mathcal{G}_P$ for a pattern of a certain length. *Right:* Percentage of the computation of $\mathcal{G}_P$ in relation to the whole running time for answering locate(P) with GCIS-nep.

We can observe in Fig. 3 that our indexes answer locate(P) fast when P is sufficiently long or has many occurrences occ in T. GCIS-uni is always slower than GCIS-nep due to the extra costs for decoding. In particular for ENGLISH.001.2, GCIS-nep is the fastest index when the pattern length reaches 10000 characters and more. At this time, the pattern grammar reached a height τ_P of almost six, which is the height τ_T. The algorithm can extend an occurrence of a core to a pattern occurrence by checking only 80–100 characters. However, when the pattern surpasses 5000 characters, the computation of $\mathcal{G}_P$ becomes the time bottleneck.

With that respect, the ESP-index shares the same characteristic. encoding make slow down the location time by about 2 to 10 times approximately. Let us have a look at the dataset FIB41, which is linearly recurrent [10], a property from which we can derive the fact that a pattern that occurs at least once in T has actually a huge number of occurrences in T. There are almost 3,000,000 occurrence of patterns with a length of 100. Here, we observe that our indexes are faster than ESP-index. ESP-index needs more time for locate than GCIS because GCIS can form a core than covers a higher percentage of the pattern than the core selected by ESP. FM-index, and ESP-index with $|P| = 10$ take 100 s or more on average – we omitted them in the graph to keep the visualization clear.

In Fig. 4, we study the maximum height $\tau_P = \mathcal{O}(\lg |P|)$ that we achieved for the patterns with $|P| = 100$ in each dataset. For this experiment, we randomly select a position j in T and extracted $P = T[j..j + 99]$. For every dataset, we could observe that τ_P is logarithmic to the pattern length, especially for the artificial datasets FIB41, TM29, and RS.13, where τ_P is empirically larger than measured in other datasets. In DNA and COMMONCRAWL, τ_P is at most 3 , but this is because $\tau_T = 3$ for these datasets.

Acknowledgements. This work was supported by JSPS KAKENHI grant numbers JP21K17701 (DK), JP21K17705 (YN), JP20H04141 (HB), JP18H04098 (MT), and JST PRESTO grant number JPMJPR1922 (SI).

References

1. Buchsbaum, A.L., Kaplan, H., Rogers, A., Westbrook, J.R.: Linear-time pointer-machine algorithms for least common ancestors, MST verification, and dominators. In: Proceedings of the STOC, pp. 279–288 (1998)
2. Burrows, M., Wheeler, D.J.: A block sorting lossless data compression algorithm. Technical report 124, Digital Equipment Corporation, Palo Alto, California (1994)
3. Christiansen, A.R., Ettienne, M.B., Kociumaka, T., Navarro, G., Prezza, N.: Optimal-time dictionary-compressed indexes. ACM Trans. Algorithms **17**(1), 8:1–8:39 (2021)
4. Claude, F., Fariña, A., Martínez-Prieto, M.A., Navarro, G.: Universal indexes for highly repetitive document collections. Inf. Syst. **61**, 1–23 (2016)
5. Claude, F., Navarro, G.: Self-indexed grammar-based compression. Fundam. Inform. **111**(3), 313–337 (2011)
6. Claude, F., Navarro, G.: Improved grammar-based compressed indexes. In: Calderón-Benavides, L., González-Caro, C., Chávez, E., Ziviani, N. (eds.) SPIRE 2012. LNCS, vol. 7608, pp. 180–192. Springer, Heidelberg (2012). https://doi.org/10.1007/978-3-642-34109-0_19
7. Cormode, G., Muthukrishnan, S.: The string edit distance matching problem with moves. ACM Trans. Algorithms **3**(1), 2:1–2:19 (2007)
8. Díaz-Domínguez, D., Navarro, G.: A grammar compressor for collections of reads with applications to the construction of the BWT. In: Proceedings of the DCC, pp. 83–92 (2021)
9. Dinklage, P., Fischer, J., Köppl, D., Löbel, M., Sadakane, K.: Compression with the tudocomp framework. In: Proceedings of the SEA. LIPIcs, vol. 75, pp. 13:1–13:22 (2017)

10. Du, C.F., Mousavi, H., Schaeffer, L., Shallit, J.O.: Decision algorithms for fibonacci-automatic words, with applications to `pattern avoidance. CoRR abs/1406.0670 (2014). http://arxiv.org/abs/1406.0670
11. Elias, P.: Efficient storage and retrieval by content and address of static files. J. ACM **21**(2), 246–260 (1974)
12. Elias, P.: Universal codeword sets and representations of the integers. IEEE Trans. Inf. Theory **21**(2), 194–203 (1975)
13. Farach-Colton, M., Ferragina, P., Muthukrishnan, S.: On the sorting-complexity of suffix tree construction. J. ACM **47**(6), 987–1011 (2000)
14. Ferragina, P., González, R., Navarro, G., Venturini, R.: Compressed text indexes: from theory to practice. ACM J. Exp. Algorithmics **13**, 1.12:1-1.123:1 (2008)
15. Ferragina, P., Manzini, G.: Opportunistic data structures with applications. In: Proceedings of the FOCS, pp. 390–398 (2000)
16. Fischer, J., I, T., Köppl, D.: Deterministic sparse suffix sorting in the restore model. ACM Trans. Algorithms **16**(4), 50:1-50:53 (2020)
17. Gagie, T., I, T., Manzini, G., Navarro, G., Sakamoto, H., Takabatake, Y.: Rpair: Rescaling RePair with Rsync. CoRR arXiv:abs/1906.00809 (2019)
18. Gagie, T., Navarro, G., Prezza, N.: Optimal-time text indexing in BWT-runs bounded space. In: Proceedings of the SODA, pp. 1459–1477 (2018)
19. Gusfield, D.: Algorithms on Strings, Trees, and Sequences: Computer Science and Computational Biology. Cambridge University Press, Cambridge (1997)
20. Kieffer, J.C., Yang, E.: Grammar-based codes: a new class of universal lossless source codes. IEEE Trans. Inf. Theory **46**(3), 737–754 (2000)
21. Knuth, D.E., Morris, J.H., Pratt, V.R.: Fast pattern matching in strings. SIAM J. Comput. **6**(2), 323–350 (1977)
22. Larsson, N.J., Moffat, A.: Offline dictionary-based compression. In: Proceedings of the DCC, pp. 296–305 (1999)
23. Manber, U., Myers, E.W.: Suffix arrays: a new method for on-line string searches. SIAM J. Comput. **22**(5), 935–948 (1993)
24. Maruyama, S., Nakahara, M., Kishiue, N., Sakamoto, H.: ESP-index: a compressed index based on edit-sensitive parsing. J. Discret. Algorithms **18**, 100–112 (2013)
25. Mehlhorn, K., Sundar, R., Uhrig, C.: Maintaining dynamic sequences under equality tests in polylogarithmic time. Algorithmica **17**(2), 183–198 (1997)
26. Nishimoto, T., I, T., Inenaga, S., Bannai, H., Takeda, M.: Dynamic index and LZ factorization in compressed space. Discret. Appl. Math. **274**, 116–129 (2020)
27. Nong, G., Zhang, S., Chan, W.H.: Two efficient algorithms for linear time suffix array construction. IEEE Trans. Comput. **60**(10), 1471–1484 (2011)
28. Nunes, D.S.N., Louza, F.A., Gog, S., Ayala-Rincn, M., Navarro, G.: Grammar compression by induced suffix sorting (2020)
29. Nunes, D.S.N., da Louza, F.A., Gog, S., Ayala-Rincón, M., Navarro, G.: A grammar compression algorithm based on induced suffix sorting. In: Proceedings of the DCC, pp. 42–51 (2018)
30. Sahinalp, S.C., Vishkin, U.: Efficient approximate and dynamic matching of patterns using a labeling paradigm (extended abstract). In: Proceedings of the FOCS, pp. 320–328 (1996)
31. Takabatake, Y., Nakashima, K., Kuboyama, T., Tabei, Y., Sakamoto, H.: siEDM: an efficient string index and search algorithm for edit distance with moves. Algorithms **9**(2), 26:1-26:18 (2016)

32. Takabatake, Y., Tabei, Y., Sakamoto, H.: Improved ESP-index: a practical self-index for highly repetitive texts. In: Gudmundsson, J., Katajainen, J. (eds.) SEA 2014. LNCS, vol. 8504, pp. 338–350. Springer, Cham (2014). https://doi.org/10.1007/978-3-319-07959-2_29

33. Tsuruta, K., Köppl, D., Nakashima, Y., Inenaga, S., Bannai, H., Takeda, M.: Grammar-compressed self-index with Lyndon words. IPSJ TOM **13**(2), 84–92 (2020)

34. Ziv, J., Lempel, A.: A universal algorithm for sequential data compression. IEEE Trans. Inf. Theory **23**(3), 337–343 (1977)

An LMS-Based Grammar Self-index
with Local Consistency Properties

Diego Díaz-Domínguez[1,2]($\boxtimes$), Gonzalo Navarro[1,2], and Alejandro Pacheco[1,2]

[1] Department of Computer Science, University of Chile, Santiago, Chile
{ddiaz,gnavarro,apacheco}@dcc.uchile.cl
[2] CeBiB—Center for Biotechnology and Bioengineering, Santiago, Chile

Abstract. A grammar self-index of a text T (Claude et al. 2012) consists of a grammar $\mathcal{G}$ that only produces T and a geometric data structure that indexes the string cuts of the right-hand sides of $\mathcal{G}$'s rules. This representation uses space proportional to G, the size of the grammar, which is small when the text is repetitive. However, the index is slow for matching long patterns; it finds the *occ* occurrences of a pattern $P[1..m]$ in $O((m^2 + occ)\log G)$ time. The most expensive part is a set of binary searches for the different cuts $P[1..j]P[j+1..m]$ in the geometric data structure. Christiansen et al. 2010 solved this problem by building a locally consistent grammar that only searches for $O(\log m)$ cuts of P. Their representation, however, requires significant extra space (tough still in $O(G)$) to store a set of permutations for the nonterminal symbols. In this work, we propose another locally consistent grammar that builds on the idea of LMS substrings (Nong et al. 2009). Our grammar also requires to try $O(\log m)$ cuts when searching for P, but it does not need to store permutations. As a result, we obtain a self-index that searches in time $O((m\log m + occ)\log G)$ and is of practical size. Our experiments showed that our index is faster than previous grammar-based indexes at the price of increasing the space by a 1.8x factor on average. Other experimental results showed that our data structure becomes convenient when the patterns to search for are long.

Keywords: Grammar compression · LMS-substrings · Locally consistent parsing

1 Introduction

Self-indexes built on dictionary compression [19] have gained increasing attention in recent years as they can reduce the space usage of massive repetitive text collections by orders of magnitude, while still supporting direct access and pattern searches on the text. Among those, grammar-based self-indexes [13] are promising as they allow to access the text with a logarithmic-time penalty [1].

Funded in part by Basal Funds FB0001, Fondecyt Grant 1-200038, Ph.D. Scholarship 21171332, ANID, Chile.

T. Lecroq and H. Touzet (Eds.): SPIRE 2021, LNCS 12944, pp. 100–113, 2021.
https://doi.org/10.1007/978-3-030-86692-1_9

Claude and Navarro [6] proposed the first self-index based on grammar compression. Its most recent version [8] offers relevant worst-case guarantees and is competitive in both time and space. Given any context-free grammar of size G representing a text $T[1..n]$, the index uses $3 + \epsilon$ words of space (i.e., $(3 + \epsilon) \log n$ bits) for any constant $\epsilon > 0$, and finds all the occ occurrences of any pattern $P[1..m]$ in time $O((m^2 + occ) \log G)$.

While this time complexity is practical for short patterns, the quadratic time complexity becomes noticeable on longer ones. There have been several attempts to decrease the m^2 term in the time complexity. In the same article [8], they show that the time can be reduced to $O(m^2 + (m + occ) \log^\epsilon G)$, for any constant $\epsilon > 0$, while maintaining the space within $O(G)$, by replacing binary searches with Patricia trees and using larger geometric data structures. More radically, Christiansen et al. [5, App. A] showed that the time complexity can be reduced to $O(m \log n + occ \log^\epsilon n)$, still within $O(G)$ space, by using new data structures called Z-fast tries instead of Patricia trees. Those improved time complexities pay a significant price, however, in the constant hidden in the $O(G)$ space term. If implemented, this index is likely to be considerably larger than the classic grammar-based index of Claude et al. [8].

The main contribution of Christiansen et al., however, is the design of a particular grammar with local consistency properties, meaning that identical text substrings are largely parsed in the same way. This feature helps to reduce the number of anchor points of P that must be tested in the text (the meaning of this concept will be made clear throughout the paper), from $m - 1$ to $O(\log m)$. Such a significant reduction enables them to search in near-optimal time $O(m + (1 + occ) \log^\epsilon n)$. This scheme, however, only works on the particular grammar they designed. Further, it is also likely that, if implemented, the heavy theoretical machinery required to achieve their result produces a large index in practice.

Our Contribution We design a practical scheme to generate a locally consistent grammar. We encode the result using the representation of Claude et al. 2021 to produce a self-index that locates the occurrences of a pattern $P[1, m]$ in $O((m \log m + occ) \log G)$ time. Our method builds on a new idea to produce a locally consistent parsing that uses induced suffix sorting [21]. We prove this parsing is locally consistent and exploit the fact that, unlike Christiansen et al. [5], it does not require us to store the alphabet permutations of the parsing levels to perform pattern matching. When querying P in the index, we use the lexicographical relations of its symbols to infer how its occurrences in T would be parsed. By not generating the random permutations, we are likely to obtain a larger grammar than Christiansen et al. We weigh this disadvantage by simplifying our grammar in a way that does not affect its locally consistent properties. Using induced suffix sorting for grammar compression is not new, nor is the idea of simplifying the grammar [9,22]. However, self-indexing on those grammar compressors, exploiting their local consistency properties, is a contribution of this paper. Our experimental results showed that our grammar is comparable in size with that of Christiansen et al., but, as explained before, does not require the permutations. Further experiments also showed that our self-index is larger than that of Claude et al. 2021 (which we built on top of RePair), but considerably faster as the pattern length grows.

2 Related Concepts

2.1 Grammar Compression

A *context-free grammar*, or just grammar, is a tuple $\mathcal{G} = (V, \Sigma, \mathcal{R}, S)$ that describes rewriting rules to produce a set of strings in Σ^*. In this tuple, V is the alphabet of *nonterminal* symbols, Σ is the alphabet of *terminal* symbols, $\mathcal{R}$ is a list of productions that maps nonterminals to strings over $\Sigma \cup V$, and $S \in V$ is the start symbol of $\mathcal{G}$. The nonterminals rewrite as strings, while terminal symbols cannot be replaced. The rules in $\mathcal{R}$ are represented as $A \rightarrow B$, where $A \in V$ and $B \in (V \cup \Sigma)^*$, meaning that A is replaced by B. The set of strings in Σ^* we can obtain from S by recursively rewriting nonterminals is the language generated by $\mathcal{G}$, $\mathcal{L}(\mathcal{G})$. The *parse tree* of a string $T \in \mathcal{L}(\mathcal{G})$ is a labeled ordinal tree that represents the recursive nonterminal replacements leading to T. The root is labeled with S, the leaves are labeled with terminals spelling out T left to right, and the internal nodes are labeled with nonterminals: the children of A are, left to right, the symbols of B for some rule $A \rightarrow B \in \mathcal{R}$.

The aim in *grammar compression* is to encode an input string $T[1..n]$ by finding a small grammar $\mathcal{G}$ whose language is $\mathcal{L}(\mathcal{G}) = \{T\}$. In this grammar there is exactly one rule $A \rightarrow B$ per $A \in V$; we call $exp(A) \in \Sigma^*$ the only string of terminals derived from A, and then $T = exp(S)$. The *size* $G = |\mathcal{G}|$ of the grammar is the sum of the lengths of all the right-hand sides of the rules. Then we significantly compress T if we manage to build a grammar of size $G \ll |T|$ that generates only T. Even approximating the smallest grammar for T within a small constant factor is NP-hard [4,25]. However, there are good heuristic that perform well in practice, RePair [16] being the most popular one.

2.2 A Grammar Self-index

A classical grammar self-index [7,8] for a string $T[1..n]$ consists of a grammar $\mathcal{G}$ generating (only) T and a geometric data structure [3] used to perform efficient pattern matching on T. Using $O(G)$ space, the index locates all the occ occurrences of a pattern $P[1..m]$ in time $O((m^2 + occ)\log G)$.

In order to use $\mathcal{G}$, it is first (easily) modified to enforce some properties:

1. For every terminal $a \in \Sigma$, there is a nonterminal rule $X_a \rightarrow a$.
2. There are no rules in $\mathcal{R}$ of the form $A \rightarrow \varepsilon$ or $A \rightarrow B$ with $B \in V$.
3. The nonterminal symbols are numbered such that, if $X < Y$, then the reverse of $exp(X)$ is lexicographically smaller than the reverse of $exp(Y)$.
4. Every nonterminal $A \in V$ appears at least twice on the right-hand sides of $\mathcal{R}$. The only exceptions are S and the nonterminals produced from Σ.

The index stores the *grammar tree* of $\mathcal{G}$ [7] (also called partial parse tree [25]), which is a pruned version of the parse tree: if a nonterminal $A \in V$ labels several parse tree nodes, we maintain only the leftmost as an internal node and convert the others to leaves. The leaves of the grammar tree induce a partition of T into *phrases*, formed by the substrings $exp(A)$ for the labels A of all the leaves. Those phrases are indexed in the geometric data structure.

When searching for $P[1..m]$ in the grammar self-index, we classify its occurrences into *primary* and *secondary*. Primary occurrences of P cross two or more phrases, while secondary occurrences are completely contained within phrases.

The pattern matching algorithm first reports the primary occurrences of P, using the geometric structure to find all the distinct sequences of consecutive phrases that contain P as a substring. For each such sequence, it obtains the lowest common ancestor v of their corresponding grammar tree leaves. Say the label of v is $A \in V$. We traverse upwards from A to the root S to find the position of the occurrence in T (for every A child of A' in the grammar tree, the index stores the offset of $exp(A)$ inside $exp(A')$). Let $A = A_0, A_1, A_2, \ldots$ be the labels of the nodes traversed in the way to S. Apart from reporting the primary occurrence, the algorithm also finds all the leaves in the grammar tree labeled A_i and reports further (secondary) occurrences of P inside $exp(A_i)$. The ancestors of those leaves labeled A_i recursively trigger further secondary occurrences. The total time amortizes to constant per occurrence thanks to the grammar transformation rules applied.

The Grid Data Structure. We use the geometric data structure to locate the primary occurrences of P. To build it, we first define two string sets; the first one, $\mathcal{Y}$, has $|\mathcal{R}|$ strings, and the second, $\mathcal{X}$, has $G - |\mathcal{R}| + |\Sigma|$ strings. The sets are built as follows; let $A \rightarrow B_1 \ldots B_t \in \mathcal{R}$ be any nonterminal rule and let v the internal node for A in the grammar tree. For every B_i, with $i \in [1..t]$, we insert the reverse sequence $exp(B_i)^r$ to $\mathcal{Y}$. Additionally, for every proper suffix $B_i \ldots B_t$, with $i \in [2,t]$, we insert the string $exp(B_i) \cdots exp(B_t)$ to $\mathcal{X}$. We build a matrix M of $|\mathcal{Y}| \times |\mathcal{X}|$ cells. Every row k is labeled with the string in $\mathcal{Y}$ with lexicographic rank k. Equivalently, every column k' is labeled with the string in $\mathcal{X}$ with lexicographic rank k'. The cell of M in the intersection of the row for $exp(B_i)^r$ and the column for $exp(B_{i+1}) \cdots exp(B_t)$ stores the identifier of the $(i+1)$th child of v in the grammar tree. This arrangement automatically gives the lowest common ancestor v of the primary occurrence and the offset inside it.

Finding the Primary Occurrences. We cut P in two halves $P[1..j]$ and $P[j+1..m]$. The idea is to locate the range of rows (y_1, y_2) in M whose labels are prefixed by $P[1..j]^r$ and the range (x_1, x_2) of columns prefixed by $P[j+1..m]$. The non-empty cells within the grid range (x_1, y_1, x_2, y_2) point to the internal nodes in the grammar tree with primary occurrences of P. We binary search for $P[1..j]^r$ in the prefixes of $\mathcal{Y}$ to define (y_1, y_2), and binary search for $P[j+1..m]$ in the prefixes of $\mathcal{X}$ to define (x_1, x_2). When comparing $P[1..j]^r$ against the row labels, we decompress from the grammar tree the last j characters of the reversed string in $\mathcal{Y}$. Similarly, when comparing $P[j+1..m]$ against the column labels, we decompress the first $m-j+2$ symbols of the string in $\mathcal{X}$. The time for both searches is then $O(m \log G)$ per partition of P, adding up to $O(m^2 \log G)$ in total. We can reduced this to $O(m^2)$, still within $O(|G|)$ space, by using Patricia trees on $\mathcal{Y}$ and $\mathcal{X}$. Once we obtain a grid range, we get the p points inside it in time $O((1+p) \log G)$, which over all the partitions adds up to $O((m^2 + occ) \log G)$.

Christiansen et al. [5, App. A] showed that one can obtain time $O(m \log n + occ \log^\epsilon n)$ for any constant $\epsilon > 0$ within space $O(G)$, but the constant multiplying their space is much higher and the resulting index is likely impractical.

2.3 Locally Consistent Parsing

Locally consistent parsing [18,26] is a method for partitioning a text $T[1..n]$ into a sequence of *phrases* in which equal substrings of T are largely parsed in the same way. Let a *phrase boundary* be a pair of positions $(j-1, j)$ such that $T[j]$ is a prefix in the kth phrase and $T[j-1]$ is a suffix in the phrase preceding it. A parsing is locally consistent if there are two integers a, b (which may depend on n) such that, for every pair of equal substrings $T[j..j+u] = T[j'..j'+u]$, only their first a and their last b phrase boundaries can differ. In general, a locally consistent parsing algorithm puts a phrase boundary in T wherever some specific symbol combination arises. The first and last phrases of $T[j..j+u]$ might be formed in a different way than those in $T[j'..j'+u]$ because they might be preceded or followed by different symbols. Note that this approach differs from other parsing algorithms such as Lempel-Ziv [27] or RePair [16], which use global information on T to define its partition.

2.4 Locally Consistent Grammars

This type of grammars is constructed by applying successive rounds of locally consistent parsing over $T[1..n]$. In every round i, we capture the distinct phrases in the input text T^i ($T^1 = T$) and create new nonterminals rewriting to them. We then build a new text T^{i+1} by replacing the phrases in T^i with their corresponding nonterminal symbols. This new text T^{i+1} is the input for the next round. The algorithm stops when T^i can no longer be partitioned. If the phrases in every T^i are of length at least 2, then the string T^{i+1} is at most half the length of T^i, and thus the number of parsing rounds is $O(\log n)$ and the total running time is of the same order as for parsing T.

The algorithm described above produces a balanced grammar $\mathcal{G}$, which is probably bigger than the one we obtain with RePair. In exchange, if a pattern P appears more than once in T, then the parse subtrees containing its occurrences will be almost identical, differing only in a few nodes at the ends of every tree level. The internal part of the subtrees remains unchanged regardless P's context. This can be exploited to speed up pattern matching.

Recently, Christiansen et al. [5] proposed a run-length locally-consistent grammar of size $G = O(\gamma \log(n/\gamma))$, where γ is the size of the smallest attractor of T [12]. In their algorithm, the parsing rounds have two steps. In the first one, they create new nonterminal rules with the equal-symbol runs of T^i. These rules are of the form $X \to x^l$, where x^l is a run of l copies of symbol x in T^i (these rules are of constant size). Then, they produce a new string $\hat{T}^i$ by replacing the runs with their generating nonterminals. In the second round step, they define a random permutation $\pi : \hat{\Sigma}^i \to [1..|\hat{\Sigma}^i|]$ for the symbols in the alphabet $\hat{\Sigma}^i$ of $\hat{T}^i$, and use this permutation to partition $\hat{T}^i$: each phrase ends in a local minima, which is a position $\hat{T}^i[j]$ such that $\pi(\hat{T}^i[j-1]) > \pi(\hat{T}^i[j]) < \pi(\hat{T}^i[j+1])$.

Christiansen et al. also showed that, if we build the self-index of Claude et al. [8] using their grammar, then we require to test only $O(\log m)$ cuts of P to find its primary occurrences in $\mathcal{G}$. Their idea consists in preprocessing P at query time with the same algorithm they used to build $\mathcal{G}$. In every round i, they obtain the symbols of $\hat{P}^i$ by querying the equal-symbol runs of P^i in a hash table storing the nonterminals assigned to the sequences of the form x^l in $\mathcal{G}$. Subsequently, they query the phrases induced by the local minima of $\hat{P}^i$ in another hash table that maps phrases (right-hand sides in $\mathcal{R}$) to their nonterminal symbols. They use the symbols returned from the lookups to compute P^{i+1}. The prefix $\hat{P}^i[1..a]$ and the suffix $\hat{P}^i[b..|P^i|]$ that are not complete phrases do not have symbols in P^{i+1}. Analogously, the first and last equal-symbol runs of P^i do not have symbols in $\hat{P}^i$ as they are (possibly) incomplete. The preprocessing yields a list Q with the positions in P that limit incomplete parsing phrases. More specifically, every position $q \in Q$ is either the rightmost symbol under the internal node for $\hat{P}^i[a]$ in P's parse tree, or the leftmost symbol under the node for $\hat{P}^i[b]$. The elements in Q denote the cuts we try in the geometric data structure. As there are $O(1)$ incomplete phrases per parsing level i, there are $O(\log m)$ cuts in total. The time obtained [5] is $O(m + (occ + 1)\log^\epsilon n)$ for any constant $\epsilon > 0$.

2.5   Induced Suffix Sorting

Induced suffix sorting (ISS) [21] is a technique to sort the suffixes of T in lexicographical order. The basic idea consists of sampling some suffixes, sort them in lexicographical order, and use the result to induce the orders of the rest. ISS is the underlying procedure in several linear-time algorithms that build the suffix array [17,20,21] and the Burrows-Wheeler transform (BWT) [2,23]. We repeat some ISS-related definitions introduced for the SA-IS algorithm [21]:

Definition 1. *Let $T[1..n]$ be a text terminated with a sentinel symbol \$, smaller than the others. A position $T[i]$ is called L-type if $T[i] > T[i+1]$ or if $T[i] = T[i+1]$ and $T[i+1]$ also L-type. Instead, $T[i]$ is said to be S-type if $T[i] < T[i+1]$ or if $T[i] = T[i+1]$ and $T[i+1]$ is also S-type. The symbol $T[n] = \$ is S-type.*

Definition 2. *A character $T[i]$ is called leftmost S-type, or LMS-type, if $T[i]$ is S-type and $i = 1$ or $T[i-1]$ is L-type.*

Definition 3. *An LMS substring is (i) a minimal substring $T[i..j]$ with both $T[i]$ and $T[j]$ being LMS characters, for $i < j$; or (ii) the sentinel itself.*

Let $D[1..n]$ be an array that stores in $D[i]$ the type (L, S, or LMS) of $T[i]$. We refer to D as the *description* of T.

SA-IS is a recursive approach. In every recursion level i, it first scans the input text T^i, with $T^1 = T$, from right to left to compute its description. As it moves through the text, SA-IS records the positions of the LMS substrings and sorts them. The idea is to use the resulting ranks to induce an ordering for the suffixes of T that are not prefixed by LMS substrings. When comparing two LMS substrings $T[a..b]$ and $T[a'..b']$, the algorithm inspects them from left to right until their symbols differ. However, if for some two positions $j \in [a..b]$ and $j' \in [a'..b']$, the type of $T[j]$ is distinct from the type of $T[j']$, then the substring with S-type gets the highest order, even if $T[j]$ and $T[j']$ are the same symbol. We refer to this ordering as $\prec_{LMS}$.

Before inducing the order of the suffixes, SA-IS recursively sorts the suffixes starting LMS substrings. For this purpose, it creates a new string T^{i+1} in which it replaces the LMS substrings with their $\prec_{LMS}$ orders, and uses T^{i+1} as input for another recursive call, of level $i+1$. The recursive call returns the suffix array of T^{i+1}, which gives the order between LMS-starting suffixes. This information is used to induce the order of the other suffixes of T^i.

Nunes et al. [22] noticed that the LMS substrings of the recursive calls of SA-IS can be used to build a grammar for T. The advantage of this construction is that it requires $O(n)$ time and space, and it is much cheaper in practice than other popular grammar heuristics, like RePair. However, it does not achieve compression ratios as good as those of RePair.

3   A Grammar Self-index Based on LMS Parsing

3.1   LMS Parsing

We define *LMS parsing* as the procedure of parsing T using its LMS substrings. The idea is similar to the method described in the SA-IS algorithm: we compute the description of T, and define a phrase $T[i..i']$ for every consecutive pair of LMS-type positions $T[i-1]$ and $T[i']$. We refer those phrases as *LMS phrases*.

The LMS parsing is locally consistent. To prove it, we demonstrate that equal substrings of T have the same descriptions, except possibly at their endpoints.

Lemma 1. *The LMS parsing is locally consistent.*

Proof. Let $T[a..b] = T[a'..b']$ be two equal substrings. Let their suffixes of length $u \geq 1$ be equal-symbol runs, and symbols $T[b-u]$ and $T[b'-u]$ be different from $T[b-u+1]$ and $T[b'-u+1]$, respectively. The symbols within the same run have the same types, by definition. However, those types might differ if $T[b]$ and $T[b']$ are followed by different symbols. In particular, if $T[b-u+1]$ is L-type and $T[b'-u+1]$ is not, then $T[b'-u+1]$ can be LMS-type, and thus a phrase may end at $T[b'-u+1]$ and not at $T[b-u+1]$ (or vice versa).

Instead, the positions $T[b-u]$ and $T[b'-u]$ preceding those runs will always have the same type because they are followed by the same symbol, $T[b-u+$

$1] = T[b' - u + 1]$. Furthermore, the equal substrings $T[a + 1..b - u - 1]$ and $T[a' + 1..b' - u - 1]$ will also have the same types because they are preceded and followed by the same symbols.

Finally, the types of $T[a]$ and $T[a']$ may differ because they may depend on the preceding symbol. Both or none can be L-type, but if they are not, then one may be LMS-type and the other be S-type, depending on the symbols at $T[a-1]$ and $T[a' - 1]$. Therefore, one substring may have an LMS phrase ending at the first position and not the other.

To conclude, there can be at most one LMS phrase boundary appearing in each extreme of one of the substrings and not in the order. □

We then produce a locally consistent grammar $\mathcal{G}$ using several rounds of LMS parsing. In every round i, we create a dictionary $\mathcal{D}^i$ with all the distinct LMS phrases of T^i. Then, for every $F \in \mathcal{D}^i$, we create a new rule $X \to F$, where X is the number of rules in $\mathcal{R}$ built before round i plus the $\prec_{LMS}$ rank of F among the strings in $\mathcal{D}^i$. After generating the new rules, we create T^{i+1} by replacing the LMS phrases in T^i with their nonterminal symbols. If there are still repeated symbols in T^{i+1}, we perform another parsing round $i + 1$ using T^{i+1} as input.

Note that this procedure is very similar to that of Christiansen et al. [5]. They randomly permute the alphabet and place a phrase boundary after every local minimum. In our LMS parsing, we place a phrase boundary after every LMS-type symbol, which is also a local minimum. The key difference is that Christiansen et al. need to store the permutations used in order to replicate the same process on the search pattern, whereas our parsing is given by the lexicographic order and thus can be applied on the pattern without further information.

To further reduce the grammar, we create a new rule $Y \to X^l$ for every maximal equal-symbol run X^l appearing on a right-hand side. The grammar tree represents rules $Y \to X^l$ as $Y \to X X^{l-1}$, where X^{l-1} is a special leaf. This unique cut is enough to detect the occurrences of any pattern, provided a special procedure is carried out to report the secondary occurrences inside X^{l-1} [5].

We also reduce space by replacing the nonterminals appearing once with the right-hand sides of their rules, unless they represent equal-symbol runs. The rules of those replaced symbols are then removed from $\mathcal{R}$.

3.2   Computing the Cuts During the Pattern Matching

We use our grammar to build the self-index of Claude et al. [8] (with the special provision for run-length rules). The main change is the way we cut the pattern.

Let us call the *projection* of $P^i[p]$ the index $q \in [1..m]$ such that $P[p]$ is the rightmost leaf under the subtree rooted at $P^i[p]$ in P's parse tree; similarly with the projection of $\hat{P}^i[p]$.

Our procedure for finding the cuts for P is analogous to that of Christiansen et al. [5]. We start with an empty set Q and apply successive rounds of LMS parsing over P. In every round i, we insert into Q the projection of $P^i[1]$ and $\hat{P}^i[1]$. We then hash the distinct LMS phrases in P^i. We discard for the hashing, however, the prefix $P^i[1..a]$ where $P^i[a]$ is the leftmost LMS-type symbol, and the suffix $P^i[b..]$ where $P^i[b-1]$ is the rightmost LMS-type symbol. These elements can be incomplete phrases, so we do not use them for the next round. Still, we do record in Q the projection of $P^i[a]$ and the projection of $P^i[b-1]$ in P's parse tree. Additionally, when the rightmost equal-symbol run of P^i (last symbol in $\hat{P}^i$) has length $u > 1$ and $P^i[|P^i| - u]$ is L-type, we also insert the projection of $P^i[|P^i| - u + 1]$ into Q. We consider this position because if the rightmost run of P^i is S-type, then $P^i[|P^i| - u + 1]$ can be LMS-type if $P^i[|P^i| - u]$ is L-type. After scanning P^i, we sort the hashed phrases in $\prec_{LMS}$ order and create a new string P^{i+1} that replaces the phrases' occurrences with their $\prec_{LMS}$ orders. The new parse P^{i+1} is the input for the next round. The processing of P stops when there are no more LMS-type symbols in P^i.

The length of P^{i+1} is at most half that of P^i, so we scan $O(m)$ symbols along the $O(\log m)$ parsing rounds. On the other hand, we sort the distinct LMS subtrings of P^i in $O(|P^i|)$ time [21], which adds up to $O(m)$ total time. Therefore, the complete preprocessing of P requires $O(m)$ time.

To find the primary occurrences, we binary search the cut $P[1..q]P[q+1..m]$ associated to every $q \in Q$. Since $|Q| \in O(\log m)$, the total time to look for the primary occurrences is $O(|Q|m \log G) \subseteq O(m \log m \log G)$, plus $O(|Q| \log G) \subseteq O(\log m \log G)$ for the geometric searches, plus $O(occ \log G)$ to extract the grid points. Our final result borrows the space figures of Claude et al. [8].

Theorem 1. *Let our grammar, built for $T[1..n]$, be of size G. Then our index uses $G \log n + (2 + \epsilon)G \log G$ bits of space, for any constant $\epsilon > 0$, and finds all the occ occurrences of $P[1..m]$ in time $O((m \log m + occ) \log G)$.*

We note that, still within $O(G)$ space, we can use Patricia trees to speed up the binary searches, obtaining $O(m \log m + (\log m + occ) \log G)$ time.

4   Experiments

We implemented our version of the grammar index in C++ on top of the SDSL-lite library [11]. The source code is available at https://github.com/ddiazdom/LPG/tree/LPG_grid. We generated two versions of our index. The regular version (**lms-ind**) implements the wavelet tree of the grid data structure using plain bit vectors. The second variant (**lms-ind-rrr**) encodes the wavelet tree using the RRR [24] data structure for compressed bit vectors. We compared our software against the state-of-the-art self-indexes for repetitive collections:

- **r-ind**[1]: The run-length compressed FM-index of [10]. It uses $O(r \log n)$ bits, where r is the number of runs in the text's BWT, and supports locate within that space.
- **lz-ind**[2]: A self-index based on Lempel-Ziv [15] that guarantees $O(z \log z)$ bits of space over a Lempel-Ziv parse of z phrases. We also included the variant that uses LZ-end parsing (**lz-end-ind**) [14].

---

[1] https://github.com/nicolaprezza/r-index.
[2] https://github.com/migumar2/uiHRDC/tree/master/uiHRDC/self-indexes/LZ.

Table 1. Datasets. The second and third columns are the number of symbols and alphabet, respectively. The rest of the columns are grammar sizes (value for G) obtained with different grammar algorithms. The column for RePair considers the postprocessing described in Claude et al. 2021 (Sect. 3). LMS is the grammar obtained with LMS parsing and LMS post is grammar resulted from the postprocessing described in Sect. 3. The last column (LC) refers to the grammar of Christiansen et al. 2020.

Dataset	n	σ	RePair	LMS	LMS post	LC
para	429,265,758	5	5,344,480	22,787,047	8,933,303	8,888,002
cere	461,286,644	5	4,069,450	37,426,507	6,802,801	4,069,450
influenza	154,808,555	15	1,957,370	4,259,746	3,304,035	4,477,322
einstein.en	467,626,544	139	212,903	643,338	427,142	601,755
kernel	257,961,616	162	1,374,650	3,769,839	2,870,350	3,795,801

- **slp-ind**: An optimized implementation of the grammar index of Claude et al. 2012 [7]. This version speeds up the binary searches by storing q-grams of the prefixes to which the nonterminals expand (grid labels). We used three q-gram values in our experiments; 4,8, and 16. We refer to these variants as **slp-ind4**, **slp-ind8** and **slp-ind16**, respectively.
- **g-ind**[3]: The grammar index of Claude et al. 2021. The first variant of this index (**g-ind-bs**) uses binary searches over the grid labels to find the primary occurrences of P. The second variant (**g-ind-pt**) speeds up the search by maintaining two Patricia trees, one for a subset of uniformly sampled column labels in the grid and the other for a subset of uniformly sampled row labels. We used three sampling rates; 1/4, 1/16, and 1/64. We refer to these variants as **g-ind-pt4**, **g-ind-pt16**, and **gt-ind-pt64**, respectively.

We used five data sets of the *Pizza&Chilli*[4] corpus for the experiments. The characteristics of these datasets are shown in Table 1. We assessed the compression ratio and the running time for locating patterns. We extracted random substrings from the datasets and then we searched them back with the different indexes. The length of these patterns ranged from 10 to 100 characters.

We also compared our grammar algorithm against RePair and the method of Christiansen et al. [5]. The metric we used for the comparison was the grammar size (G value). The algorithm of Christiansen et al. has no formal implementation, so we produced one ourselves.

All the experiments were carried out on a machine with eight Intel(R) Xeon(R) CPU E5-2407 processors at 2.40 GHz and 250 GB RAM. We compiled our source code using full compiler optimizations and we do not use multi threading.

---

[3] https://github.com/apachecom/grammar_improved_index.
[4] http://pizzachili.dcc.uchile.cl.

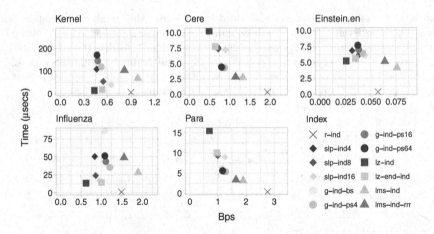

Fig. 1. Time-space tradeoffs for locating 1000 random pattern of length 100 on different collections and indexes. The time (y-axis) is given in μ secs per occurrence and the index space (x-axis) in bits per symbol (bps).

5   Results and Discussion

The grammars produced with our method were, on average, 4.2 times bigger than the RePair grammars (see columns 4 and 5 of Table 1). This considerable difference is expected as our grammar algorithm prioritizes consistency over compression. By further processing the grammars produced with our method (see Sect. 3), we reduced their sizes by 41% on average. However, their final sizes were still far from those of RePair; they were 1.82 times bigger on average. Interestingly, the sizes of our post-processed grammars were similar to those of Christiansen et al. (columns 6 and 7 of Table 1), even though we are not using random permutations. It is important to note that the grammar of Christiansen et al. cannot be further simplified without losing local consistency.

Figure 1 shows the trade-offs between index space usage and time for locating patterns of length 100. The results varied widely depending on the dataset. For instance, in *cere* and *para*, the index that used the most space was **r-ind** (1.93 and 2.76 bps, respectively), but it was also the fastest (0.34 and 0.37 μ secs). The second-largest index was **lms-ind** (1.32 and 1.89 bps), and the second-fastest after **r-ind**. In both datasets, the variant **lms-ind-rrr** reduced the space usage and stayed competitive for locating, but in the other datasets, **lms-ind-rrr** reduced the space at the cost of becoming slower. The smallest representation in *cere* and *para* was **lz-ind** (0.49 and 0.70 bps), but it was the slowest at locating (10.2 and 15.5 μ secs). In *einstein.en*, **lms-ind** was the biggest index (0.076 bps), even bigger than **r-ind**, which remained the fastest. However, **lms-ind** was the fastest dictionary-based data structure. The **lms-ind-rrr** variant reduced the space, but it did not outperform **r-ind**.

Things went differently with *influenza* and *kernel*. The Lempel-Ziv data structures (**lz-ind** and **lz-end-ind**) were competitive with **r-ind** for locating

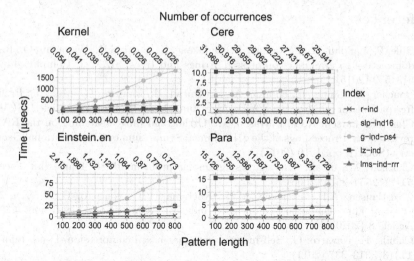

Fig. 2. Locating time for increasing pattern lengths. The time is given in μ secs per occurrence. The upper x-axes indicate the number of occurrences (in millions) searched in every combination of document and pattern length.

(14.16 and 18.05 μ secs versus 5.89 μ secs, respectively). Nevertheless, they used less space. This result is unexpected as *influenza* and *kernel* are not as repetitive as *einstein.en*. Our index performed poorly in these datasets. Both variants (**lms-ind** and **lms-ind-rrr**) were the biggest dictionary-based data structures, and they were not the fastest ones. However, they were competitive with **r-ind** in terms of space, with **lms-ind-rrr** using less space than **r-ind** in *kernel* (0.81 bps versus 0.89 bps, respectively), although they were significantly slower.

Figure 2 shows the performance of the indexes for the locate operation using different pattern lengths (from 100 to 800). In *para* and *cere*, **lms-ind-rrr** greatly outperformed the other dictionary-based indexes as the pattern length increased. This was not the case in *einstein.en* and *kernel*, where the performance of **lms-ind-rrr** was not different from that of **slp-ind16**. We also noted that in those datasets, the performance of **lz-ind** was very close to that of **r-ind**. Interestingly, the performance of **r-ind** remained steady as the pattern length increased.

6 Concluding Remarks

We presented a locally consistent grammar that allows us to produce a self-index using $G \log n + (2 + \epsilon)\, G \log G$ bits of space and that performs pattern matching in $O((m \log m + occ) \log G)$ time. Our experimental results showed that our method is a practical alternative to the technique of Christiansen et al. as we obtain a locally consistent grammar of comparable size without storing the symbol permutations. The resulting self-index is thus not much larger than the other popular dictionary-based indexes but generally faster at locating patterns, especially long ones.

References

1. Bille, P., Landau, G.M., Raman, R., Sadakane, K., Satti, S.R., Weimann, O.: Random access to grammar-compressed strings and trees. SIAM J. Comput. **44**(3), 513–539 (2015)
2. Boucher, C., Gagie, T., Kuhnle, A., Langmead, B., Manzini, G., Mun, T.: Prefix-free parsing for building big BWTs. Algorithms Mole. Biol. **14**(1), Article 13 (2019)
3. Chan, T., Larsen, K.G., Pătrașcu, M.: Orthogonal range searching on the RAM, revisited. In: Proceedings of the 27th Annual Symposium on Computational Geometry (SoCG), pp. 1–10 (2011)
4. Charikar, M., et al.: The smallest grammar problem. IEEE Trans. Inf. Theory **51**(7), 2554–2576 (2005)
5. Christiansen, A.R., Ettienne, M.B., Kociumaka, T., Navarro, G., Prezza, N.: Optimal-Time Dictionary-compressed indexes. ACM Trans. Algorithms **17**(1), Article 8 (2020)
6. Claude, F., Navarro, G.: Self-indexed grammar-based compression. Fund. Inform. **111**(3), 313–337 (2011)
7. Claude, F., Navarro, G.: Improved grammar-based compressed indexes. In: Proceedings of the 19th International Symposium on String Processing and Information Retrieval (SPIRE), pp. 180–192 (2012)
8. Claude, F., Navarro, G., Pacheco, A.: Grammar-compressed indexes with logarithmic search time. J. Comput. Syst. Sci. **118**, 53–74 (2021)
9. Díaz-Domínguez, D., Navarro, G.: A grammar compressor for collections of reads with applications to the construction of the BWT. In: Proceedings of the 31st Data Compression Conference (DCC) (2021)
10. Gagie, T., Navarro, G., Prezza, N.: Fully functional suffix trees and optimal text searching in BWT-runs bounded space. J. ACM **67**(1), 1–54 (2020)
11. Gog, S., Beller, T., Moffat, A., Petri, M.: From theory to practice: plug and play with succinct data structures. In: Gudmundsson, J., Katajainen, J. (eds.) SEA 2014. LNCS, vol. 8504, pp. 326–337. Springer, Cham (2014). https://doi.org/10.1007/978-3-319-07959-2_28
12. Kempa, D., Prezza, N.: At the roots of dictionary compression: string attractors. In: Proceedings of the 50th Annual ACM SIGACT Symposium on Theory of Computing (STOC), pp. 827–840 (2018)
13. Kieffer, J.C., Yang, E.H.: Grammar-based codes: a new class of universal lossless source codes. IEEE Trans. Inf. Theory **46**(3), 737–754 (2000)
14. Kreft, S., Navarro, G.: LZ77-Like compression with fast random access. In: Proceedings of the 10th Data Compression Conference (DCC), pp. 239–248 (2010)
15. Kreft, S., Navarro, G.: On compressing and indexing repetitive sequences. Theor. Comput. Sci. **483**, 115–133 (2013)
16. Larsson, J., Moffat, A.: Off-line dictionary-based compression. Proc. IEEE **88**(11), 1722–1732 (2000)
17. Louza, F., Gog, S., Telles, G.P.: Inducing enhanced suffix arrays for string collections. Theor. Comput. Sci. **678**(1), 22–39 (2017)
18. Mehlhorn, K., Sundar, R., Uhrig, C.: Maintaining dynamic sequences under equality tests in polylogarithmic time. Algorithmica **17**(2), 183–198 (1997)
19. Navarro, G.: Indexing highly repetitive string collections, Part II : compressed indexes. ACM Comput. Surv. **54**(2), Article 26 (2021)
20. Nong, G.: Practical linear-time O(1)-workspace suffix sorting for constant alphabets. ACM Trans. Inf. Syst. **31**(3), 1–15 (2013)

21. Nong, G., Zhang, S., Chan, W.H.: Linear suffix array construction by almost pure induced-sorting. In: Proceedings of the 19th Data Compression Conference (DCC), pp. 193–202 (2009)

22. Nunes, D.S.N., Louza, F.A., Gog, S., Ayala-Rincón, M., Navarro, G.: A grammar compression algorithm based on induced suffix sorting. In: Proceedings of the 28th Data Compression Conference (DCC), pp. 42–51 (2018)

23. Okanohara, D., Sadakane, K.: A linear-time Burrows-Wheeler transform using induced sorting. In: Proceedings of the 16th International Symposium on String Processing and Information Retrieval (SPIRE), pp. 90–101 (2009)

24. Raman, R., Raman, V., Satti, S.R.: Succinct indexable dictionaries with applications to encoding k-ary trees, prefix sums and multisets. ACM Trans. Algorithms **3**(4), Article 43 (2007)

25. Rytter, W.: Application of Lempel-Ziv factorization to the approximation of grammar-based compression. Theor. Comput. Sci. **302**(1–3), 211–222 (2003)

26. Sahinalp, C., Vishkin, U.: Data compression using locally consistent parsing. Technical report, UMIACS Technical report (1995)

27. Ziv, J., Lempel, A.: A universal algorithm for sequential data compression. IEEE Trans. Inf. Theory **23**(3), 337–343 (1977)

On the Approximation Ratio of LZ-End to LZ77

Takumi Ideue[1]($\boxtimes$), Takuya Mieno[2,3], Mitsuru Funakoshi[2,3],
Yuto Nakashima[2], Shunsuke Inenaga[2,4], and Masayuki Takeda[2]

[1] Department of Information Science and Technology, Kyushu University,
Fukuoka, Japan
ideue.takumi.274@s.kyushu-u.ac.jp
[2] Department of Informatics, Kyushu University, Fukuoka, Japan
{takuya.mieno,mitsuru.funakoshi,yuto.nakashima,
inenaga,takeda}@inf.kyushu-u.ac.jp
[3] Japan Society for the Promotion of Science, Tokyo, Japan
[4] PRESTO, Japan Science and Technology Agency, Kawaguchi, Japan

Abstract. A family of Lempel-Ziv factorizations is a well-studied string structure. The LZ-End factorization is a member of the family that achieved faster extraction of any substrings (Kreft & Navarro, TCS 2013). One of the interests for LZ-End factorizations is the possible difference between the size of LZ-End and LZ77 factorizations. They also showed families of strings where the approximation ratio of the number of LZ-End phrases to the number of LZ77 phrases asymptotically approaches 2. However, the alphabet size of these strings is unbounded. In this paper, we analyze the LZ-End factorization of the period-doubling sequence. We also show that the approximation ratio for the period-doubling sequence asymptotically approaches 2 for the binary alphabet.

Keywords: Lempel-Ziv 77 factorization · LZ-End factorization · Period-doubling sequence

1 Introduction

The *Lempel-Ziv 77 compression* (*LZ77*) [33] is one of the most successful lossless compression algorithms to date. On the practical side, LZ77 and its variants have been used as a core of compression software such as zip, gzip, rar, and compressed formats such as PNG, JPEG, PDF. In addition to these real world applications, compressed self-indexing structures based on LZ77 have been proposed [10–12, 22]. An LZ77-based compressed representation of a string allowing for fast access, rank, and select queries also exists [2].

On the (more) theoretical side, the left-to-right greedy factorization in LZ77, a.k.a. the *LZ77-factorization*, has widely been considered for decades. It parses a given input string w into a sequence $p_1, \ldots, p_z$ of non-empty substrings such that $p_1 = w[1]$ and p_i for $i \geq 2$ is the shortest prefix of $p_i \cdots p_z$ that does not

© Springer Nature Switzerland AG 2021
T. Lecroq and H. Touzet (Eds.): SPIRE 2021, LNCS 12944, pp. 114–126, 2021.
https://doi.org/10.1007/978-3-030-86692-1_10

occur in $p_1 \cdots p_{i-1}$. This implies that the prefix $p_i[1..|p_i|-1]$ occurs in $p_1 \cdots p_{i-1}$, and such an occurrence is called a *source* of p_i[1].

Among many versions of LZ77 (c.f. [9,13,20,21,23,29,34]), this paper focuses on the *LZ-End* compressor proposed by Kreft and Navarro [21]. It is also based on a greedy parsing $q_1, \ldots, q_{z'}$ of an input string, with a restriction that for each phrase q_i there has to be a source which ends at the right-end of a phrase in $q_1, \ldots, q_{i-1}$. This constraint permits fast substring extraction without expanding the whole input string. It is known that the LZ-End compression can be computed in linear time in the input string length [17], or in compressed space with slight slow-down on compression time [16].

One can regard LZ-End as a mix of LZ77 and LZ78 [34], since in the LZ78 factorization the source of each phrase has to begin and end at boundaries of previous phrases. Since LZ78 belongs to the class of grammar compression [6], LZ-End can be seen as a new bridge between grammar compression and LZ77.

Now, a natural question arises. How good is the compression performance of LZ-End? Practical evaluation in the literature [21] has revealed that the compression ratio of LZ-End is quite close to that of LZ77 (at most 20% worse), but very little is understood in theory. As in the literature, we measure and compare the sizes of LZ-End and LZ77 by the numbers z' and z of their phrases in the factorizations, i.e., "z' versus z".

Since LZ77 is an optimal greedy unidirectional parsing, $z' \geq z$ always holds. Thus we are concerned with *the approximation ratio* of LZ-End to LZ77, which is defined by z'/z. Kreft and Navarro [21] presented a simple family of strings for which z'/z is asymptotically 2 over an alphabet of size $n/3$, where n is the length of the string. Kreft and Navarro [21] conjectured that the upper bound for z'/z is also 2, but to our knowledge no non-trivial upper bound is known.

In this paper, we show that the same lower bound for z'/z can be obtained on a binary alphabet, thus significantly reducing the number of distinct characters used in the analysis from $n/3$ to 2. In particular, we prove that z'/z is asymptotically 2 for the *period-doubling sequences*, an interesting family of recursive strings. While the LZ77-factorization of the period-doubling sequences has an obvious structure (Proposition 1), the LZ-End factorization of the period-doubling sequences has a non-trivial structure and needs careful analysis (see our extensive discussions in Sect. 4 for detail).

Since the LZ77 factorization (without self-references) and the LZ-End factorization for the unary string a^n are the same, our result uses a minimum possible number of distinct characters to achieve such a lower bound for z'/z.

Related Work. A famous variant of the LZ77 factorization, which is called the *C-factorization* [9] and is denoted by $w = c_1 \cdots c_x$, differs from the LZ77 in that each phrase c_i is either a fresh character or the longest prefix of $c_i \cdots c_x$ that occurs in $c_1 \cdots c_{i-1}$. The size x of the C-factorization is known to be a lower bound for the size of the smallest grammar which generates only the input string [30]. A comparison of the LZ77 factorization and the C-factorization was

---

[1] This version of LZ77 is often called *non-overlapping LZ77* or *LZ77 without self-references*, since each phrase p_i never overlaps with any of its sources.

also considered in the literature [3, 26]. The structure of the C-factorization of the period-doubling sequences was investigated in [3]. We emphasize that our analysis of the LZ-End factorization of the period-doubling sequences is independent and is quite different from this existing work [3].

Relative LZ (RLZ) is a practical modification of LZ77 which efficiently compresses a collection of highly repetitive sequences [23]. In [20] an RLZ-based factorization of a string, called the *ReLZ-factorization*, was proposed. The approximation ratio of ReLZ to LZ77 was shown to be $\Omega(\log n)$ [20], where n denotes the length of the input string. On the other hand, in practice ReLZ was larger than LZ77 by at most a factor of two in all the tested cases in [20].

2   Preliminaries

2.1   Strings

Let Σ be the binary *alphabet*. An element of Σ^* is called a *string*. The length of a string w is denoted by $|w|$. The empty string ε is the string of length 0. Let Σ^+ be the set of non-empty strings, i.e., $\Sigma^+ = \Sigma^* \setminus \{\varepsilon\}$. For a string $w = xyz$, x, y and z are called a *prefix*, *substring*, and *suffix* of w, respectively. They are called a *proper prefix*, a *proper substring*, and a *proper suffix* of w if $x \neq w$, $y \neq w$, and $z \neq w$, respectively. Further, we say that w has an *internal occurrence* of y if y occurs in w as a proper substring which is neither a prefix nor a suffix. The i-th character of a string w is denoted by $w[i]$, where $1 \leq i \leq |w|$. For a string w and two integers $1 \leq i \leq j \leq |w|$, let $w[i..j]$ denote the substring of w that begins at position i and ends at position j. For convenience, let $w[i..j] = \varepsilon$ when $i > j$. For any $1 \leq i \leq |w|$, $w[i..|w|] \cdot w[1..i - 1]$ is called a *cyclic rotation* of w. If a cyclic rotation of w is not equal to w, the cyclic rotation is said to be proper. For any string w, let $w^1 = w$ and let $w^k = ww^{k-1}$ for any integer $k \geq 2$, i.e., w^k is the k-times repetition of w. A string w is said to be *primitive* if w cannot be written as x^k for any $x \in \Sigma^*$ and $k \geq 2$. Let $\bar{c}$ be the opposite character of c in a binary alphabet (e.g., $\bar{a} = b, \bar{b} = a$ for alphabet $\{a, b\}$). For any non-empty binary string w, $\widehat{w}$ denotes the string $w[1..|w| - 1] \cdot \overline{w[|w|]}$. We sometimes use $\mathsf{b}(x)$ and $\mathsf{e}(x)$ as the beginning position and the ending position of a substring x of a given string w, if the occurrence of x in w is clear from a discussion.

2.2   Lempel-Ziv Factorizations

We introduce the Lempel-Ziv 77 and LZ-End factorizations.

Definition 1 (LZ77 [33][2]). *The Lempel-Ziv 77 factorization (LZ77 factorization for short) of a string w is the factorization $\mathsf{LZ}_{77}(w) = p_1, \ldots, p_z$ of w such that $p_i[1..|p_i| - 1]$ is the longest prefix of $p_i \cdots p_z$ which occurs in $p_1 \cdots p_{i-1}$. As an exception, the last phrase p_z can be a suffix of w which occurs in $p_1 \cdots p_{z-1}$.*

---

[2] This definition of LZ77 is different from the original one [33] (see [21] for more information).

Definition 2 (LZ-End [21]). *The* LZ-End *factorization of a string w is the factorization* $\mathsf{LZ}_{\mathsf{end}}(w) = q_1, \ldots, q_{z'}$ *of w such that* $q_i[1..|q_i| - 1]$ *is the longest prefix of* $q_i \cdots q_{z'}$ *which occurs as a suffix of* $q_1 \cdots q_j$ *for some* $j < i$. *As an exception, the last phrase* $q_{z'}$ *can be a suffix of w which occurs as a suffix of* $q_1 \cdots q_j$ *for some* $j < z'$.

We refer to each p_i and q_i as an *LZ phrase* and *LZ-End phrase*, respectively. For each phrase, associated longest substring is called a *source* of the phrase. $z_{77}(w)$ and $z_{\mathsf{end}}(w)$ denote the number of the LZ phrases and the LZ-End phrases of a string w, respectively. For each $1 \leq i \leq z_{\mathsf{end}}(w)$, $\mathsf{LZ}_{\mathsf{end}}(w)[i]$ denotes the i-th LZ-End phrase of $\mathsf{LZ}_{\mathsf{end}}(w)$. Let $\mathsf{LZ}_{\mathsf{end}}(w).\mathsf{last}$ be the last LZ-End phrase of a string w, i.e., $\mathsf{LZ}_{\mathsf{end}}(w).\mathsf{last} = \mathsf{LZ}_{\mathsf{end}}(w)[z_{\mathsf{end}}(w)]$. Figure 1 shows examples of two factorizations.

$$\mathsf{LZ}_{77}(w) = a\,|\,b\,|\,a\ a\,|\,a\ b\ a\ b\,|\,a\ b\ a\ a\ a\ b\ a\ a\,|\,a\ b\ a\ a\ a\ b\ a\ b\ a\ b\ a\ a\ a\ b\ a\ b$$

$$\mathsf{LZ}_{\mathsf{end}}(w) = a\,|\,b\,|\,a\ a\,|\,a\ b\ a\,|\,b\ a\ b\,|\,a\ a\ a\ b\ a\ a\,|\,a\ b\ a\ a\ a\ b\ a\ b\ a\ b\ a\,|\,a\ a\ b\ a\ b$$

Fig. 1. The upper one shows the LZ77 factorization of w and the lower one shows the LZ-End factorization of w, where $w = abaaabababaaabaaabaaabababaaabab$. This w is the fifth period-doubling sequence S_5 which will be defined later.

2.3 Period-Doubling Sequence

The *period-doubling sequence* (cf. [1]) is one of the prominent automatic sequences. Let S_k be the k-th period-doubling sequence for any $k \geq 0$. The following two definitions are equivalent:

Definition 3. $S_0 = a$ *and* $S_k = \phi(S_{k-1})$ *for* $k \geq 1$ *where ϕ is the morphism such that* $\phi(a) = ab, \phi(b) = aa$.

Definition 4. $S_0 = a$ *and* $S_k = S_{k-1} \cdot \widehat{S_{k-1}}$ *for* $k \geq 1$.

Let n_k be the length of the k-th period-doubling sequence, i.e., $n_k = 2^k$.

3 Properties on Period-Doubling Sequence

The period-doubling sequences have many good combinatorial properties (see cf. [1]). In this section, we introduce helpful properties for our results on the period-doubling sequences.

Lemma 1. *For any* $k \geq 0$, S_k *is primitive.*

Proof. If S_k is not primitive, S_k has a period 2^i for some i. This implies that $S_k[n_k/2] = S_k[n_k]$, which contradicts Definition 4. □

Lemma 2 (Proposition 8.1.5 of [25]). *If a string w is primitive, ww has no internal occurrence of w.*

Lemma 3. *For any $k \geq 2$, $S_k = A_k B_k A_k A_k$ where $A_k = S_{k-2}$ and $B_k = \widehat{A_k}$. Moreover, $A_k = A_{k-1} B_{k-1}$ and $B_k = A_{k-1} A_{k-1}$ for any $k \geq 3$.*

Proof. Straightforward from Definition 3. $\square$

Lemma 4. *For any $k \geq 2$, $A_k A_k, A_k B_k$, and $B_k A_k$ have no internal occurrence of A_k. Hence the number of occurrences of A_k in $S_k = A_k B_k A_k A_k$ is 3.*

Proof. If $k = 2$, the lemma clearly holds. We assume $k \geq 3$. Since $A_k = S_{k-2}$, A_k is primitive. By Lemma 2, $A_k A_k$ has no internal occurrence of A_k. Since $A_k B_k = \widehat{A_k A_k}$, $A_k B_k$ also has no internal occurrence of A_k. Similarly, $A_{k-1} A_{k-1}$ and $A_{k-1} B_{k-1}$ have no internal occurrence of A_{k-1}. Also, by Lemma 3, $B_k A_k$ can be written as $A_{k-1} A_{k-1} A_{k-1} B_{k-1}$. These imply that $B_k A_k$ have no internal occurrence of $A_k = A_{k-1} B_{k-1}$. $\square$

Lemma 5. *For any $k \geq 3$ and any proper cyclic rotation α of A_k, the number of occurrences of α in $A_k A_k A_k$, $A_k B_k$, and $B_k A_k$ are 2, 1, and 0, respectively.*

Proof. Since $A_k = S_{k-2}$ and Lemma 1, A_k is primitive. This implies that α is also primitive. Thus, $A_k A_k$ has exactly one (internal) occurrences of α. Namely, α occurs in $A_k A_k A_k$ exactly two times. Since $A_k B_k = \widehat{A_k A_k}$, $A_k B_k$ also has exactly one (internal) occurrence of α. Finally, let us consider $B_k A_k = A_{k-1} A_{k-1} A_{k-1} B_{k-1}$. In a similar way of the proof of Lemma 4, we can show that both $A_{k-1} A_{k-1}$ and $A_{k-1} B_{k-1}$ have no internal occurrence of B_{k-1}. From this facts and Lemma 4, A_{k-1} occurs exactly three times and B_{k-1} occurs exactly once in $B_k A_k$. If $\alpha = B_{k-1} A_{k-1}$, α cannot occur in $B_k A_k$. Otherwise, α can be written as either $x B_{k-1} y$ or $x' A_{k-1} y'$ where x (resp. y) is a non-empty suffix (resp. prefix) of A_{k-1}, and x' (resp. y') is a non-empty suffix (resp. prefix) of B_{k-1}. If $\alpha = x B_{k-1} y$, α cannot occur in $B_k A_k$ due to the constraint of B_{k-1}. If $\alpha = x' A_{k-1} y'$, α cannot occur in $B_k A_k$ due to the constraint of A_{k-1} and the difference between the last characters of A_{k-1} and x'. Therefore α cannot occur in $B_k A_k$ in all cases. $\square$

4   Factorizations of Period-Doubling Sequence

By the definition of LZ77, the following proposition immediately holds:

Proposition 1. $\mathsf{LZ}_{77}(S_k) = (S_0, \widehat{S_0}, \widehat{S_1}, \ldots, \widehat{S_{k-1}})$ and thus $\mathsf{z}_{77}(S_k) = k + 1$.

In this section, we mainly discuss the LZ-End factorization of the period-doubling sequence, and give the following result.

Theorem 1. $\mathsf{z}_{\mathsf{end}}(S_k) = 2k - f(k)$ where $f(k) = O(\log^* k)$.

By Proposition 1 and Theorem 1, we can reach our goal of this paper:

Corollary 1. *There exists a family of* binary *strings w such that the ratio* $z_{end}(w)/z_{77}(w)$ *asymptotically approaches* 2.

In the rest of this paper, we show Theorem 1. The next lemma gives the LZ-End factorization of the period-doubling sequence. Notice that statement (I) in the lemma is not an immediate property for the LZ-End factorization due to the next example. Let $S = abaababaabbabbaababa$. Then,

$$\mathsf{LZ_{end}}(S) = a|b|aa|ba|baab|bab|baabab|a,$$
$$\mathsf{LZ_{end}}(Saba) = a|b|aa|ba|baab|bab|baababaaba.$$

Lemma 6. *For any* $k \geq 5$, *the following statements (I)-(IV) hold.*

(I) $\mathsf{LZ_{end}}(S_k)[i] = \mathsf{LZ_{end}}(S_{k-1})[i]$ *for every* $1 \leq i \leq z_{end}(S_{k-1}) - 1$.
(II) $z_{end}(S_k) \geq z_{end}(S_{k-1}) + 1$.

Let

$$w_k = \mathsf{LZ_{end}}(S_k)[z_{end}(S_{k-1})],$$
$$x_k = \mathsf{LZ_{end}}(S_k)[z_{end}(S_{k-1}) + 1],$$
$$y_k = S_k[e(x_k) + 1..n_k] \ (possibly \ empty).$$

(III) If $w_k \neq \mathsf{LZ_{end}}(S_{k-1}).last$,

$$|w_k| = \frac{1}{8}n_k + 1, |x_k| = \frac{3}{8}n_k, |y_k| = \frac{3}{16}n_{\ell(k)} - (k - \ell(k)) - 1,$$

where $\ell(k) = \max\{i \mid i \leq k, w_i = \mathsf{LZ_{end}}(S_{i-1}).last\}$.
Otherwise (if $w_k = \mathsf{LZ_{end}}(S_{k-1}).last$),

$$|w_k| = \frac{3}{16}n_k, |x_k| = \frac{5}{16}n_k + 1, |y_k| = \frac{3}{16}n_k - 1.$$

(IV) If $|y_k| \geq 2$, $y_k[1..|y_k| - 1]$ *has another occurrence to the left which ends with some LZ-End phrase of* S_k. *Namely,* y_k *is the last LZ-End phrase of* S_k *if* y_k *is not empty.*

Proof. In this proof, we use $Z'_k = \mathsf{LZ_{end}}(S_k)$ and $z'_k = z_{end}(S_k)$ for simplicity. We prove this lemma by induction on k.

Suppose that $k = 5$. The LZ-End factorizations of S_4, S_5 are

$$Z'_4 = a|b|aa|aba|bab|aaabaa,$$
$$Z'_5 = a|b|aa|aba|bab|aaabaa|abaaababababa|aabab.$$

Statements (I) and (II) clearly hold. Then, $w_5 = aaabaa, x_5 = abaaababababa, y_5 = aabab$. Hence, statement (III) holds since $n_5 = 32$ and $w_5 = Z'_4.last$ (i.e., the latter case). Statement (IV) also holds since $y_5[1..4] = aaba$ has an occurrence which ends with the fourth phrase aba.

Suppose that all the statements hold for any $k \in [5, k' - 1]$ for some $k' > 5$. We show that all the statements hold for k'. Firstly, suppose on the contrary

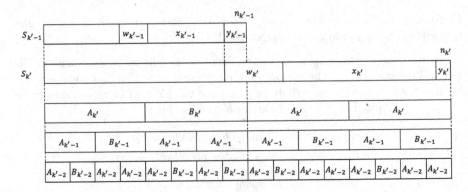

Fig. 2. Illustration for the LZ-End factorization when $w_{k'} \neq Z'_{k'-1}.\mathsf{last}$.

that statement (I) does not hold for k'. This implies that there exists a phrase $T = S_{k'}[\mathsf{b}(Z'_{k-1}[i])..j]$ for some $i < z'_{k'-1}$ and $j > n_{k'-1}$. Since $|x_{k'-1}y_{k'-1}| \geq \frac{3}{8}n_{k'-1} > \frac{1}{4}n_{k'-1}$ and $x_{k'-1}y_{k'-1}$ is a substring of T, T has an internal occurrence of the length-$\frac{1}{4}n_{k'-1}$ suffix $A_{k'-1}$ of $S_{k'-1}$. By Lemma 4 (showing the occurrences of A_{k-1} in S_{k-1}), $A_{k'-1}$ occurs exactly three times in $S_{k'}[1..n_{k'-1}]$. The first occurrence of $A_{k'-1}$ cannot be included by a source of T since $A_{k'-1}$ is not a prefix of $T[1..|T|-1]$. In addition, the second occurrence of $A_{k'-1}$ also cannot be included by a source of T since the source overlaps phrase T. Thus, $T[1..|T|-1]$ cannot have another occurrence to the left as a source of T. This contradicts that T is an LZ-End phrase of $S_{k'}$ at that position. Hence, statement (I) holds for k'. Due to statement (I), $w_{k'}$ must have $y_{k'-1}$ as a prefix. On the other hand, $w_{k'}$ cannot reach the end of $S_{k'}$. Hence, statement (II) also holds. Thanks to statements (I) and (II) for k', three substrings $w_{k'}$, $x_{k'}$, and $y_{k'}$ are well-defined (see Fig. 2 and 5 for illustrations).

Next, we show statements (III) and (IV).

- Assume that $\ell(k'-1) = \ell(k')$ (i.e., $w_{k'} \neq Z'_{k'-1}.\mathsf{last}$). We consider a phrase $w_{k'}$. If $|y_{k'-1}| = 0$, $x_{k'-1}$ is the suffix of length $\frac{3}{8}n_{k'-1}$ of $S_{k'-1}$, i.e., $x_{k'-1} = B_{k'-2}A_{k'-1}$. From Lemma 4, $x_{k'-1}$ does not have other occurrences to the left. This implies that $w_{k'} = x_{k'-1}$. This contradicts to $w_{k'} \neq Z'_{k'-1}.\mathsf{last}$. Thus, $|y_{k'-1}| > 0$ holds. Namely, $x_{k'-1} = Z'_{k'-1}[z'_{k'-1}-1]$ and $y_{k'-1} = Z'_{k'-1}.\mathsf{last}$ (see also Fig. 2). Let W be the string of length $\frac{1}{8}n_{k'}$ which begins at $\mathsf{b}(Z'_{k'-1}.\mathsf{last})$. $\ell(k'-1) = \ell(k')$ also implies that $\ell(k'-1) < k'$. Hence, $|y_{k'-1}| < \frac{3}{16}n_{\ell(k'-1)} \leq \frac{3}{32}n_{k'} < \frac{1}{8}n_{k'}$. This fact means that W is a proper cyclic rotation of $A_{k'-1}$. By Lemma 5, W occurs twice to the left (one is in $A_{k'-1}B_{k'-1}$, the other is in $A_{k'-1}A_{k'-1}$). Since the second occurrence ends with phrase $Z'_{k'}[z'_{k'-1}-1]$, Wc_W is a candidate of phrase $w_{k'}$ where c_W is the character preceded by W. Assume on the contrary that a source of phrase $w_{k'}$ is Wu for some $u \in \Sigma^+$ (see Fig. 3). The second occurrence of W cannot be the beginning position of a source of $w_{k'}$ since Wu overlaps $w_{k'}$. Hence, the only candidate of the beginning position of source Wu is in the first $A_{k'-1}B_{k'-1}$. Moreover, Wu cannot contain $B_{k'-1}$ since the original

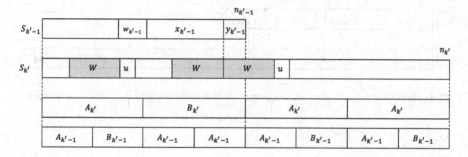

Fig. 3. Illustration for a part of the proof. W is a candidate of a source of phrase w'_k.

Wu occurs in $A_{k'-1}A_{k'-1}\cdots$. Thus, Wu is a proper substring of $A_{k'-1}A_{k'-1}$ and $A_{k'-1}B_{k'-1}$. In other words, $u'Wu$ is a proper prefix of $A_{k'-1}A_{k'-1}$ and $A_{k'-1}B_{k'-1}$ for some u'. Since $x_{k'-1}$ is a proper substring of $A_{k'-1}A_{k'-1}$, $x_{k'-1}$ also occurs in $u'Wu$. Hence, this contradicts that phrase $x_{k'-1}$ ends with W (i.e., $x_{k'-1}$ has to be a longer phrase.), and then, $w_{k'} = Wc_W$. Next, we consider a phrase $x_{k'}$. By the definition of the period-doubling sequence, there exists a clear candidate X of a source which ends at $\mathsf{e}(x_{k'-1})$ (see Fig. 4). Then, an equation $|y_{k'-1}| + \frac{1}{2}n_{k'} = |w_{k'}| + |X| + |y_{k'-1}|$ stands w.r.t. the length of suffix $S_{k'}[\mathsf{b}(y_{k'-1})..n_{k'}]$. Thus, $|X| = \frac{3}{8}n_{k'} - 1$ holds since $|w_{k'}| = \frac{1}{8}n_{k'} + 1$. This implies that X has $B_{k'-1}A_{k'-1}$ as a substring. There does not exist a longer candidate since $B_{k'-1}A_{k'-1}$ has only one occurrence to the left. Hence, $x_{k'} = Xc_X$ where c_X is the character preceded by X. Finally, we consider the suffix $y_{k'}$ of $S_{k'}$. If $|y_{k'}| \geq 2$, from the above discussion, $y_{k'-1}[2..|y_{k'-1}| - 1] = y_{k'}[1..|y_{k'}| - 1]$ holds. Since $y_{k'-1}[2..|y_{k'-1}| - 1]$ has an occurrence to the left which ends with some phrase ($\because$ statement (IV) for $k' - 1$), $y_{k'}[1..|y_{k'}| - 1]$ too. Therefore, statements (III) and (IV) also hold.

- Assume that $\ell(k'-1) \neq \ell(k')$ (i.e., $w_{k'} = Z'_{k'-1}.\mathsf{last}$). We can show that all the statements also hold for this case in a similar way. If we assume $|y_{k'-1}| > 0$, then $|w_{k'}| > |y_{k'-1}|$ holds by the above discussions. This contradicts that $w_{k'} = Z'_{k'-1}.\mathsf{last}$, and hence, $|y_{k'-1}| = 0$ and $w_{k'} = x_{k'-1}$ hold (see Fig. 5). Hence, $|w_{k'}| = |x_{k'-1}| = \frac{3}{8}n_{k'-1} = \frac{3}{16}n_{k'}$. We consider a phrase $x_{k'}$ that begins at position $\frac{1}{2}n_{k'} + 1$. Let $X' = S_{k'}[1..\mathsf{e}(w_{k'-1})]$ be a clear candidate of a source of $x_{k'}$. Since $|X'| = \frac{1}{2}n_{k'} - \frac{3}{16}n_{k'} = \frac{5}{16}n_{k'}$, X' has A'_k as a prefix. From Lemma 4, X' is the only candidate of a source, and thus $x_{k'} = X'c_{X'}$ where $c_{X'} = S_{k'}[\frac{13}{16}n_{k'} + 1]$ is the character preceded by X'. Moreover, the length of $y_{k'}$ is $\frac{1}{2}n_{k'} - (\frac{5}{16}n_{k'} + 1) = \frac{3}{16}n_{k'} - 1$. Since $|y_{k'}| = |w_{k'}| - 1$ and phrase $w_{k'}$ is a suffix of $S_{k'-1}$, a source of $w_{k'}$ can be also a source of $y_{k'}$. Namely, $y_{k'}$ is the last phrase. Thus, all the statements also hold for this case.

Therefore, this lemma holds. $\square$

We have just finished showing the form of the LZ-End factorization of S_k. Now, we will analyze the number of phrases of the factorization. Let $\mathcal{K}$ be the

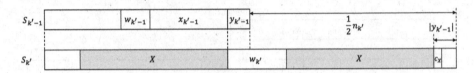

Fig. 4. Illustration for a part of the proof. X is a candidate of a source of phrase x'_k.

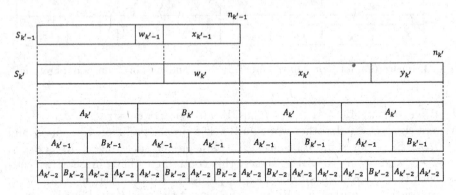

Fig. 5. Illustration for the LZ-End factorization when $w_{k'} = Z'_{k'-1}.\text{last}$.

sequence of integers k which satisfies $\ell(k) = k$. Let k_m^* denotes the m-th smallest integer in $\mathcal{K}$. Each k_m^* can be represented by the following recurrence formula:

Lemma 7

$$k_1^* = 5 \text{ and } k_m^* = k_{m-1}^* + \frac{3}{16} \cdot 2^{k_{m-1}^*} \text{ for } m \geq 2.$$

Proof. Let m be an integer greater than one. By the discussion of the proof for the previous lemma, $|y_{i-1}| - 1 = |y_i|$ holds for any integer $i \in [k_{m-1}^* + 1, k_m^* - 1]$. In addition, $|y_{k_m^* - 1}| = 0$. Hence,

$$k_m^* = k_{m-1}^* + |y_{k_{m-1}^*}| + 1 = k_{m-1}^* + \frac{3}{16} n_{k_{m-1}^*} = k_{m-1}^* + \frac{3}{16} \cdot 2^{k_{m-1}^*}.$$

$\square$

Lemma 8. *For any $k \geq 5$,*

$$\mathsf{z_{end}}(S_k) = 2k - f(k),$$

where $f(k)$ is a function such that $f(k) = m + 1$ if $k \in [k_m^* - 1, k_{m+1}^* - 2]$.

Proof. By Lemma 6, if $|y_k| = 0$ (i.e., $k + 1 \in \mathcal{K}$), then $\mathsf{z_{end}}(S_k) = \mathsf{z_{end}}(S_{k-1}) + 1$ holds, otherwise, $\mathsf{z_{end}}(S_k) = \mathsf{z_{end}}(S_{k-1}) + 2$ holds. Hence, for any $k \in [k_m^* - 1, k_{m+1}^* - 2]$,

$$\mathsf{z_{end}}(S_k) = \mathsf{z_{end}}(S_5) + 2(k - 5) - (m - 1) = 2k - (m + 1) = 2k - f(k).$$

$\square$

Lemma 9. $f(k) = O(\log^* k)$.

Proof. By Lemma 7,

$$k_m^* = O(2^{k_{m-1}^*}) \subseteq O\left(2^{2^{\cdot^{\cdot^{\cdot^{2^{k_1^*}}}}}}\right).$$

Thus, $m = O(\log^* k)$ holds. This implies that $f(k) = O(\log^* k)$ by Lemma 8. $\square$

By Lemmas 8 and 9, Theorem 1 holds.

5   Conclusions and Further Work

Let z' and z be the number of phrases in the LZ-End and LZ77 factorizations in a string. In this paper, we proved that the approximation ratio z'/z of LZ-End to LZ77 is asymptotically 2 for the period-doubling sequences. This significantly reduces the number of distinct characters needed to achieve such a lower bound from $n/3$ (in the existing work [21]) to 2 (in this work). We believe that our work initiates analysis of theoretical performance of LZ-End compression.

A lot of interesting further work remains for LZ-End, including the following:

- Is our lower bound for the approximation ratio tight? Kreft and Navarro [21] conjectured that $z'/z \le 2$ holds for *any* string. We performed some exhaustive experiments on binary strings and the result supports their conjecture.
- Is the size z' of the LZ-End factorization a lower bound for the size g of the smallest grammar generating the input string? It is known that the size of the C-factorization [9], a variant of LZ77, is a lower bound of g [6,30]. In particular case of the period-doubling sequences, there exists the following small SLP (i.e., grammar in the Chomsky normal form) generating the k-th period-doubling sequence: $S_k = S_{k-1}T_k$, $T_k = S_{k-2}S_{k-2}$, ..., $S_1 = ab$, $S_0 = a$. Following [30], the size of an SLP is evaluated by the number of productions and thus the above grammar is of size $2k + 1$. It is quite close to the size of the LZ-End factorization which is $2k - O(\log^* k)$ but is slightly larger.
- Interesting relationships between the size of the C-factorization and other string repetitive measures such as the size r of the run-length BWT [5], the size s of the smallest run-length SLP [28], the size ℓ of the Lyndon factorization [7], the size b of the smallest bidirectional scheme [31], the size γ of the smallest string attractor [18], the substring complexity δ [8], have been considered in the literature [4,14,15,19,24,27,32]. Can we extend these results to the LZ-End?

Acknowledgments. This work was supported by JSPS KAKENHI Grant Numbers JP20J11983 (TM), JP20J21147 (MF), JP18K18002 (YN), JP21K17705 (YN), JP18H04098 (MT), JP20H05964 (MT), and by JST PRESTO Grant Number JPMJPR1922 (SI).

References

1. Allouche, J.P., Shallit, J.: Automatic Sequences: Theory, Applications, Generalizations. Cambridge University Press, Cambridge (2003). https://doi.org/10.1017/CBO9780511546563
2. Belazzougui, D., et al.: Queries on LZ-bounded encodings. In: Bilgin, A., Marcellin, M.W., Serra-Sagristà, J., Storer, J.A. (eds.) 2015 Data Compression Conference, DCC 2015, Snowbird, UT, USA, 7-9 April 2015, pp. 83-92. IEEE (2015). https://doi.org/10.1109/DCC.2015.69
3. Berstel, J., Savelli, A.: Crochemore factorization of sturmian and other infinite words. In: Královič, R., Urzyczyn, P. (eds.) MFCS 2006. LNCS, vol. 4162, pp. 157-166. Springer, Heidelberg (2006). https://doi.org/10.1007/11821069_14
4. Bille, P., Gagie, T., Gørtz, I.L., Prezza, N.: A separation between RLSLPs and LZ77. J. Discret. Algorithms 50, 36-39 (2018). https://doi.org/10.1016/j.jda.2018.09.002
5. Burrows, M., Wheeler, D.: A block-sorting lossless data compression algorithm. Technical report, Digital SRC Research Report (1994)
6. Charikar, M., et al.: The smallest grammar problem. IEEE Trans. Inf. Theory 51(7), 2554-2576 (2005). https://doi.org/10.1109/TIT.2005.850116
7. Chen, K.T., Fox, R.H., Lyndon, R.C.: Free differential calculus, IV. The quotient groups of the lower central series. Ann. Math. 68(1), 81-95 (1958). http://www.jstor.org/stable/1970044
8. Christiansen, A.R., Ettienne, M.B., Kociumaka, T., Navarro, G., Prezza, N.: Optimal-time dictionary-compressed indexes. ACM Trans. Algorithms 17(1), 8:1-8:39 (2021). https://doi.org/10.1145/3426473
9. Crochemore, M.: An optimal algorithm for computing the repetitions in a word. Inf. Process. Lett. 12(5), 244-250 (1981). https://doi.org/10.1016/0020-0190(81)90024-7
10. Do, H.H., Jansson, J., Sadakane, K., Sung, W.: Fast relative Lempel-Ziv self-index for similar sequences. Theor. Comput. Sci. 532, 14-30 (2014). https://doi.org/10.1016/j.tcs.2013.07.024
11. Gagie, T., Gawrychowski, P., Kärkkäinen, J., Nekrich, Y., Puglisi, S.J.: A faster grammar-based self-index. In: Dediu, A.-H., Martín-Vide, C. (eds.) LATA 2012. LNCS, vol. 7183, pp. 240-251. Springer, Heidelberg (2012). https://doi.org/10.1007/978-3-642-28332-1_21
12. Gagie, T., Gawrychowski, P., Kärkkäinen, J., Nekrich, Y., Puglisi, S.J.: LZ77-based self-indexing with faster pattern matching. In: Pardo, A., Viola, A. (eds.) LATIN 2014. LNCS, vol. 8392, pp. 731-742. Springer, Heidelberg (2014). https://doi.org/10.1007/978-3-642-54423-1_63
13. Goto, K., Bannai, H., Inenaga, S., Takeda, M.: *LZD Factorization*: simple and practical online grammar compression with variable-to-fixed encoding. In: Cicalese, F., Porat, E., Vaccaro, U. (eds.) CPM 2015. LNCS, vol. 9133, pp. 219-230. Springer, Cham (2015). https://doi.org/10.1007/978-3-319-19929-0_19
14. Kärkkäinen, J., Kempa, D., Nakashima, Y., Puglisi, S.J., Shur, A.M.: On the size of Lempel-Ziv and Lyndon factorizations. In: Vollmer, H., Vallée, B. (eds.) 34th Symposium on Theoretical Aspects of Computer Science, STACS 2017. LIPIcs, Hannover, Germany, 8-11 March 2017, vol. 66, pp. 45:1-45:13. Schloss Dagstuhl - Leibniz-Zentrum für Informatik (2017). https://doi.org/10.4230/LIPIcs.STACS.2017.45

15. Kempa, D., Kociumaka, T.: Resolution of the Burrows-Wheeler transform conjecture. In: 61st IEEE Annual Symposium on Foundations of Computer Science, FOCS 2020, Durham, NC, USA, 16–19 November 2020, pp. 1002–1013. IEEE (2020). https://doi.org/10.1109/FOCS46700.2020.00097

16. Kempa, D., Kosolobov, D.: LZ-End parsing in compressed space. In: Bilgin, A., Marcellin, M.W., Serra-Sagristà, J., Storer, J.A. (eds.) 2017 Data Compression Conference, DCC 2017, Snowbird, UT, USA, 4–7 April 2017, pp. 350–359. IEEE (2017). https://doi.org/10.1109/DCC.2017.73

17. Kempa, D., Kosolobov, D.: LZ-End parsing in linear time. In: Pruhs, K., Sohler, C. (eds.) 25th Annual European Symposium on Algorithms, ESA 2017. LIPIcs, Vienna, Austria, 4–6 September 2017, vol. 87, pp. 53:1–53:14. Schloss Dagstuhl - Leibniz-Zentrum für Informatik (2017). https://doi.org/10.4230/LIPIcs.ESA.2017.53

18. Kempa, D., Prezza, N.: At the roots of dictionary compression: string attractors. In: Diakonikolas, I., Kempe, D., Henzinger, M. (eds.) Proceedings of the 50th Annual ACM SIGACT Symposium on Theory of Computing, STOC 2018, Los Angeles, CA, USA, 25–29 June 2018, pp. 827–840. ACM (2018). https://doi.org/10.1145/3188745.3188814

19. Kociumaka, T., Navarro, G., Prezza, N.: Towards a definitive measure of repetitiveness. In: Kohayakawa, Y., Miyazawa, F.K. (eds.) LATIN 2021. LNCS, vol. 12118, pp. 207–219. Springer, Cham (2020). https://doi.org/10.1007/978-3-030-61792-9_17

20. Kosolobov, D., Valenzuela, D., Navarro, G., Puglisi, S.J.: Lempel–Ziv-Like Parsing in Small Space. Algorithmica 82(11), 3195–3215 (2020). https://doi.org/10.1007/s00453-020-00722-6

21. Kreft, S., Navarro, G.: On compressing and indexing repetitive sequences. Theor. Comput. Sci. 483, 115–133 (2013). https://doi.org/10.1016/j.tcs.2012.02.006

22. Kärkkäinen, J., Ukkonen, E.: Lempel-Ziv parsing and sublinear-size index structures for string matching (extended abstract). In: Proceedings of the 3rd South American Workshop on String Processing, WSP 1996, pp. 141–155. Carleton University Press (1996)

23. Kuruppu, S., Puglisi, S.J., Zobel, J.: Relative Lempel-Ziv compression of genomes for large-scale storage and retrieval. In: Chavez, E., Lonardi, S. (eds.) SPIRE 2010. LNCS, vol. 6393, pp. 201–206. Springer, Heidelberg (2010). https://doi.org/10.1007/978-3-642-16321-0_20

24. Kutsukake, K., Matsumoto, T., Nakashima, Y., Inenaga, S., Bannai, H., Takeda, M.: On repetitiveness measures of Thue-Morse words. In: Boucher, C., Thankachan, S.V. (eds.) SPIRE 2020. LNCS, vol. 12303, pp. 213–220. Springer, Cham (2020). https://doi.org/10.1007/978-3-030-59212-7_15

25. Lothaire, M.: Applied Combinatorics on Words, vol. 105. Cambridge University Press, Cambridge (2005)

26. Mitsuya, S., Nakashima, Y., Inenaga, S., Bannai, H., Takeda, M.: Compressed communication complexity of Hamming distance. Algorithms 14(4), 116 (2021). https://doi.org/10.3390/a14040116

27. Navarro, G., Ochoa, C., Prezza, N.: On the approximation ratio of ordered parsings. IEEE Trans. Inf. Theory 67(2), 1008–1026 (2021). https://doi.org/10.1109/TIT.2020.3042746

28. Nishimoto, T., Tomohiro, I., Inenaga, S., Bannai, H., Takeda, M.: Dynamic index and LZ factorization in compressed space. Discret. Appl. Math. 274, 116–129 (2020). https://doi.org/10.1016/j.dam.2019.01.014

126   T. Ideue et al.

29. Nishimoto, T., Tabei, Y.: LZRR: LZ77 parsing with right reference. In: Bilgin, A., Marcellin, M.W., Serra-Sagristà, J., Storer, J.A. (eds.) Data Compression Conference, DCC 2019, Snowbird, UT, USA, 26–29 March 2019, pp. 211–220. IEEE (2019). https://doi.org/10.1109/DCC.2019.00029
30. Rytter, W.: Application of Lempel-Ziv factorization to the approximation of grammar-based compression. Theor. Comput. Sci. **302**(1–3), 211–222 (2003). https://doi.org/10.1016/S0304-3975(02)00777-6
31. Storer, J.A., Szymanski, T.G.: Data compression via textual substitution. J. ACM **29**(4), 928–951 (1982). https://doi.org/10.1145/322344.322346
32. Urabe, Y., Nakashima, Y., Inenaga, S., Bannai, H., Takeda, M.: On the size of overlapping Lempel-Ziv and Lyndon factorizations. In: Pisanti, N., Pissis, S.P. (eds.) 30th Annual Symposium on Combinatorial Pattern Matching, CPM 2019. LIPIcs, Pisa, Italy, 18–20 June 2019, vol. 128, pp. 29:1–29:11. Schloss Dagstuhl - Leibniz-Zentrum für Informatik (2019). https://doi.org/10.4230/LIPIcs.CPM.2019.29
33. Ziv, J., Lempel, A.: A universal algorithm for sequential data compression. IEEE Trans. Inf. Theory **23**(3), 337–343 (1977). https://doi.org/10.1109/TIT.1977.1055714
34. Ziv, J., Lempel, A.: Compression of individual sequences via variable-rate coding. IEEE Trans. Inf. Theory **24**(5), 530–536 (1978). https://doi.org/10.1109/TIT.1978.1055934

Data Structures

Computing the Original eBWT Faster, Simpler, and with Less Memory

Christina Boucher[1], Davide Cenzato[2], Zsuzsanna Lipták[2],
Massimiliano Rossi[1]([✉]), and Marinella Sciortino[3]

[1] Department of Computer and Information Science and Engineering,
University of Florida, Gainesville, FL, USA
`{cboucher,rossi.m}@cise.ufl.edu`
[2] Department of Computer Science, University of Verona, Verona, Italy
`{davide.cenzato,zsuzsanna.liptak}@univr.it`
[3] Department of Computer Science, University of Palermo, Palermo, Italy
`marinella.sciortino@unipa.it`

Abstract. Mantaci et al. [TCS 2007] defined the eBWT to extend the definition of the BWT to a collection of strings. However, since this introduction, it has been used more generally to describe any BWT of a collection of strings, and the fundamental property of the original definition (i.e., the independence from the input order) is frequently disregarded. In this paper, we propose a simple linear-time algorithm for the construction of the original eBWT, which does not require the preprocessing of Bannai et al. [CPM 2021]. As a byproduct, we obtain the first linear-time algorithm for computing the BWT of a single string that uses neither an end-of-string symbol nor Lyndon rotations. We combine our new eBWT construction with a variation of prefix-free parsing to allow for scalable construction of the eBWT. We evaluate our algorithm (`pfpebwt`) on sets of human chromosomes 19, *Salmonella*, and SARS-CoV2 genomes, and demonstrate that it is the fastest method for all collections, with a maximum speedup of $7.6\times$ on the second best method. The peak memory is at most $2\times$ larger than the second best method. Comparing with methods that are also, as our algorithm, able to report suffix array samples, we obtain a $57.1\times$ improvement in peak memory. The source code is publicly available at https://github.com/davidecenzato/PFP-eBWT.

1 Introduction

In the last several decades, the number of sequenced human genomes has been growing at unprecedented pace. In 2015 the number of sequenced genomes was doubling every 7 months [35]. Now the amount of data for some species is large enough that it poses challenges with respect to storage and analysis. One of the most widely-used methods for indexing data in bioinformatics is the Burrows-Wheeler Transform (BWT), which is a text transformation that compresses the input in a manner that allows for efficient substring queries. Not only can it be constructed in linear-time, it is also reversible—meaning the original input can

© Springer Nature Switzerland AG 2021
T. Lecroq and H. Touzet (Eds.): SPIRE 2021, LNCS 12944, pp. 129–142, 2021.
https://doi.org/10.1007/978-3-030-86692-1_11

be constructed from its compressed form. The BWT is formally defined over a single input string; thus, in order to define and construct it for one or more strings, the input strings need to be concatenated or modified in some way. In 2007 Mantaci et al. [26] presented a formal definition of the BWT for a multiset of strings, which they call the *extended Burrows-Wheeler Transform* (eBWT). It is a bijective transformation that sorts the cyclic rotations of the strings of the multiset according to the ω-order relation, an order, defined by considering infinite iterations of each string, which is different from the lexicographic order.

Since its introduction, several algorithms have been developed that construct the BWT of collections of strings for various types of biological data, including short sequence reads [1,3,4,9–11,16,23,32,33], protein sequences [39], metagenomic data [15], and longer DNA sequences such as long sequence reads and whole chromosomes [21]. However, we note that in the development of some of these methods the underlying definition of eBWT was loosened. For example, ropebwt2 [21] tackles a similar problem of building what they describe as the FM-index for a multiset of long sequence reads, however, they do not construct the suffix array (SA) or SA samples, and also, require that the sequences are delimited by separator symbols. Similarly, gsufsort [23] and egap [11] construct the BWT for a collection of strings but do not construct the eBWT according to its original definition. gsufsort [23] requires the collection for strings to be concatenated in a manner that the strings are deliminated by separator symbols that have an augmented relative order among them. egap [11], which was developed to construct the BWT and LCP for a collection of strings in external memory, uses the gSACA-K algorithm to construct the suffix array of the concatenated input and then constructs the BWT from the suffix array, using an additional $(\sigma + 1) \log N$ bits, where σ is the size of the alphabet and N the total length of the strings in the collection. Lastly, we note that there exists a number of methods for construction of the BWT for a collection of short sequence reads, including ble [4], BCR [3], G2BWT [10], egsa [24]; however, these methods make implicit or explicit use of end-of-string symbols appended to strings in the collection. For an example of the effects of these manipulations, see Sect. 2, and [8] for a more detailed study.

We present an efficient algorithm for constructing the eBWT that preserves the original definition of Mantaci et al. [26]—thus, it does not impose any ordering of the input strings or delimiter symbols. It is an adaptation of the well-known Suffix Array Induced Sorting (SAIS) algorithm of Nong et al. [29], which computes the suffix array of a single string T ending with an end-of-string character $. Our adaptation is similar to the algorithm proposed by Bannai et al. [2] for computing the BBWT (bijective BWT), which can also be used for computing the eBWT after additional linear-time preprocessing of the input strings. The key change in our approach is based on the insight that the properties necessary for applying Induced Sorting are valid also for the ω-order between different strings. As a result, it is not necessary that the input be Lyndon words, or that their relative order be known at the beginning. Furthermore, our algorithmic strategy, when applied to a single string, provides the first linear-time algorithm for computing the BWT of the string that uses neither an end-of-string symbol nor Lyndon rotations.

We then combine our new eBWT construction with a variation of a pre-processing technique called *prefix-free parsing* (PFP). PFP was introduced by Boucher et al. [6] for building the (run-length encoded) BWT of large and highly repetitive input text. Since its original introduction, it has been extended to construct the *r*-index [20], applied as a preprocessing step for building grammars [12], and used as a data structure itself [5]. Briefly, PFP is a one-pass algorithm that divides the input into overlapping variable length phrases with delimiting prefixes and suffixes; which in effect, leads to the construction of what is referred to as the dictionary and parse of the input. It follows that the BWT can be constructed in space proportional to the size of the dictionary and parse, which is expected to be significantly smaller than linear for repetitive text.

In our approach, PFP is applied to obtain a parse that is a multiset of cyclic strings (*cyclic prefix-free parse*) to which our eBWT construction is applied. We implement our approach (called pfpebwt), measure the time and memory required to build the eBWT for sets of increasing size of chromosome 19, *Salmonella*, and SARS-CoV2 genomes, and compare this to that required by gsufsort, ropebwt2, and egap. We show that pfpebwt is consistently faster and uses less memory than gsufsort and egap on reasonably large input (≥ 4 copies of chromosome 19, ≥ 50 *Salmonella* genomes, and $\geq 25,000$ SARS-CoV2 genomes). Although ropebwt2 uses less memory than pfpebwt on large input, pfpebwt is 7× more efficient in terms of wall clock time, and 2.8× in terms of CPU time. Moreover, pfpebwt is capable of reporting SA samples in addition to the eBWT with a negligible increase in time and memory [20], whereas ropebwt2 does not have that ability. If we compare pfpebwt only with methods that are able to report SA samples in addition to the eBWT (e.g., egap and gsufsort), we obtain a 57.1× improvement in peak memory.

Due to space limitations, proofs are deferred to the full version of the paper.

2   Preliminaries

A string $T = T[1..n]$ is a sequence of characters $T[1] \cdots T[n]$ drawn from an ordered alphabet Σ of size σ. We denote by $|T|$ the length n of T, and by ε the empty string, the only string of length 0. Given two integers $1 \leq i, j \leq n$, we denote by $T[i..j]$ the string $T[i] \cdots T[j]$, if $i \leq j$, while $T[i..j] = \varepsilon$ if $i > j$. We refer to $T[i..j]$ as a *substring* (or *factor*) of T, to $T[1..j]$ as the j-th *prefix* of T, and to $T[i..n] = T[i..]$ as the i-th *suffix* of T. A substring S of T is called *proper* if $T \neq S$. Given two strings S and T, we denote by $\mathrm{lcp}(S, T)$ the length of the *longest common prefix* of S and T, i.e., $\mathrm{lcp}(S, T) = \max\{i \mid S[1..i] = T[1..i]\}$.

Given a string $T = T[1..n]$ and an integer k, we denote by T^k the kn-length string $TT \cdots T$ (k-fold concatenation of T), and by T^ω the infinite string $TT \cdots$ obtained by concatenating an infinite number of copies of T. A string T is called *primitive* if $T = S^k$ implies $T = S$ and $k = 1$. For any string T, there exists a unique primitive word S and a unique integer k such that $T = S^k$. We refer to $S = S[1..\frac{n}{k}]$ as root(T) and to k as exp(T). Thus, $T = \mathrm{root}(T)^{\exp(T)}$.

We denote by $<_{\mathrm{lex}}$ the lexicographic order: for two strings $S[1..n]$ and $T[1..m]$, $S <_{\mathrm{lex}} T$ if S is a proper prefix of T, or there exists an index $1 \leq i \leq n, m$ such that $S[1..i-1] = T[1..i-1]$ and $S[i] < T[i]$. Given a string $T[1..n]$, the *suffix array* [25], denoted by $\mathrm{SA} = \mathrm{SA}_T$, is the permutation of $\{1,\ldots,n\}$ such that $T[\mathrm{SA}[i]..]$ is the i-th lexicographically smallest suffix of T.

We denote by $\prec_\omega$ the ω-order [13,26], defined as follows: for two strings S and T, $S \prec_\omega T$ if $\mathrm{root}(S) = \mathrm{root}(T)$ and $\exp(S) < \exp(T)$, or $S^\omega <_{\mathrm{lex}} T^\omega$ (this implies $\mathrm{root}(S) \neq \mathrm{root}(T)$). One can verify that the ω-order relation is different from the lexicographic one. For instance, $CG <_{\mathrm{lex}} CGA$ but $CGA \prec_\omega CG$.

The string S is a *conjugate* of the string T if $S = T[i..n]T[1..i-1]$, for some $i \in \{1,\ldots,n\}$ (also called the *i-th rotation* of T). The conjugate S is also denoted $\mathrm{conj}_i(T)$. It is easy to see that T is primitive if and only if it has n distinct conjugates. A *Lyndon word* is a primitive string which is lexicographically smaller than all of its conjugates. For a string T, the *conjugate array*[1] $\mathrm{CA} = \mathrm{CA}_T$ of T is the permutation of $\{1,\ldots,n\}$ such that $\mathrm{CA}[i] = j$ if $\mathrm{conj}_j(T)$ is the i-th conjugate of T with respect to the lexicographic order, with ties broken according to string order, i.e. if $\mathrm{CA}[i] = j$ and $\mathrm{CA}[i'] = j'$ for some $i < i'$, then either $\mathrm{conj}_j(T) <_{\mathrm{lex}} \mathrm{conj}_{j'}(T)$, or $\mathrm{conj}_j(T) = \mathrm{conj}_{j'}(T)$ and $j < j'$. Note that if T is a Lyndon word, then $\mathrm{CA}[i] = \mathrm{SA}[i]$ for all $1 \leq i \leq n$ [14].

Given a string T, U is a *circular* or *cyclic substring* of T if it is a factor of TT of length at most $|T|$, or equivalently, if it is the prefix of some conjugate of T. For instance, ATA is a cyclic substring of $AGCAT$. It is sometimes also convenient to regard a given string $T[1..n]$ itself as *circular* (or *cyclic*); in this case we set $T[0] = T[n]$ and $T[n+1] = T[1]$.

Burrows-Wheeler-Transform. Given a string T, $\mathrm{BWT}(T)$ [7] is a permutation of the letters of T which equals the last column of the matrix of the lexicographically sorted conjugates of T. The mapping $T \mapsto \mathrm{BWT}(T)$ is reversible, up to rotation. It can be made uniquely reversible by adding to $\mathrm{BWT}(T)$ an index indicating the rank of T in the lexicographic order of all of its conjugates. Given $\mathrm{BWT}(T)$ and an index i, the original string T can be computed in linear time [7]. The BWT itself can be computed from the conjugate array, since for all $i = 1,\ldots,n$, $\mathrm{BWT}(T)[i] = T[\mathrm{CA}[i] - 1]$, where T is considered to be cyclic.

It should be noted that in many applications, it is assumed that an end-of-string-character (usually denoted \$), which is not element of Σ, is appended to the string; this character is assumed to be smaller than all characters from Σ. Computing the conjugate array becomes equivalent to computing the suffix array, since $\mathrm{CA}_{T\$}[i] = \mathrm{SA}_{T\$}[i]$. Thus, applying one of the linear-time suffix-array computation algorithms [28] leads to linear-time computation of the BWT.

When no \$-character is appended to the string, the situation is slightly more complex. For primitive strings T, first the Lyndon conjugate of T has to be computed (in linear time, see [34]) and then a linear-time suffix array algorithm can be employed [14]. For strings T which are not primitive, one can take

---

[1] Our conjugate array CA is called *circular suffix array* and denoted $\mathrm{SA}_\circ$ in [2,17], and *BW-array* in [19,31], but in both cases defined for primitive strings only.

advantage of the following well-known property of the BWT: if $T = S^k$ and $BWT(S) = U[1..m]$, then $BWT(T) = U[1]^k U[2]^k \cdots U[m]^k$ (Prop. 2 in [27]).

Generalized Conjugate Array and Extended BWT. Given a multiset of strings $\mathcal{M} = \{T_1[1..n_1], \ldots, T_m[1..n_m]\}$, the *generalized conjugate array* of $\mathcal{M}$, denoted by $GCA_{\mathcal{M}}$ or just by GCA, contains the list of the conjugates of all strings in $\mathcal{M}$, sorted according to the ω-order relation. More formally, $GCA[i] = (j, d)$ if $conj_j(T_d)$ is the i-th string in the $\preceq_\omega$-sorted list of the conjugates of all strings of $\mathcal{M}$, with ties broken first w.r.t. the index of the string (in case of identical strings), and then w.r.t. the index in the string itself.

The *extended Burrows-Wheeler Transform* (eBWT) is an extension of the BWT to a multiset of strings [26]. It is a bijective transformation that, given a multiset of strings $\mathcal{M} = \{T_1, \ldots, T_m\}$, produces a permutation of the characters on the strings in the multiset $\mathcal{M}$. Formally, $eBWT(\mathcal{M})$ can be computed by sorting all the conjugates of the strings in the multiset according to the $\preceq_\omega$-order, and the output is the string obtained by concatenating the last character of each conjugate in the sorted list, together with the set of indices representing the positions of the original strings of $\mathcal{M}$ in the list. Similarly to the BWT, the eBWT is thus uniquely reversible. The $eBWT(\mathcal{M})$ can be computed from the generalized conjugate array of $\mathcal{M}$ in linear time, since $eBWT(\mathcal{M})[i] = T_d[j - 1]$ if $GCA[i] = (j, d)$, where again, the strings in $\mathcal{M}$ are considered to be cyclic. It is easy to see that when $\mathcal{M}$ consists of only one string, i.e. $\mathcal{M} = \{T\}$, then $eBWT(\mathcal{M}) = BWT(T)$.

Example 1. Let $\mathcal{M} = \{GTACAACG, CGGCACACACGT, C\}$. Then $GCA(\mathcal{M})$ is as follows, where we give the pair (j, d) vertically, i.e. the first row contains the position in the string, and the second row the index of the string:

5 3 5 7 6 9 4 4 6 8 1 1 7 10 3 2 8 1 11 2 12
1 1 2 2 1 2 1 2 2 2 3 2 1 2 2 2 1 1 2 1 2

From the GCA we can compute $eBWT(\mathcal{M}) = CTCCACAGAACTAAGCCG$ CGG, with index set $\{11, 12, 18\}$. Note that $conj_8(T_2)$ comes before $conj_1(T_3)$, since $CACGTCGGCACA \prec_\omega C$. In fact, $(CACGTCGGCACA)^\omega <_{\text{lex}} C^\omega = CCCC \ldots$.

Remark 1. Note that if end-of-string symbols are appended to the string of the collection then the output of eBWT could be quite different. For instance, if $\mathcal{M} = \{GTACAACG\$_1, CGGCACACACGT\$_2, C\$_3\}$, then $eBWT(\mathcal{M}) = GTCCTCCAC\$_3AGAAA\$_2ACGCC\$_1GG$.

While in the original definition of eBWT [26], the multiset $\mathcal{M}$ was assumed to contain only primitive strings, our definition is more general and allows also for non-primitive strings. For example, $eBWT(\{ATA, TATA\}) = TATTAAA$, with index set $\{2, 6\}$, while $eBWT(\{ATA, TA, TA\}) = TATTAAA$, with index set $\{2, 6, 7\}$. Also the linear-time algorithm for recovering the original multiset can be straightforwardly extended.

The following lemma shows how to construct the generalized conjugate array $GCA_{\mathcal{M}}$ of a multiset $\mathcal{M}$ of strings (not necessarily primitive), once we know the generalized conjugate array $GCA_\mathcal{R}$ of the multiset $\mathcal{R}$ of the roots of the strings in $\mathcal{M}$. It follows straightforwardly from the fact that equal conjugates will end up consecutively in the GCA.

Lemma 1. *Let* $\mathcal{M} = \{T_1, \dots, T_m\}$ *be a multiset of strings and let* $\mathcal{R}$ *the multiset of the roots of the strings in* $\mathcal{M}$, *i.e.* $\mathcal{R} = \{S_1, \dots, S_m\}$, *where* $T_i = (S_i^{r_i})$, *with* $r_i \geq 1$ *for* $1 \leq i \leq m$. *Let* $GCA_\mathcal{R}[1..K] = [(j_1, i_1), (j_2, i_2), \dots, (j_K, i_K)]$, *where* $K = \sum_{i=1}^m |S_i|$. $GCA_\mathcal{M}$ *is then given by (where* $N = \sum_{i=1}^m |S_i| \cdot r_i$):

$$GCA_\mathcal{M}[1..N] = [(j_1, i_1), (j_1 + |S_{i_1}|, i_1), \dots, (j_1 + (r_{i_1} - 1) \cdot |S_{i_1}|, i_1), \quad \dots,$$
$$(j_K, i_K), (j_K + |S_{i_K}|, i_K), \dots, (j_K + (r_{i_K} - 1) \cdot |S_{i_K}|, i_K)].$$

3   Computing the eBWT and GCA

Here, we describe our algorithm to compute the eBWT of a multiset of strings $\mathcal{M}$. We will assume that all strings in $\mathcal{M}$ are primitive, since we can use Lemma 1 to compute the eBWT of $\mathcal{M}$ otherwise. Our algorithm is an adaptation of the well-known SAIS algorithm of Nong et al. [29], which computes the suffix array of a single string T ending with an end-of-string character $. Our adaptation is similar to that of Bannai et al. [2] for computing the BBWT, which can also be used for computing the eBWT. Even though our algorithm does not improve the latter asymptotically (both are linear time), it is significantly simpler, since it does not require first computing and sorting the Lyndon rotations of the input strings. We assume some familiarity with the SAIS algorithm, focusing on the differences between our algorithm and the original SAIS. Detailed explanations of SAIS can be found in [22, 29, 30].

The main differences between our algorithm and the original SAIS algorithm are: (1) we are comparing conjugates rather than suffixes, (2) we have a multiset of strings rather than just one string, (3) the comparison is done w.r.t. the omega-order rather than the lexicographic order, and (4) the strings are not terminated by an end-of-string symbol.

We need the following definition, which is the cyclic version of the definition in [29] (where S stands for smaller, L for larger, and *LMS* for leftmost-S):

Definition 1. (Cyclic types, *LMS*-substrings). *Let* T *be a primitive string of length at least 2, and* $1 \leq i \leq |T|$. *Position* i *of* T *is called (cyclic) S-type if* $conj_i(T) <_{lex} conj_{i+1}(T)$, *and (cyclic) L-type if* $conj_i(T) >_{lex} conj_{i+1}(T)$. *An S-type position* i *is called (cyclic) LMS-position if* $i - 1$ *is L-type (where we view* T *as a cyclic string). An LMS-substring is a cyclic substring* $T[i, j]$ *of* T *such that both* i *and* j *are LMS-positions, but there is no LMS-position between* i *and* j. *Given a conjugate* $conj_i(T)$, *its LMS-prefix is the cyclic substring from* i *to the first LMS-position strictly greater than* i *(viewed cyclically).*

Since T is primitive, no two conjugates are equal, and in particular, no two adjacent conjugates are equal. Thus, the type of every position is defined.

Example 2. Continuing Example 1 (where we mark *LMS*-positions with a ∗),

$$G\,T\,A\,C\,A\,A\,C\,G \quad C\,G\,G\,C\,A\,C\,A\,C\,A\,C\,G\,T$$
$$S\,L\,S\,L\,S\,S\,S \quad S\,L\,L\,L\,S\,L\,S\,L\,S\,S\,S\,L$$
$$\quad *\quad *\quad\quad\quad *\quad\quad *\quad *\quad *$$

The *LMS*-substrings are ACA, $AACGGTA$, $CGGCA$, and $ACGTC$. The *LMS*-prefix of the conjugate $\mathrm{conj}_7(T_1) = CGGTACAA$ is $CGGTA$.

Lemma 2. (Cyclic type properties). *Let T be primitive string of length at least 2. Let a_1 be the smallest and a_σ the largest character of the alphabet. Then the following hold, where T is viewed cyclically:*

1. *if $T[i] < T[i+1]$, then i is type S, and if $T[i] > T[i+1]$, then i is type L,*
2. *if $T[i] = T[i+1]$, then the type of i is the same as the type of $i+1$,*
3. *i is of type S iff $T[i'] > T[i]$, where $i' = \min\{i+j \mid j > 0, T[i+j] \neq T[i]\}$,*
4. *if $T[i] = a_1$, then i is type S, and if $T[i] = a_\sigma$, then i is type L.*

Corollary 1. (Linear-time cyclic type assignment). *Let T be a primitive string of length at least 2. Then all positions can be assigned a type in altogether at most $2|T|$ steps.*

Let N be the total length of the strings in $\mathcal{M}$. The algorithm constructs an initially empty array A of size N, which, at termination, will contain the GCA of $\mathcal{M}$. The algorithm also returns the set $\mathcal{I}$ containing the set of indices in A representing the positions of the strings of $\mathcal{M}$. The overall procedure consists of the following steps (each step will be explained below):

Algorithm SAIS-for-eBWT
Step 1 remove strings of length 1 from $\mathcal{M}$ (to be added back at the end)
Step 2 assign cyclic types to all positions of strings from $\mathcal{M}$
Step 3 use procedure Induced Sorting to sort cyclic *LMS*-substrings
Step 4 assign names to cyclic *LMS*-substrings; if all distinct, go to Step 6
Step 5 recurse on new multiset $\mathcal{M}'$, returning array A', map A' back to A
Step 6 use procedure Induced Sorting to sort all positions in $\mathcal{M}$, add
       length-1 strings in their respective positions, return $(A, \mathcal{I})$

At the heart of the algorithm is the procedure Induced Sorting of [29] (Algorithms 3.3 and 3.4), which is used once to sort the *LMS*-substrings (Step 3), and once to induce the order of all conjugates from the correct order of the *LMS*-positions (Step 6), as in the original SAIS. Before sketching this procedure, we need to define the order according to which the *LMS*-substrings are sorted in Step 2. Our definition of *LMS*-order extends the *LMS*-order of [29] to *LMS*-prefixes. It can be proved that these definitions coincide for *LMS*-substrings.

Definition 2. (*LMS*-order). *Given two strings S and T, let U resp. V be their LMS-prefixes. We define $U <_{LMS} V$ if either V is a proper prefix of U, or neither is a proper prefix of the other and $U <_{\mathrm{lex}} V$.*

The procedure **Induced Sorting** for the conjugates of the multiset is analogous to the original one, except that strings are viewed cyclically. First, the array A is subdivided into so-called *buckets*, one for each character. For $c \in \Sigma$, let n_c denote the total number of occurrences of the character c in the strings in $\mathcal{M}$. Then the buckets are $[1, n_{a_1}], [n_{a_1} + 1, n_{a_1} + n_{a_2}], \ldots, [N - n_{a_\sigma} + 1, N]$, i.e., the k-th bucket will contain all conjugates starting with character a_k. The procedure **Induced Sorting** first inserts all *LMS*-positions at the end of their respective buckets, then induces the L-type positions in a left-to-right scan of A, and finally, induces the S-type positions in a right-to-left scan of A, possibly overwriting previously inserted positions. We need two pointers for each bucket **b**, *head*(**b**) and *tail*(**b**), pointing to the current first resp. last free position.

Procedure **Induced Sorting** [29]
1. insert all *LMS*-positions at the end of their respective buckets; for all buckets **b**, initialize *head*(**b**), *tail*(**b**) to the first resp. last position
2. induce the L-type positions in a left-to-right scan of A: for i from 1 to $N - 1$, if $A[i] = (j, d)$ then $A[head(bucket(T_d[j - 1]))] \leftarrow (j - 1, d)$; increment $head(bucket(T_d[j - 1]))$
3. induce the S-type positions in a right-to-left scan of A: for i from N to 2, if $A[i] = (j, d)$ then $A[tail(bucket(T_d[j - 1]))] \leftarrow (j - 1, d)$; decrement $tail(bucket(T_d[j - 1]))$

At the end of this procedure, the *LMS*-substrings are listed in correct relative *LMS*-order (see Lemma 4), and they can be named according to their rank. For the recursive step, we define, for $i = 1, \ldots, m$, a new string T_i', where each *LMS*-substring of T_i is replaced by its name (i.e. its rank). The algorithm is called recursively on $\mathcal{M}' = \{T_1', \ldots, T_m'\}$ (Step 5).

Finally (Step 6), the array $A' = \text{GCA}(\mathcal{M}')$ from the recursive step is mapped back into the original array, resulting in the placement of the *LMS*-substrings in their correct relative order. This is then used to induce the full array A. All length-1 strings T_i, which were removed in Step 1, can now be inserted between the L- and S-type positions in their bucket (Lemma 3).

Correctness and Running Time. The following lemma shows that the individual steps of **Induced Sorting** are applicable for the ω-order on conjugates of a multiset (part 1), that L-type conjugates (of all strings) come before the S-type conjugates within the same bucket (part 2), and that length-1 strings are placed between S-type and L-type conjugates (part 3). The second property was originally proved for the lexicographic order between suffixes in [18]:

Lemma 3. (Induced sorting for multisets). *Let $U, V \in \Sigma^*$.*

1. *If $U \prec_\omega V$, then for all $c \in \Sigma$, $cU \prec_\omega cV$.*
2. *If $U[i] = V[j]$, i is an L-type position, and j an S-type position, then $conj_i(U) \prec_\omega conj_j(V)$.*
3. *If $U[i] = V[j] = c$, i is an L-type position, and j an S-type position, then $conj_i(U) \prec_\omega c \prec_\omega conj_j(V)$.*

Next, we show that after applying procedure Induced Sorting, the conjugates will appear in A such that they are correctly sorted w.r.t. to the LMS-order of their LMS-prefixes, while the order in which conjugates with identical LMS-prefixes appear in A is determined by the input order of the LMS-positions.

Lemma 4. (Extension of Thm. 3.12 of [29]). *Let $T_1, T_2 \in \mathcal{M}$, let U be the LMS-prefix of $conj_i(T_1)$, with i' the last position of U; let V be the LMS-prefix of $conj_j(T_2)$, and j' the last position of V. Let k_1 be the position of $conj_i(T_1)$ in array A after the procedure Induced Sorting, and k_2 that of $conj_j(T_2)$.*

1. *If $U <_{LMS} V$, then $k_1 < k_2$.*
2. *If $U = V$, then $k_1 < k_2$ if and only if $conj_{i'}(T_1)$ was placed before $conj_{j'}(T_2)$ at the start of the procedure.*

Lemma 5. *Let $S, T \in \Sigma^*$, let U be the LMS-prefix of S and V the LMS-prefix of T. If $U <_{LMS} V$ then $S \prec_\omega T$.*

Theorem 1. *Algorithm SAIS-for-eBWT correctly computes GCA and eBWT of multiset of strings $\mathcal{M}$ in time $O(N)$, with N the total length of strings in $\mathcal{M}$.*

Computing the BWT for One Single String. The special case where $\mathcal{M}$ consists of one single string leads to a new algorithm for computing the BWT, since for a singleton set, the eBWT coincides with the BWT. To the best of our knowledge, this is the first linear-time algorithm for computing the BWT *of a string without an end-of-string character* that uses neither Lyndon rotations nor end-of-string characters.

We demonstrate the algorithm on a well-known example, $T = banana$. We get the following types, from left to right: $LSLSLS$, and all three S-type positions are LMS. We insert $2, 4, 6$ into the array A; after the left-to-right pass, indices are in the order $2, 4, 6, 1, 3, 5$, and after the right-to-left pass, in the order $6, 2, 4, 1, 3, 5$. The LMS-substring aba (pos. 6) gets the name A, and the LMS-substring ana (pos. 2,4) gets the name B. In the recursive step, the new string $T' = ABB$, with types SLL and only one LMS-position 1, the GCA gets induced in just one pass: $1, 3, 2$. This maps back to the original string: $6, 2, 4$, and one more pass over the array A results in $6, 4, 2, 1, 5, 3$ and the BWT $nnbaaa$.

4   eBWT and PFP

We show how to extend the PFP to build the eBWT. We define the *cyclic prefix-free parse* for a multiset of strings $\mathcal{M} = \{T_1, T_2, \ldots, T_m\}$ (with $|T_i| = n_i$, $1 \le i \le m$) as the multiset of parses $\mathcal{P} = \{P_1, P_2, \ldots, P_m\}$ with dictionary D, where we consider T_i as circular, and P_i is the parse of T_i. We denote by p_i the length of the parse P_i. In the following we show how the multiset of parses $\mathcal{P}$ and the dictionary D are constructed. Given a positive integer w, let E be a set of strings of length w called *trigger strings*. We assume that each string $T_h \in \mathcal{M}$ has length at least w and at least one cyclic factor in E. We divide each string

$T_h \in \mathcal{M}$ into overlapping phrases as follows: a phrase is a circular factor of T_h of length $> w$ that starts and ends with a trigger string and has no internal occurrences of a trigger string. The set of phrases obtained from strings in $\mathcal{M}$ is the dictionary D. The parse P_h can be computed from the string T_h by replacing each occurrence of a phrase in T_h with its lexicographic rank in D.

We denote by $\mathcal{S}$ the set of suffixes of D having length greater than w. The first important property of the dictionary D is that the set $\mathcal{S}$ is *prefix-free*, i.e., no string in $\mathcal{S}$ is prefix of another string of $\mathcal{S}$. This follows directly from [6].

The computation of eBWT from the prefix-free parse consists of three steps: computing the cyclic prefix-free parse of $\mathcal{M}$ (denoted as $\mathcal{P}$), computing the eBWT of $\mathcal{P}$ by using the algorithm described in Sect. 3; and computing the eBWT of $\mathcal{M}$ from the eBWT of $\mathcal{P}$ using the lexicographically sorted dictionary $D = \{D_1, D_2, \ldots, D_{|D|}\}$ and its prefix-free suffix set $\mathcal{S}$. We now describe the last step as follows. We define δ as the function that uniquely maps each character of $T_h[j]$ to the pair (i, k), where with $1 \le i \le p_h$, $k > w$, and $T_h[j]$ appears as the k-th character of the $P_h[i]$-th phrase of D. We call i and k the *position* and the *offset* of $T_h[j]$, respectively. Furthermore, we define α as the function that uniquely associates to each conjugate $conj_j(T_h)$ the element $s \in \mathcal{S}$ such that s is the k-th suffix of the $P_h[i]$-th element of D, where $(i, k) = \delta(T_h[j])$. By extension, i and k are also called the *position* and the *offset* of the suffix $\alpha(conj_j(T_h))$.

Lemma 6. *Given two strings* $T_g, T_h \in \mathcal{M}$, *if* $\alpha(conj_i(T_g)) <_{\text{lex}} \alpha(conj_i(T_h))$ *it follows that* $conj_i(T_g) \prec_w conj_j(T_h)$.

Proposition 1. *Given two strings* $T_g, T_h \in \mathcal{M}$. *Let* $conj_i(T_g)$ *and* $conj_j(T_h)$ *be the* i-th *and* j-th *conjugates of* T_g *and* T_h, *respectively, and let* $(i', g') = \delta(T_g[i])$ *and* $(j', h') = \delta(T_h[j])$. *Then* $conj_i(T_g) \prec_w conj_j(T_h)$ *if and only if either* $\alpha(conj_i(T_g)) <_{\text{lex}} \alpha(conj_j(T_h))$, *or* $conj_{i'+1}(P_g) \prec_w conj_{j'+1}(P_h)$, *i.e.,* $P_g[i']$ *precedes* $P_h[j']$ *in* eBWT$(\mathcal{P})$.

Next, using Proposition 1, we define how to build the eBWT of the multiset of strings $\mathcal{M}$ from $\mathcal{P}$ and D. We consider the eBWT partitioned into $|\mathcal{S}|$ consecutive blocks, each block associated to one $s \in \mathcal{S}$, such that each block consists of the last characters of all conjugates of the strings in $\mathcal{M}$ prefixed by s. Hence, it follows that we only need to describe how to build an eBWT block corresponding to a suffix $s \in \mathcal{S}$. We will iterate through all the suffixes in $\mathcal{S}$ in lexicographic order. Given $s \in \mathcal{S}$, we let $\mathcal{S}_s$ be the set of the lexicographic ranks of the phrases of D that have s as a suffix, i.e., $\mathcal{S}_s = \{i \mid 1 \le i \le |D|, s \text{ is a suffix of } D_i \in D\}$. Moreover, given the string $T_h \in \mathcal{M}$, we let $conj_i(T_h)$ be the i-th conjugate of T_h, let j and k be the position and offset of $T_h[i]$, and lastly, let p be the position of $P_h[j]$ in eBWT$(\mathcal{P})$. We define $f(p, k) = D_{P_h[j]}[k-1]$ if $k > 1$, otherwise $f(p, k) = D_{P_h[j-1]}[|D_{P_h[j-1]}| - w]$ where we view P_h as a cyclic string.

Finally, we let $\mathcal{O}_s$ be the set of pairs (p, c) such that for all $d \in \mathcal{S}_s$, p is the position of an occurrence of d in eBWT$(\mathcal{P})$, and c is the character resulting the application of the f function considering as k the offset of s in D_d, i.e., $c = f(p, |D_d| - |s| + 1)$. Formally, $\mathcal{O}_s = \{(p, f(p, |D_{\text{eBWT}(\mathcal{P})[p]}| - |s| + 1) \mid$ eBWT$(\mathcal{P})[p] \in \mathcal{S}_s\}$.

To build the eBWT block corresponding to $s \in \mathcal{S}$, we scan the set $\mathcal{O}_s$ in increasing order of the first element of the pair, i.e., the position of the occurrence in $eBWT(\mathcal{P})$, and concatenate the values of the second element of the pair, i.e., the character preceding the occurrence of s in T_h. Note that if all the occurrences in $\mathcal{O}_s$ are preceded by the same character c, we do not need to iterate through all the occurrences but rather concatenate $|\mathcal{O}_s|$ copies of the character c.

More details about the implementation will be given in the full version.

5 Results and Discussion

We implemented the algorithm for building the eBWT and measured its performance on real biological data. We performed the experiments on a server with Intel(R) Xeon(R) CPU E5-2620 v4 @ 2.10 GHz with 16 cores and 62 gigabytes of RAM running Ubuntu 16.04 (64bit, kernel 4.4.0). The compiler was g++ version 9.4.0 with -O3 -DNDEBUG -funroll-loops -msse4.2 options. We recorded the runtime and memory usage using the wall clock time, CPU time, and maximum resident set size from /usr/bin/time. The source code is available online at: https://github.com/davidecenzato/PFP-eBWT.

Competing Methods. We compared our method (pfpebwt) with the BCR algorithm implementation of [21] (ropebwt2), gsufsort [23], and egap [11]. We did not compare against G2BWT [10], lba [4], and BCR [3] since they are currently implemented only for short reads[2]. We did not compare against egsa [24] since it is the predecessor of egap. Note that all tested implementations produce different outputs due to the different rules used for the end of string symbols, see [8].

Datasets. We evaluated our method using 2,048 copies of human chromosomes 19 from the 1000 Genomes Project [37]; 10,000 *Salmonella* genomes taken from the GenomeTrakr project [36], and 400,000 SARS-CoV2 genomes from EBI's COVID-19 data portal [38]. We used 12 sets of variants of human chromosome 19 (chr19), containing 2^i variants for $i = 0, \ldots, 11$ respectively. We used 6 collections of *Salmonella* genomes (salmonella) containing 50, 100, 500, 1,000, 5,000, and 10,000 genomes respectively. We used 5 sets of SARS-CoV2 genomes (sars-cov2) containing 25,000, 50,000, 100,000, 200,000, 400,000 genomes respectively. Each collection is a superset of the previous one. We ran pfpebwt and ropebwt2 with 16 threads, and gsufsort and egap with a single threads since they do not support multi-threading. We used ropebwt2 with the -R flag. The experiments that exceeded 48 h of wall clock time or exceeded 62 GB of memory were omitted from further consideration.

Experiments. In Fig. 1 we illustrate the construction time and memory usage to build the eBWT and the BWT of collections of strings for the chromosome 19 dataset. pfpebwt was the fastest method to build the eBWT of 4 or more sequences of chromosomes 19, with a maximum speedup of 7.6× of wall-clock

---

[2] G2BWT crashed and BCR did not terminate within 48 h with the smallest of each dataset; lba works only with sequences of length up to 255.

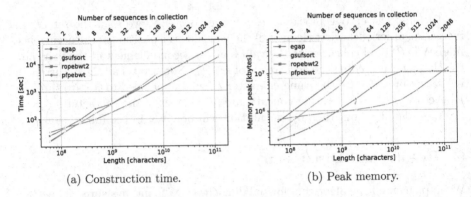

(a) Construction time.   (b) Peak memory.

Fig. 1. Chromosome 19 dataset construction CPU time and peak memory usage. We compare `pfpebwt` with `ropebwt2`, `gsufsort`, and `egap`. Note that for time, `egap` has almost the same performance as `gsufsort`, therefore the two plots are indistinguishable.

time and 2.9× of CPU time over `ropebwt2` on 256 sequences of chromosomes 19, 2.7× of CPU time over `egap` on 64 sequences, and 3.8× of CPU time over `gsufsort` on 128 sequences. Considering the peak memory, on the chromosomes 19 dataset, `ropebwt2` used the smallest amount of memory for 1, 2, 4, 8, and 2,048 sequences, while `pfpebwt` used the smallest amount of memory in all other cases. `pfpebwt` used a maximum of 5.6× less memory than `ropebwt2` on 256 sequences of chromosomes 19, 28.0× less than `egap` on 64 sequences, and 45.3× less than `gsufsort` on 128 sequences. Note that the memory increase in `gsufsort` from 16 to 32 sequences is due to the switch from the 32 bit implementation to the 64 bit implementation.

Similar results were obtained also on *Salmonella* and SARS-CoV2 sequences, for which we provide an in-depth discussion in the full version of the paper.

The memory peak of `ropebwt2` is given by the default buffer size of 10 GB, and the size of the run-length encoded BWT stored in the rope data structure. This explains the memory plateau on 10.5 GB of `ropebwt2` on the chromosomes 19 dataset. However, `ropebwt2` is able only to produce the BWT of the input sequence collection, while `pfpebwt` can be trivially extended to produce also the samples of the conjugate array at the run boundaries with negligible additional costs in terms of time and peak memory.

Acknowledgements. CB and MR are funded by National Science Foundation NSF IIBR (Grant No. 2029552), NSF SCH (Grant No. 2013998), National Institutes of Health (NIH) NIAID (Grant No. HG011392) and NIH NIAID (Grant No. R01AI141810).

References

1. Ander, C., Schulz-Trieglaff, O., Stoye, J., Cox, A.: metaBEETL: high-throughput analysis of heterogeneous microbial populations from shotgun DNA sequences. BMC Bioinf. **14**(5), S2 (2013)

2. Bannai, H., Kärkkäinen, J., Köppl, D., Piatkowski, M.: Constructing the bijective and the extended Burrows-Wheeler Transform in linear time. In: Proceedings of the 32nd Annual Symposium on Combinatorial Pattern Matching, CPM 2021. LIPIcs, vol. 191, pp. 7:1–7:16 (2021)
3. Bauer, M., Cox, A., Rosone, G.: Lightweight algorithms for constructing and inverting the BWT of string collections. Theor. Comput. Sci. **483**, 134–148 (2013)
4. Bonizzoni, P., Vedova, G.D., Pirola, Y., Previtali, M., Rizzi, R.: Computing the multi-string BWT and LCP array in external memory. Theor. Comput. Sci. **862**, 42–58 (2021)
5. Boucher, C., et al.: PFP compressed suffix trees. In: Proceedings of the Symposium on Algorithm Engineering and Experiments, ALENEX 2021, pp. 60–72. SIAM (2021)
6. Boucher, C., Gagie, T., Kuhnle, A., Langmead, B., Manzini, G., Mun, T.: Prefix-free parsing for building big BWTs. Algorithms Mol. Biol. **14**(1), 13:1–13:15 (2019)
7. Burrows, M., Wheeler, D.: A block sorting lossless data compression algorithm. Technical report 124, Digital Equipment Corporation (1994)
8. Cenzato, D., Lipták, Zs.: On different variants of the extended Burrows-Wheeler-Transform. Unpublished manuscript (2021)
9. Cox, A., Bauer, M., Jakobi, T., Rosone, G.: Large-scale compression of genomic sequence databases with the Burrows-Wheeler Transform. Bioinformatics **28**(11), 1415–1419 (2012)
10. Díaz-Domínguez, D., Navarro, G.: Efficient construction of the extended BWT from grammar-compressed DNA sequencing reads. CoRR, arXiv:2102.03961 (2021)
11. Egidi, L., Louza, F., Manzini, G., Telles, G.: External memory BWT and LCP computation for sequence collections with applications. Algorithms Mol. Biol. **14**(1), 1–15 (2019)
12. Gagie, T., Tomohiro, I., Manzini, G., Navarro, G., Sakamoto, H., Takabatake, Y.: Rpair: rescaling RePair with Rsync. In: Brisaboa, N.R., Puglisi, S.J. (eds.) SPIRE 2019. LNCS, vol. 11811, pp. 35–44. Springer, Cham (2019). https://doi.org/10.1007/978-3-030-32686-9_3
13. Gessel, I.M., Reutenauer, C.: Counting permutations with given cycle structure and descent set. J. Combin. Theory Ser. A **64**(2), 189–215 (1993)
14. Giancarlo, R., Restivo, A., Sciortino, M.: From first principles to the Burrows and Wheeler Transform and beyond, via combinatorial optimization. Theor. Comput. Sci. **387**, 236–248 (2007)
15. Guerrini, V., Louza, F., Rosone, G.: Metagenomic analysis through the extended Burrows-Wheeler Transform. BMC Bioinf. **21**(299) (2020)
16. Guerrini, V., Rosone, G.: Lightweight metagenomic classification via eBWT. In: Holmes, I., Martín-Vide, C., Vega-Rodríguez, M.A. (eds.) AlCoB 2019. LNCS, vol. 11488, pp. 112–124. Springer, Cham (2019). https://doi.org/10.1007/978-3-030-18174-1_8
17. Hon, W.-K., Ku, T.-H., Lu, C.-H., Shah, R., Thankachan, S.V.: Efficient algorithm for circular Burrows-Wheeler Transform. In: Kärkkäinen, J., Stoye, J. (eds.) CPM 2012. LNCS, vol. 7354, pp. 257–268. Springer, Heidelberg (2012). https://doi.org/10.1007/978-3-642-31265-6_21
18. Ko, P., Aluru, S.: Space efficient linear time construction of suffix arrays. J. Discret. Algorithms **3**(2), 143–156 (2005)
19. Kucherov, G., Tóthmérész, L., Vialette, S.: On the combinatorics of suffix arrays. Inf. Process. Lett. **113**(22–24), 915–920 (2013)

20. Kuhnle, A., Mun, T., Boucher, C., Gagie, T., Langmead, B., Manzini, G.: Efficient construction of a complete index for pan-genomics read alignment. In: Cowen, L.J. (ed.) RECOMB 2019. LNCS, vol. 11467, pp. 158–173. Springer, Cham (2019). https://doi.org/10.1007/978-3-030-17083-7_10

21. Li, H.: Fast construction of FM-index for long sequence reads. Bioinformatics **30**(22), 3274–3275 (2014)

22. Louza, F.A., Gog, S., Telles, G.P.: Construction of Fundamental Data Structures for Strings. SCS, Springer, Cham (2020). https://doi.org/10.1007/978-3-030-55108-7

23. Louza, F., Telles, G., Gog, S., Prezza, N., Rosone, G.: gsufsort: constructing suffix arrays, LCP arrays and BWTs for string collections. Algorithms Mol. Biol. **15**(1), 1–5 (2020)

24. Louza, F.A., Telles, G.P., Hoffmann, S., de Aguiar Ciferri, C.D.: Generalized enhanced suffix array construction in external memory. Algorithms Mol. Biol. **12**(1), 26:1–26:16 (2017)

25. Manber, U., Myers, G.W.: Suffix arrays: a new method for on-line string searches. SIAM J. Comput. **22**(5), 935–948 (1993)

26. Mantaci, S., Restivo, A., Rosone, G., Sciortino, M.: An extension of the Burrows-Wheeler Transform. Theor. Comput. Sci. **387**(3), 298–312 (2007)

27. Mantaci, S., Restivo, A., Sciortino, M.: Burrows-Wheeler Transform and Sturmian words. Inf. Process. Lett. **86**(5), 241–246 (2003)

28. Navarro, G.: Compact Data Structures: A Practical Approach. Cambridge University Press, Cambridge (2016)

29. Nong, G., Zhang, S., Chan, W.H.: Two efficient algorithms for linear time suffix array construction. IEEE Trans. Comput. **60**(10), 1471–1484 (2011)

30. Ohlebusch, E.: Bioinformatics Algorithms: Sequence Analysis, Genome Rearrangements, and Phylogenetic Reconstruction. Oldenbusch Verlag (2013)

31. Perrin, D., Restivo, A.: Enumerative combinatorics on words. In: Bona, M. (ed.) Handbook of Enumerative Combinatorics (2015)

32. Prezza, N., Pisanti, N., Sciortino, M., Rosone, G.: SNPs detection by eBWT positional clustering. Algorithms Mol. Biol. **14**(1), 1–13 (2019)

33. Prezza, N., Pisanti, N., Sciortino, M., Rosone, G.: Variable-order reference-free variant discovery with the Burrows-Wheeler Transform. BMC Bioinform. **21**–**S**(8), 260 (2020)

34. Shiloach, Y.: Fast canonization of circular strings. J. Algorithms **2**(2), 107–121 (1981)

35. Stephens, Z.D., et al.: Big data: astronomical or genomical? PLoS Biol. **13**(7), e1002195 (2015)

36. Stevens, E., et al.: The public health impact of a publically available, environmental database of microbial genomes. Front. Microbiol. **8**, 808 (2017)

37. The 1000 Genomes Project Consortium. A global reference for human genetic variation. Nature **526**, 68–74 (2015)

38. The COVID-19 Data Portal. https://www.covid19dataportal.org/. Accessed 17 May 2021

39. Yang, L., Zhang, X., Wang, T.: The Burrows-Wheeler similarity distribution between biological sequences based on Burrows-Wheeler Transform. J. Theor. Biol. **262**(4), 742–749 (2010)

Extracting the Sparse Longest Common Prefix Array from the Suffix Binary Search Tree

Tomohiro I[1], Robert W. Irving[2], Dominik Köppl[3(⊠)], and Lorna Love[2]

[1] Department of Artificial Intelligence, Kyushu Institute of Technology,
Iizuka, Japan
tomohiro@ai.kyutech.ac.jp
[2] School of Computing Science, University of Glasgow, Glasgow, UK
Rob.Irving@glasgow.ac.uk
[3] M&D Data Science Center, Tokyo Medical and Dental University, Tokyo, Japan
koeppl.dsc@tmd.ac.jp

Abstract. Given a text T of length n, the sparse suffix sorting problem asks for the lexicographic order of suffixes starting at m selectable text positions P. The suffix binary search tree [Irving and Love, JDA'03] is a dynamic data structure that can answer this problem dynamically in the sense that insertions and deletions of positions in P are allowed. While a standard binary search tree on strings needs to store two longest-common prefix (LCP) values per node for providing the same query bounds, each suffix binary search tree node only stores a single LCP value and a bit flag. Its tree topology induces the sorting of the m suffixes by an Euler tour in $\mathcal{O}(m)$ time. However, it has not been addressed how to compute the lengths of the longest common prefixes of two suffixes with neighboring ranks with this data structure. We show that we can compute these lengths again by an Euler tour in $\mathcal{O}(m)$ time.

Keywords: Suffix binary search tree · Sparse suffix sorting · Longest common prefixes · Euler tour

1 Introduction

While common full-text indexing data structures provide interfaces answering pattern matching queries for all positions in the underlying text, the sparse variation of such data structures index the text only at certain m positions with the aim to improve the space and construction time bounds to a complexity related to m. Such a technique can make sense if we work with large data sets for which the maintenance of a full-text data structure is prohibitive with respect to its space or construction time. In such a case, when only certain positions are of interest such as word beginnings in natural language texts, we can resort to a sparse text index. One of the most well-studied sparse text indices is the sparse suffix array. Given a text $T[1..n]$ of length n, and a set of text

© Springer Nature Switzerland AG 2021
T. Lecroq and H. Touzet (Eds.): SPIRE 2021, LNCS 12944, pp. 143–150, 2021.
https://doi.org/10.1007/978-3-030-86692-1_12

positions $P = \{p_1, \ldots, p_m\}$ with $p_i \in [1..n]$, the sparse suffix array SSA determines the lexicographic order of all suffixes starting with the positions of P. Formally, the sparse suffix array $\text{SSA}[1..m]$ is a ranking of P with respect to the suffixes starting at the respective positions, i.e., $T[\text{SSA}[i]..] \prec T[\text{SSA}[i+1]..]$ for $i \in [1..m-1]$ with $\{\text{SSA}[1], \ldots, \text{SSA}[m]\} = P$. To support pattern matching efficiently, the sparse suffix array can be augmented with the sparse longest common prefix (LCP) array SLCP to support the classic suffix array pattern matching algorithm of Manber and Myers [19]. The sparse LCP array SLCP stores the length of the LCP of each suffix stored in SSA with its preceding entry, i.e., $\text{SLCP}[1] = 0$ and $\text{SLCP}[i] := \text{lcp}(T[\text{SSA}[i-1]..], T[\text{SSA}[i]..])$ for all integers $i \in [2..|\text{SSA}|]$.

Other applications of SSA beyond plain pattern matching include the computation of the LCP array in external memory [12], a Burrows-Wheeler transform [4] variation [5], or finding maximal exact matches [14, 22]. There are algorithms that can compute SSA and SLCP for a given set P efficiently (e.g., [2, 3, 6–8, 13, 15, 17, 20, 21]), where most approaches resort to a data structure answering longest common extension (LCE) queries, i.e., a query that asks for the length of the LCP of two suffixes of the underlying text. As pointed out by Fischer et al. [7, Observation 1.2], given a data structure that answers LCE queries in $\mathcal{O}t_{\text{LCE}}(n)$ time for $t_{\text{LCE}}(n) > 0$, we can solve the sparse suffix sorting problem for m positions in $\mathcal{O}t_{\text{LCE}}(n) \cdot m \cdot t_{\mathcal{D}}(m, n)$ time by inserting the m respective suffixes into a dynamic dictionary $\mathcal{D}$ with an insertion operation taking $t_{\mathcal{D}}(m, n)$ time when $\mathcal{D}$ stores m suffix starting positions. In particular, this observation generalizes the suffix sorting problem to be dynamic. By *dynamic* we mean that we allow insertions or deletions of positions in P, and hence have the additional need for updating $\mathcal{D}$. This problem is a fundamental task for dynamic pattern matching, where the user can (a) change P and (b) query an indexing data structure built on P and T, both in an arbitrary order. A straightforward approach is to use a binary search tree (BST) as the dictionary representation since it supports the necessary insertion operations. However, the time complexity can be a problem, since a suffix has $\mathcal{O}(n)$ characters and hence, we need $t_{\mathcal{D}}(m, n) = \mathcal{O}(nh)$ time for an insertion into a BST of height h. We can use a balanced representation such as the AVL-tree [1] to obtain $h = \mathcal{O}(\lg m)$, but the insertion time complexity has still a linear factor in n. Exactly for this use case scenario, Irving and Love [11] provided an augmentation of the standard BST to obtain $\mathcal{O}(m + h)$ time, which they called suffix BST. In the suffix BST, they augmented each BST node with the length of the LCEs with an ancestor node. They also proposed a variant with the virtues of the AVL-tree, which they called suffix AVL-tree, having $h = \mathcal{O}(\lg m)$. In the following we write SBST for the suffix BST or its variants, the suffix AVL-tree, built upon the m suffixes of our text. Although we can retrieve SSA from SBST with a simple Euler tour, we show in the following that retrieving SLCP is also possible:[1]

---

[1] A precursor of this research is the technical report [10, Section 5] and a PhD thesis [18].

Theorem 1. *We can compute* SLCP *from* SBST *in time linear in the number of nodes stored in* SBST.

Theorem 1 is important in use cases where we need to compute the order of the suffixes dynamically, but then need SSA and SLCP for one of the aforementioned applications. The problem tackled by Theorem 1 can be made trivial with a variation of SBST that stores two LCP values per node [16, Sect. 4.6], and thus this variant is a constant factor larger than SBST. Another way to solve the problem naively would be to use an LCE data structure to compute the LCPs of two neighboring entries in SSA, after extracting SSA from SBST, as done in [7, Corollary 4.2].

2   Preliminaries

We work in the pointer machine model. Let $T[1..n]$ be a text of length n whose characters are drawn from an ordered alphabet Σ. We assume that we can compare two characters of Σ in constant time. We write $T[i]$ for the i-th character of T, for $i \in [1..n]$. For $X, Y, Z \in \Sigma^*$ with $T = XYZ$, X, Y, and Z are called a *prefix*, *substring*, and *suffix* of S, respectively. Since X, Y, or Z may be empty, X and Z are also substrings of S at the same time by this definition.

Let SSA and SLCP be defined as in the introduction. The sparse inverse suffix array SISA$[1..n]$ is (partially) defined by SISA$[\text{SSA}[i]] = i$. We only need SISA conceptually, and only care about the entries determined by this equation. The idea is that SISA$[i]$ for $i \in P$ is the rank of the suffix $T[i..]$ among all suffixes starting with a position in P. See Fig. 1 for an example of the defined arrays for $n = m$. There, the entries of rules are sorted in suffix order, i.e., rules$[i]$ is the rule for the node representing SSA$[i]$.

i	1	2	3	4	5	6	7	8	9	10	11	12	13	14	15
T	c	a	a	t	c	a	c	g	g	t	c	g	g	a	c
SSA	2	14	6	3	15	1	5	11	7	13	12	8	9	4	10
SISA	6	1	4	14	7	3	9	12	13	15	8	11	10	2	5
SLCP	0	1	2	1	0	1	2	1	3	0	1	2	1	0	2
rules	E	A	L	L	A	D	R	A	L	A	L	L	R	D	A

Fig. 1. The example string $T = \texttt{caatcacggtcggac}$ used in [11, Fig. 2]. The row rules shows from which rule or scenario (cf. Sect. 4) the SLCP value was obtained.

3   The Suffix AVL Tree

Given a set of text positions P, the *suffix AVL tree* represents each suffix $T[p..]$ starting at a text position $p \in P$ by a node. The nodes are arranged in a binary

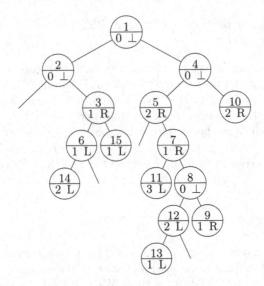

Fig. 2. The unbalanced SBST of the string $T = \text{caatcacggtcggac}$ defined in Fig. 1 when inserting all text positions in increasing order. A node consists of its position v (top), m_v (bottom left) and d_v (bottom right) abbreviated to L and R for left and right, respectively.

search tree topology such that reading the nodes with an in-order traversal gives the sparse suffix array. To support fast operations, each node is augmented with the following extra information:

Given a node v of SBST, cla_v (resp. cra_v) is the lowest node having v as a descendant in its left (resp. right) subtree. We write $T[v..]$ for the suffix represented by the node v, i.e., we identify nodes with their respective suffix starting positions. Each node v stores a tuple (d_v, m_v), where m_v is

- $\text{lcp}(T[v..], T[\text{cla}_v..])$ if $d_v = \text{left}$ and cla_v exists,
- $\text{lcp}(T[v..], T[\text{cra}_v..])$ if $d_v = \text{right}$ and cra_v exists, or
- 0 if $d_v = \bot$.

The value of $d_v \in \{\text{left}, \text{right}, \bot\}$ is set such that m_v is maximized. Let ca_v be cla_v (resp. cra_v) if $d_v = \text{left}$ (resp. $d_v = \text{right}$). If $d_v = \bot$, then ca_v as well as cla_v and cra_v are not defined. See Fig. 2 for an example.

4   Computing the Sparse LCP Array

Since an SBST node does not necessarily store the LCP with the lexicographically preceding suffix, it is not obvious how to compute SLCP from SBST. For computing SLCP from SBST, we use the following two facts and a lemma:

Fact 1: We have $T[\text{cra}_v..] \prec T[v..] \prec T[\text{cla}_v..]$ in case that cla_v and cra_v exist.

Fact 2: During an Euler tour (an in-order traversal), we can compute SSA by reading the text positions represented by the nodes in the order we visit the nodes for the first time. We can additionally keep track of the in-order rank $\mathsf{SISA}[v]$ of each node v.

Lemma 1 ([11, Lemma 1]). *Given three strings X, Y, Z with the lexicographic order $X \prec Y \prec Z$, we have $\mathrm{lcp}(X, Z) = \min(\mathrm{lcp}(X, Y), \mathrm{lcp}(Y, Z))$.*

With the following three rules Rule L, R, and E (as abbreviations for left, right and non-existing), we can partially compute SLCP:

Rule L: If $d_v = \texttt{left}$ and the right sub-tree of v is empty, then $\mathsf{SLCP}[\mathsf{SISA}[v] + 1] = m_v$. That is because $\mathrm{ca}_v = \mathrm{cla}_v = \mathsf{SSA}[\mathsf{SISA}[v]+1]$ is the starting position of the lexicographically next larger suffix with respect to the suffix starting with v.

Rule R: If $d_v = \texttt{right}$ then $\mathsf{SLCP}[\mathsf{SISA}[v]] \geq m_v$ since v shares a prefix of at least m_v characters with the lexicographically (not necessarily next) smaller suffix cra_v. In particular, if v does not have a left child, then $\mathsf{SLCP}[\mathsf{SISA}[v]] = m_v$ since $\mathrm{cra}_v = \mathsf{SSA}[\mathsf{SISA}[v] - 1]$ in this case.

Rule E: If v does not have a left child and cra_v does not exist, then $\mathsf{SLCP}[\mathsf{SISA}[v]] = 0$. This is the case when $T[v..]$ is the smallest suffix stored in SBST.

To compute all SLCP values, there remain the two scenarios Scenario D and A (the letters D and A have the meaning to compare with a descendent or ancestor node):

Scenario D: If a node v has a left child, then we have to compare v with the rightmost leaf in v's left subtree because this leaf corresponds to the lexicographically preceding suffix of the suffix starting with v.

Scenario A: Otherwise, this lexicographically preceding suffix corresponds to cra_v, such that we have to compare cra_v with v. If $d_v = \texttt{right}$, we are already done due to Rule R since $\mathrm{ca}_v = \mathrm{cra}_v$ in this case (such that the answer is already stored in m_v).

Figure 1 lists which rule or scenario has been applied to compute the SLCP value of a specific node of the SBST instance shown in Fig. 2. We cope with both scenarios by an Euler tour on SBST. For Scenario D, we want to know $\mathrm{lcp}(T[v..], T[\mathrm{cla}_v..])$ for each leaf v regardless of whether $d_v = \texttt{left}$ or not. For Scenario A, we want to know $\mathrm{lcp}(T[v..], T[\mathrm{cra}_v..])$ for each node v regardless of whether $d_v = \texttt{right}$ or not. We can obtain this LCP information by the following lemma:

Lemma 2. *Given $\mathrm{lcp}(T[\mathrm{cla}_v..], T[\mathrm{cra}_v..])$, $\mathrm{lcp}(T[v..], T[\mathrm{ca}_v..])$, and d_v, we can compute $\mathrm{lcp}(T[v..], T[\mathrm{cla}_v..])$ and $\mathrm{lcp}(T[v..], T[\mathrm{cra}_v..])$ in constant time.*

Proof. If $d_v = \texttt{left}$,

- $\mathrm{lcp}(T[v..], T[\mathrm{cla}_v..]) = \mathrm{lcp}(T[v..], T[\mathrm{ca}_v..])$ since $\mathrm{cla}_v = \mathrm{ca}_v$, and

- $\mathrm{lcp}(T[v..], T[\mathrm{cra}_v..]) = \mathrm{lcp}(T[\mathrm{cla}_v..], T[\mathrm{cra}_v..])$.

The latter is because of Fact 1 (assuming cla_v and cra_v exist) and

$$\mathrm{lcp}(T[\mathrm{cra}_v..], T[\mathrm{cla}_v..]) = \min(\mathrm{lcp}(T[\mathrm{cra}_v..], T[v..]), \mathrm{lcp}(T[v..], T[\mathrm{cla}_v..]))$$
$$= \mathrm{lcp}(T[v..], T[\mathrm{cra}_v..]) \leq \mathrm{lcp}(T[v..], T[\mathrm{cla}_v..])$$

according to Lemma 1. The case $d_v = \texttt{right}$ is symmetric:

- $\mathrm{lcp}(T[v..], T[\mathrm{cla}_v..]) = \mathrm{lcp}(T[\mathrm{cla}_v..], T[\mathrm{cra}_v..])$, and
- $\mathrm{lcp}(T[v..], T[\mathrm{cra}_v..]) = \mathrm{lcp}(T[v..], T[\mathrm{ca}_v..])$.   □

With Lemma 2, we can keep track of $\mathrm{lcp}(T[\mathrm{cla}_v..], T[\mathrm{cra}_v..])$ while descending the tree from the root: Suppose that we know $\mathrm{lcp}(T[\mathrm{cla}_v..], T[\mathrm{cra}_v..])$ and v's left and right children are x and y, respectively. Then the following holds for x and y:

- Since $\mathrm{cla}_x = v$ and $\mathrm{cra}_x = \mathrm{cra}_v$, $\mathrm{lcp}(T[\mathrm{cla}_x..], T[\mathrm{cra}_x..]) = \mathrm{lcp}(T[v..], T[\mathrm{cra}_v..])$.
- Since $\mathrm{cra}_y = v$ and $\mathrm{cla}_y = \mathrm{cla}_v$, $\mathrm{lcp}(T[\mathrm{cla}_y..], T[\mathrm{cra}_y..]) = \mathrm{lcp}(T[v..], T[\mathrm{cla}_v..])$.

During an Euler tour, we keep the values $\mathrm{lcp}(T[\mathrm{cla}_u..], T[\mathrm{cra}_u..])$ in a stack for the ancestors u of the current node. By applying the above rules and using the LCP information of Lemma 2 for both scenarios, we can compute SLCP during a single Euler tour. Since we did not make any assumptions on the height of SBST, this algorithm runs in linear time, regardless of whether the tree is balanced or not. Altogether, we obtain SSA and SLCP with a single Euler tour with constant time operations per node, and thus could prove the claim of Theorem 1.

Finally, one might be interested not in the complete arrays, but in the recovery of certain parts. By augmenting each node v with the size of the subtree rooted at v, we can answer SSA[i] in $\mathcal{O}(h)$ time by a top-down traversal in SBST. We select the left child if its subtree size s is at most i, otherwise we exchange i with $i - s - 1$ and either select the right child if $i > 1$, or stop ($i = 1$) since we arrived at a node representing the suffix starting position in question. Like with classic AVL trees, the subtree sizes can be maintained dynamically without additionally costs to the asymptotic time bounds.

For computing SLCP[i], we need to locate the nodes representing SSA[i] and SSA[$i+1$], which is done by two top-down traversals as above. If we additionally keep track of $\mathrm{lcp}(T[\mathrm{cla}_v..], T[\mathrm{cra}_v..])$ while descending the tree as in the aforementioned Euler tour, we gain information of all the above described cases for computing SLCP[i]. By then continuing the Euler tour as described above we obtain:

Corollary 1. *SBST augmented by the subtree sizes can retrieve* SSA[$i..i+\ell$] *and* SLCP[$i..i+\ell$] *in* $\mathcal{O}(h+\ell)$ *time for any* $\ell \in [0..m-i]$.

5 Future Work

Future directions of research include the analysis of space efficient data structures that have the same capabilities as the suffix BST, but work in compressed or succinct space. For instance, it is possible to use the B tree of [9] whose topology can be succinctly represented in $o(nk)$ bits. Another benefit of this B tree would be that it can read SLCP from the satellite values stored in leaves from left to right. Finally, for B+ variants, the data can be practically faster accessed than in classic binary search trees due to memory locality.

Acknowledgments. This work was supported by JSPS KAKENHI Grant Numbers JP21K17701 (DK) and JP19K20213 (TI). We thank the four anonymous reviewers of SPIRE'21 for their valuable comments on our manuscript. They give additional inspiration for Corollary 1 and proposed the problem of how to efficiently merge two suffix binary search tree instances. Tackling this problem could indeed be useful for building and updating FM-indexes and other related indexing data structures.

References

1. Adelson-Velsky, G.M., Landis, E.M.: An algorithm for organization of information. Dokl. Akad. Nauk SSSR **146**, 263–266 (1962)
2. Bille, P., Fischer, J., Gørtz, I.L., Kopelowitz, T., Sach, B., Vildhøj, H.W.: Sparse text indexing in small space. ACM Trans. Algorithms **12**(3), 39:1–39:19 (2016)
3. Birenzwige, O., Golan, S., Porat, E.: Locally consistent parsing for text indexing in small space. In: Proceedings of SODA, pp. 607–626 (2020)
4. Burrows, M., Wheeler, D.J.: A block sorting lossless data compression algorithm. Technical report 124, Digital Equipment Corporation, Palo Alto, California (1994)
5. Chien, Y., Hon, W., Shah, R., Thankachan, S.V., Vitter, J.S.: Geometric BWT: compressed text indexing via sparse suffixes and range searching. Algorithmica **71**(2), 258–278 (2015)
6. Ferragina, P., Fischer, J.: Suffix arrays on words. In: Ma, B., Zhang, K. (eds.) CPM 2007. LNCS, vol. 4580, pp. 328–339. Springer, Heidelberg (2007). https://doi.org/10.1007/978-3-540-73437-6_33
7. Fischer, J., I, T., Köppl, D.: Deterministic sparse suffix sorting in the restore model. ACM Trans. Algorithms **16**(4), 50:1–50:53 (2020)
8. I, T., Kärkkäinen, J., Kempa, D.: Faster sparse suffix sorting. In: Proceedings of STACS. LIPIcs, vol. 25, pp. 386–396 (2014)
9. I, T., Köppl, D.: Load-balancing succinct B trees. arXiv CoRR abs/2104.08751 (2021)
10. Irving, R.W., Love, L.: Suffix binary search trees and suffix arrays. University of Glasgow, Technical report (2001)
11. Irving, R.W., Love, L.: The suffix binary search tree and suffix AVL tree. J. Discret. Algorithms **1**(5–6), 387–408 (2003)
12. Kärkkäinen, J., Kempa, D.: LCP array construction using O(sort(n)) (or Less) I/Os. In: Inenaga, S., Sadakane, K., Sakai, T. (eds.) SPIRE 2016. LNCS, vol. 9954, pp. 204–217. Springer, Cham (2016). https://doi.org/10.1007/978-3-319-46049-9_20
13. Kärkkäinen, J., Ukkonen, E.: Sparse suffix trees. In: Cai, J.-Y., Wong, C.K. (eds.) COCOON 1996. LNCS, vol. 1090, pp. 219–230. Springer, Heidelberg (1996). https://doi.org/10.1007/3-540-61332-3_155
14. Khan, Z., Bloom, J.S., Kruglyak, L., Singh, M.: A practical algorithm for finding maximal exact matches in large sequence datasets using sparse suffix arrays. Bioinform. **25**(13), 1609–1616 (2009)

15. Kolpakov, R., Kucherov, G., Starikovskaya, T.A.: Pattern matching on sparse suffix trees. In: Proceedings of CCP, pp. 92–97 (2011)
16. Köppl, D.: Exploring regular structures in strings. Ph.D. thesis, TU Dortmund (2018)
17. Kosolobov, D., Sivukhin, N.: Construction of sparse suffix trees and LCE indexes in optimal time and space. arXiv CoRR abs/2105.03782 (2021)
18. Love, L.: The suffix binary search tree. Ph.D. thesis, University of Glasgow, UK (2001)
19. Manber, U., Myers, E.W.: Suffix arrays: a new method for on-line string searches. SIAM J. Comput. 22(5), 935–948 (1993)
20. Prezza, N.: In-place sparse suffix sorting. In: Proceedings of SODA, pp. 1496–1508 (2018)
21. Uemura, T., Arimura, H.: Sparse and truncated suffix trees on variable-length codes. In: Giancarlo, R., Manzini, G. (eds.) CPM 2011. LNCS, vol. 6661, pp. 246–260. Springer, Heidelberg (2011). https://doi.org/10.1007/978-3-642-21458-5_22
22. Vyverman, M., Baets, B.D., Fack, V., Dawyndt, P.: essaMEM: finding maximal exact matches using enhanced sparse suffix arrays. Bioinform. 29(6), 802–804 (2013)

findere: Fast and Precise Approximate Membership Query

Lucas Robidou$^{(\boxtimes)}$ and Pierre Peterlongo$^{(\boxtimes)}$

Univ. Rennes, Inria, CNRS, IRISA, Rennes, France
{lucas.robidou,pierre.peterlongo}@inria.fr

Abstract. Motivation: Approximate membership query (AMQ) structures such as Cuckoo filters or Bloom filters are widely used for representing large sets of elements. Their lightweight space usage explains their success, mainly as they are the only way to scale hundreds of billions or trillions of elements. However, they suffer by nature from non-avoidable false-positive calls that bias downstream analyses of methods using these data structures.

Results: In this work we propose a simple strategy and its implementation for reducing the false-positive rate of any AMQ data structure indexing k-mers (words of length k). The method we propose, called findere, enables to speed-up the queries by a factor two and to decrease the false-positive rate by two order of magnitudes. This achievement is done on the fly at query time, without modifying the original indexing data-structure, without generating false-negative calls and with no memory overhead.

This method yields so-called "construction false positives", but the amount of such false positives is negligible when the method is used within classical parameter ranges. This method, as simple as effective, reduces either the false-positive rate or the space required to represent a set given a user-defined false-positive rate.

Availability: https://github.com/lrobidou/findere.

Keywords: Approximate membership query · Data structure · Indexation · k-mers · Bloom filters · Sequence data

1 Introduction

Genomic studies generate a "data deluge" [13]. Public data banks providing sequencing data or assembled genome sequences are growing at an exponential rate [1]. Alongside, a fundamental need consists in comparing sequences at large scale, for instance for species identification [15], metagenome similarity estimation [3,11], or any generic need for estimating the presence of a sequence among available datasets [10].

Electronic supplementary material The online version of this chapter (https://doi.org/10.1007/978-3-030-86692-1_13) contains supplementary material, which is available to authorized users.

© Springer Nature Switzerland AG 2021
T. Lecroq and H. Touzet (Eds.): SPIRE 2021, LNCS 12944, pp. 151–163, 2021.
https://doi.org/10.1007/978-3-030-86692-1_13

Given the overwhelming amount of data to compare, these large scale sequence comparisons are made through alignment-free methods [16] mainly based on the number of words of fixed length, usually called k-mers for words of length k, shared between the two (set of) sequences to compare. So then, indexing k-mers for fast and low memory membership queries is a fundamental need. This last decade, an intense research activity was carried out in order to optimize such indexes (see [9] and [6] reviewing these efforts). Most of those indexes use approximate membership query (AMQ) structures such as Bloom filters [4] (called "BF" in this manuscript). With low space needed per represented element (usually less than 10 bits), these data structures are widely used although they suffer by nature from the existence of false-positive calls.

In genomics, BFs are the simplest and the most employed AMQ data structure used for representing the set of k-mers from a set of sequences. However, more sophisticated AMQ data structures enable under certain conditions to improve the BF features (cuckoo filters [7], SAT filters [14], quotient filters [2]). However, these improvements are marginal and/or at the expense of important limitations such as an important query time. For instance with SAT filters, despite they need $\approx 22\%$ less space, their queries are roughly 14 times slower than BFs' ones. Also, Cuckoo filters may show marginal smaller space cost per element for false-positive rates below 3% (space gain of 16% for instance for a false-positive rate of 0.1%, at the expense of longer query time).

In this work, we do not propose yet another AMQ data structure. Instead, we propose a downstream analysis of results returned by such data structures when used to query k-mers from a sequence. To the best of our knowledge, the only work in the same spirit is kBF [12]. kBF is designed to work on genomic sequences with the alphabet $\{A, C, G, T\}$ and indexes k-mers using a BF. With kBF if a queried k-mer is positive in the BF, the presence in the BF of at least one of the four (the alphabet size) potential previous and incoming k-mers is also checked. If none of them is positive, then the original queried k-mer is considered a negative. This leads to a lower false-positive rate (up to 30x lower than a raw BF) at the expense of longer query time and higher memory usage. kBF presents interesting features that can be further improved as 1/ its query times are 1.3 to 1.6x longer than a classical BF, and 2/ its strategy is limited to the previous and the k-mer that comes just after a queried k-mer. Extending this approach to n neighbor increases the query time by (4^n) fold. Finally, this approach applies only for small alphabets (e.g. of size four) as the number of queries depends on the size of the alphabet. The work we propose aims to overcome these limitations.

In this paper, we propose a method for improving AMQ results when used for querying k-mers from a sequence. Our strategy is based on the observation that false-positive k-mers from AMQ are not likely to occur consecutively on a queried sequence. Hence, in a procedure that we call the "Query Time Filtration" (QTF), small stretches of positive calls surrounded by negative calls are considered as false-positives and are filtered out. This simple strategy leads to an unprecedented decrease of the AMQ false-positive rate. However, this leads to the introduction of false-negative calls that are a barrier for many downstream applications. Nevertheless, we show that the QTF strategy can be used for querying K-mers (with $K > k$), with no false-negative.

We implemented this approach in a tool called findere. Used on results from any original AMQ data structure, findere presents only advantages when querying K-mers: it does not necessitate any change to the original AMQ structure, it does not use any additional memory or disk, it has no false-negative calls, and it has a false-positive rate two orders of magnitude lower than original AMQ. Moreover, findere does not entail any additional query time penalty and even enables faster query of K-mers (in average >2 times faster with recommended parameters) with no negative impact on result quality.

2 Method

2.1 Background

Preliminary Definitions
A k-mer is a word of length k over an alphabet Σ. Given a sequence S, $|S|$ denotes the length of S.

In the current framework, we consider a dataset as composed of one sequence or a set of sequences. We consider that a k-mer occurs in a dataset if it occurs at least once in any of the sequences composing the set.

An AMQ data structure represents a set of elements $\mathcal{D}$. It can be queried with any element d. If $d \in \mathcal{D}$, then the AMQ answer is positive (there is no false-negative). If $d \notin \mathcal{D}$, the AMQ answer may be negative or positive, in this last case it is a false-positive call. The false-positive rate, denoted by FPR_{AMQ}, is defined by $FPR_{AMQ} = \frac{\#FP}{\#FP + \#TN}$ with $\#FP$ and $\#TN$ denoting respectively the number of false-positive calls and the number of true negative calls. FPR_{AMQ} depends on the used AMQ strategy and mainly on the amount of space used by this AMQ.

Sequence Similarity Estimated by the Number of Shared k-mers
Given two sequences, a k-mer occurring in both sequences is called "shared". The number of shared k-mers between a queried sequence q and a dataset B provides an insight of the presence of q in B, or at least of a sequence similar to q in B [5].

In practice, for scaling Terabyte-sized sets B, a static AMQ indexes all (overlapping) k-mers from B. At query time, all (overlapping) k-mers from the query q are read on the fly and, for each of them, the k-mer is queried using the AMQ. Note that for each position p on $q \in [0, |q| - k + 1]$ the k-mer starting at this position is queried. Thus, it overlaps by $k - 1$ characters with the previously queried k-mer if it exists (if $p > 0$).

2.2 Decreasing the AMQ False-Positive Rate with "Query Time Filtration"

In the context of computing the k-mer similarity between a bank B represented by its set of k-mers indexed in an AMQ and a query sequence q, we propose a surprisingly simple approach for drastically decreasing the AMQ false-positive calls.

Observation About the False-Positive Calls

Given the AMQ properties, one can assume that false-positive calls appear at random when querying elements from Σ^*. In particular, when querying negative k-mers from a queried sequence q, the probability to query one false-positive is FPR_{AMQ}, the probability to query two successive false-positives is FPR_{AMQ}^2, and so on. Overall, the probability to query z successive false-positives is FPR_{AMQ}^z. Given a classical FPR_{AMQ} value of 1% (usual expected false-positive rate with BFs for instance), with $z = 3$, the chances to call three consecutive false-positives k-mers is 0.0001%.

Intuition About the True Positive Calls

When approximating the sequence similarity between q and B using the k-mer similarity, the underlying idea is to choose k (usually chosen higher than 20 and lower than 40) such that q and sequences of B share large ($\geq k$) sub-sequences. An intuitive consequence of this choice stands in the fact that, when querying successive k-mers from q, it is unlikely that only a low number (*e.g.* less than 3) of successive k-mers are true positives.

Query Time Filtration

Motivated by the observation and the intuition presented in the two previous sections, we propose a method that we call the "Query Time Filtration" (QTF in short), designed for lowering the FPR_{AMQ} at the expense of the introduction of false-negative calls.

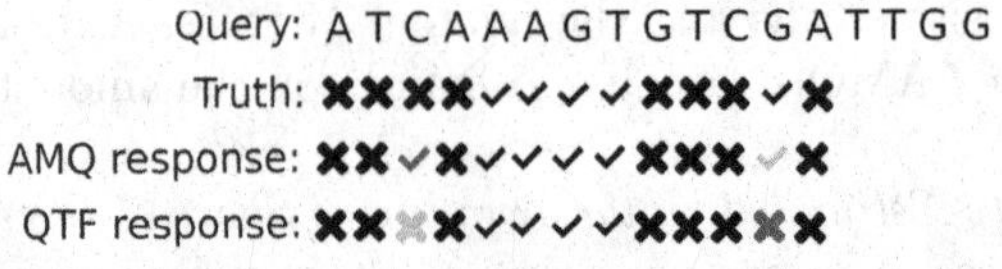

Query: A T C A A A G T G T C G A T T G G
Truth: ✗✗✗✗✓✓✓✓✗✗✗✓✗
AMQ response: ✗✗✓✗✓✓✓✓✗✗✗✓✗
QTF response: ✗✗✗✗✓✓✓✓✗✗✗✗✗

Fig. 1. Example with $k = 5$ and $z = 2$, showing a query sequence (first line) and the various answers in the three next lines while querying 5-mers (indexed truth, AMQ answer and QTF answer). For each k-mer, a cross mark (resp. a check mark) indicates that this k-mer is absent from (resp. present in) the queried set. False answers are shown in red. This AMQ false-positive response (5-mer $CAAAG$) is filtered out by QTF as it generates a positive stretch of size one, lower than $z = 2$. Hence, one the last line showing the result after QTF, this false-positive does not occur. However, by applying this strategy, QTF may also remove the true positive stretch of size $\leq z$ (example shown with a red cross), leading to a false-negative QTF answer for the 5-mer $GATTG$.

As illustrated Fig. 1, the fundamental idea is to filter the AMQ answers depending on the context of the queried k-mers when those are queried successively. Given a parameter z, QTF answers "positive" only for k-mers having at least z consecutive neighbors indexed in the AMQ, and it answers "negative" else. Said differently, QTF considers as negative any k-mer that does not belong to a set of at least $z + 1$ successive k-mers positive for the AMQ.

In the following, a set of x consecutive k-mers positive for the AMQ is called a "positive stretch of length x", or simply an "x-stretch".

QTF Algorithm

The QTF algorithm is straightforward. However, given its designed use cases (querying billions or trillions of k-mers) it has to be as much optimized as possible. We propose a simple yet efficient algorithm (see Algorithm 1) not using any extra disk space or RAM. Compared to a usual usage of an AMQ querying consecutive k-mers, it only requires a single additional integer "$currentStretchLength$" that represents the number of consecutive positives k-mers being read on q and a single test, with no impact on query time complexity that is $O(|q|)$.

Algorithm 1: QTF

Data: $q \in \Sigma^*$; AMQ indexing k-mers; k and z in $\mathbb{N}^+$
Result: Prints k-mers from q shared with AMQ, after QTF

1   $currentStretchLength = 0$;
2   **for** each position i in $[0, |q| - k + 1]$ **do**
3     $kmer$ = kmer starting position i in q;
4   **if** AMQ contains $kmer$ **then**
5       $currentStretchLength = currentStretchLength + 1$;
6   **else**
7   **if** $currentStretchLength > z$ **then**
8   print all overlapping k-mers occurring on q from $i - currentStretchLength$ to $i - 1$;
9       $currentStretchLength = 0$

10 **if** $currentStretchLength > z$ **then**
11   print all overlapping k-mers occurring on q from $|q| - k + 1 - currentStretchLength$ to $|q| - k$;

Considerations About QTF False-Positive and False-Negative Rates

The QTF strategy drawback is the introduction of false-negative calls. A false-negative call occurs when a true-positive k-mer belongs to a x-stretch with $x \le z$. This situation happens when q and B share a (small) sub-sequence of length $< k + z$. This may happen in practice.

As shown by intermediate results (see Supplementary Materials, Sect. S1.1), the QTF strategy enables to reduce the original AMQ false-positive rate by several orders of magnitude at the expense of non-null false-negative rate. For instance, when applied to an AMQ composed of a BF indexing 31-mers with a false-positive rate of 5%, the QTF filter, using $z = 3$, enables to decrease the false-positive rate from 5% to $\approx 0.02\%$ but increases the false-negative rate from 0%

to $\approx 1.37\%$. With no impact on query time or memory, these results are highly satisfying for applications in which a small false-negative rate is acceptable.

By the way, in the following section we propose a second main contribution. We show how to take advantage of the QTF strategy to query K-mers (with $K > k$) with no false-negative call, with low false-positive rate and with query time being faster than usual AMQ queries.

2.3   Querying K-mers with findere

As exposed previously, after QTF, a false-negative call arises only when a positive k-mer belongs to a positive stretch of length $\leq z$. Conversely, if a sequence of length K (with $K > k$) is shared between q and B, then it generates $K - k + 1$ successive true positive k-mers, thus defining a $(K - k + 1)$-stretch.

Conceptually, we take advantage of this remark by proposing an algorithm that we call findere in which we query K-mers based on indexed k-mers filtered by the QTF algorithm, using $z = K - k$, hence looking for stretches of length $\geq z + 1$.

More precisely, given two integer values K and k, with $K > k > 0$, a bank dataset B and a query sequence q, the findere strategy consists in indexing all the k-mers from B using an AMQ. In contrast to QTF that calls k-mers, findere calls K-mers based on their k-mer content. Given a position $i \in [0, |q| - K + 1]$ on q, the K-mer starting at this position is considered as "present" by findere if the k-mer starting at position i and the $K - k$ next successive k-mers are considered as present by QTF. This explains the "findere" name that comes from Latin and means "*divide*".

The findere algorithm is obtained from a straightforward modification of the QTF algorithm (Algorithm 1). It is sufficient to define $z = K - k$ and to print K-mers instead of k-mers lines 8 and 11, taking care to avoid printing the last $K - k$ K-mers of each stretch. The complete algorithm is presented in Algorithm 2, including an additional time optimization as presented in the following section.

findere Time Optimization

While walking a sequence q searching for $(z + 1)$-stretches, it is possible to skip some k-mer queries. Indeed, if two negative k-mers start positions i and $i + z + 1$ on q, it is impossible to have a $(z+1)$-positive stretch starting from any position between i and $i + z + 1$. This is because at most z positive k-mers can start between i and $i + z + 1$ both excluded, hence no $(z+1)$-stretch can contain any position in $[i, i + z + 1]$.

Thus, when a negative k-mer is found position i on q, we check directly whether the k-mer starting at position $i + z + 1$ is positive or negative for the AMQ.

- If it is negative, we know that no z-stretch can exists including any position in $[i, i + z + 1]$. There is in this case no need to query k-mers starting at positions in $[i, i + z]$, and we can repeat the process position $i + 2(z + 1)$, and so on.

Algorithm 2: findere

Data: $q \in \Sigma^*$; AMQ indexing k-mers; K and k in $\mathbb{N}^+$, $K \geq k$
Result: Prints k-mers from q shared with AMQ, after QTF

1 $z = K - k$;
2 $currentStretchLength = 0$;
3 $extendingStretch = True$;
4 $i = 0$;
5 **while** $i \leq |q| - k + 1$ **do**
6      $kmer$ = kmer starting position i in q;
7   **if** *AMQ contains kmer* **then**
8          $currentStretchLength = currentStretchLength + 1$;
9   **if** *extendingStretch* **then**
10             $i = i + 1$;
11   **else**
12             $extendingStretch = True$;
13             $i = i - z$;
14   **else**
15   **if** $currentStretchLength \geq z$ **then**
16   print all overlapping K-mers occurring on q from
               $i - currentStretchLength$ to $i - 1 - (K - k)$;
17         $currentStretchLength = 0$;
18         $extendingStretch = False$;
19         $i = i + z + 1$;
20 **if** $currentStretchLength \geq z$ **then**
21   print all overlapping K-mers occurring on q from
       $|q| - k + 1 - currentStretchLength$ to $|q| - K$;

- If it is positive, k-mers starting from position $i+1$ have to be queried following the same process.

Note that even if we report this optimisation specifically for the findere algorithm, it also applies for the QTF strategy.

findere Algorithm
Algorithm 2 proposes an overview of the findere algorithm. This includes the time optimization described in the previous section. This optimisation mainly takes effects line 19 where z positions are not queried unless the next checked position contains a positive k-mer, leading to rewind the query z positions back (line 13).

findere False-Positives and "construction False-Positives" (cFP)
findere reduces the false-positive rate of an AMQ by detecting and filtering out stretches of length $\leq z$ $(= K - k)$. The greater the length of the stretches, the fewer false-positives stretches pass through the filtration, as illustrated by results presented in Supp. Mat, Sect. S1.1.

However, detecting K-mers based on their k-mer content leads to the appari-
tion of a novel kind of false-positives for `findere`. It may appear that for a negative K-mer, all its k-mers are true-positives. In this case, `findere` generates
a false-positive call for this K-mer. We name those false-positive K-mers "construction false-positives" (cFP). The higher z, the higher the number of cFP, as
when k gets too small, more k-mers occur by chance.

`findere` Implementation

We propose an implementation of `findere`, available at https://github.com/
lrobidou/findere. This implementation uses a Bloom filter as its inner AMQ.
However, any other AMQ implementation can be used through a simple wrap-
per (provided with the `findere` implementation). The BF chosen for this implementation is a fork of the original https://github.com/mavam/libbf, which was
modified to add the support of serialization. Although `findere` can index and
query any alphabet, its implementation proposes a specialisation for genomic
sequences: as such, one can index not only natural language, but also fasta and
fastq files (gzipped or not) representing genomes or any sequencing read files.
In this genomic context, a function to index and query canonical K-mers is also
available.

3   Results

We propose results on real biological data and on natural texts. The aim is
to show the practical advantages offered by `findere`, both in terms of query
precision, index size, and query time. Being developed to be used on top of any
AMQ, we do not compare `findere` with any of those. Remind that `findere` may
be used on filtering the results from any such data structure, including Cuckoo
Filters for instance. To the best of our knowledge, the only tool comparable to
`findere` is kBF. We compared `findere` and kBF on biological data only as kBF
is not designed to work on a generic alphabet.

Executions were performed on the GenOuest platform on a node with 4×8-
cores Xeon E5-2660 2,20 GHz with 200 Go of memory. A complete description
of tool versions, data acquisition, command lines, and numerical results are
provided in the Github repository https://github.com/lrobidou/findere/tree/
master/paper_companion.

3.1   Experimental Data

Metagenomic Data. In order to measure the impacts of the `findere` algorithm
on real genomic data, we used two HMP [8] fastq files, indexing reads1 from sample SRS014107 and querying reads1 from sample SRS016349, both downloaded
from the NCBI Sequence Read Archive. Theses samples contain respectively 4.2
million reads of average size 92 characters and 2.3 million reads of average size
96 characters. We simply refer to this dataset as the "hmp" dataset.

Natural Language Data. In order to test the `findere` implementation on natural
language, we used a dump of Wikipedia, from which we extracted two subsets

overlapping with 10 Mo. They have each a size of 105 Mo, leading to about 10^8 31-mers each. We refer to this dataset as the "natural language dataset".

3.2 Results on Genomics Data

In this Section we propose results using $K = 31$. As shown in Supp. mat. (Sect. S1.2), findere results are robust with this main parameter.

False-Positive Analyses

We obtained results with a classical BF, indexing SRS014107 and querying SRS016349 with K-mers of length $K = 31$. For all experiences the size of the used BF is ≈ 2.6 billion bits, leading to 5% FPR when indexing 31-mers. With the same BF size, findere was run varying the z value, with $K = 31$. As shown Fig. 2, with $z = 0$, findere obtains as expected exactly the same results as those obtained with the original used AMQ. With low z values (e.g. lower than 5), the findere FPR quickly drops close to zero. For instance, with $z = 3$, the findere FPR is equal to 0.056%, which is two orders of magnitudes smaller than the original BF FPR. Also, with such low z values, k-mers are large enough to limit the findere "construction FP" (cFP) to negligible values. For instance, $z = 3$ leads to $k = 28$ and the cFP rate is 0.025%.

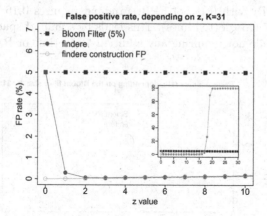

Fig. 2. Comparative false-positive rate obtained on the hmp dataset. A BF indexes 31-mers, with a pratical FP rate of $\approx 5\%$ (blue filled squares, independent of the z value). With the same BF size, findere was run varying the z value (and thus the k value), leading to FP rates as shown in red filled circles. Orange empty circles show the amount of findere "construction FP", the rest of findere FP being due to stretches of length $\geq z+1$ containing BF FP. The full figure zooms on recommended z values (in particular $z \in [2,5]$). The small frame shows results including higher but discouraged z values. (Color figure online)

When using large values of z (here > 15), k-mers get too small to be specific enough. Hence they have high chances to appear at random, leading to a dramatic increase of the cFP rate. This happens with z values leading to k-mers of size < 13. For instance, with $z = 19$, one has $k = 12$, which results for findere in an FPR of 89.27%, being almost only composed of cFP (curves overlap on the figure, cFP representing 99.95% of the FP).

Fortunately, fixing z is easily anticipated by choosing a value small enough so that the indexed k-mers in the AMQ have a low chance (e.g.. lower than 0.01%) to appear by chance. The probability for a k-mer to appear by chance in a random text of length n can be roughly approximated as $1 - \left(1 - \frac{1}{|\Sigma|^k}\right)^n$, with $|\Sigma|$ being the size of the alphabet. For instance on the indexed SRS014107 sample, where $|\Sigma| = 4$ and $n = 386$ millions, the probability to appear by chance for a k-mer is respectively equal to 0.99% for $k = 13$ ($z = 18$), to 0.30% for $k = 15$ ($z = 16$), to 0.02% for $k = 17$ ($z = 14$), and to 0.005% for $k = 18$ ($z = 13$). Hence choosing $z < 14$ is acceptable. By default findere uses $z = 3$.

Space Gain

We recall first that findere memory usage has no overhead compared to the size of the used AMQ. For a given AMQ size, we can deduce the FPR for a BF and findere. Results are presented Fig. 3 (with the default $z = 3$ value). This can also be used for deducing the amount of space needed for a given FPR. For instance, to obtain a usual value of 1% FPR, findere requires 0.05 Gio of space while a BF requires 1.06 Go. The findere advantage gets even more important with a lower FPR: with 0.1% of FPR, findere requires 0.16 Go while a BF requires about 17 Go (dotted lines). This leads to a gain of space of two orders of magnitude, while not requiring any additional run time or RAM.

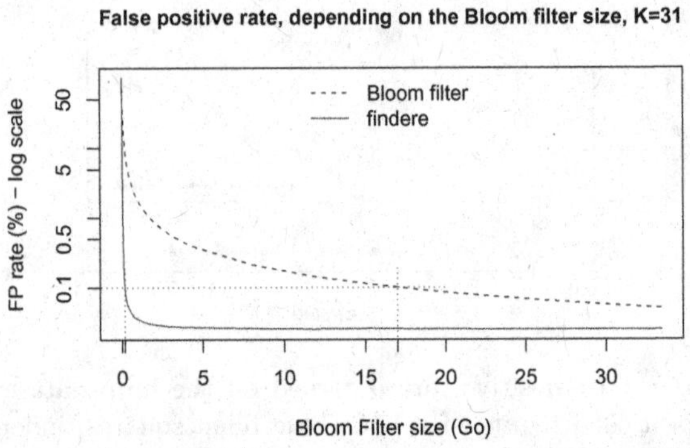

False positive rate, depending on the Bloom filter size, K=31

Fig. 3. findere and BF FPR depending on the space used, on the hmp dataset. Dotted line segment corresponds to 0.1% false-positive rate

Query Time

Thanks to the optimization detailed Sect. 2.3, the query time decreases when z increases. As shown Table 1, with discouraged values $z = 0$ or $z = 1$, the query time of findere is slightly higher than querying the original BF. This is due to additional conditional tests. With recommended z values ($z = 2$ to 5), compared to the query time of the BF, the `findere` query time is divided by a factor 2 to 3. With the default $z = 3$ value, query time is divided by 2.4, while query time still decreases when z increases: with $z = 10$, the query time is divided by ≈ 5.

Table 1. BF and `findere` query time in seconds on the hmp dataset, depending on the z value. BF result does not depend on z and is reported only for $z = 0$.

z	0	1	2	3	4	5	10
BF	42.4						
`findere`	42.9	43.7	24.3	17.5	14.1	12.0	8.6

Comparisons with kBF

To compare `findere` and kBF, we created a fork of kBF, available at http://github.com/lrobidou/kbf/. This fork enables to specify the amount of memory to be used by kBF and more importantly, it enables to index a set of reads and query another set (the original implementation does not allow that). kBF comes with two versions: 1-kBF and 2-kBF. 1-kBF uses less space at higher FPR than 2-kBF. The kBF strategy imposes to query up to nine times the BF when asking for the membership of a single k-mer. At the same time, `findere` is ≈ 2.4 times faster than a Bloom filter with the recommended value $z = 3$. Moreover, kBF dumps all queried k-mers in RAM, and 2-kBF uses an additional hash set. Hence, kBF is much slower than `findere`. For instance, on the hmp data, with a BF FP of 5%, 1-kBF (resp. 2-kBF) query needs $\approx 300s$ (resp. $\approx 1450s$) while `findere` needs $\approx 17s$.

Moreover, as it shows a higher FPR for a fixed amount of space, kBF uses more space than `findere` for an equivalent FPR. For instance with for $\approx 1\%$ FPR, `findere` requires 0.05 Go of space while 1-kBF requires ≈ 0.40 Go. Comparisons with 2-kBF are not fair as it computes a hash set of every k-mer, leading to unreasonable space usage (*e.g.* 7.78 Go for an FPR of $\approx 1\%$ when `findere` requires 0.05 Go). Full kBF results are proposed in Supp. Mat, Sect. S1.4.

3.3 Results on Natural Languages

We applied `findere` and BF on the natural language corpus. Full results are provided in Supp. Mat, Sect. S1.3.

Memory Gain. As in Sect. 3.2, we computed the FPR in function of the space used. From those results, we can deduce that, also on natural languages, for an FPR of 0.1%, the `findere` space usage is two orders of magnitude less than BF. Indeed, `findere` needs 0.023 Go, while a BF requires 3.38 Go.

Query Time. As described in Sect. 2.3, the query time decreases when z increases. It holds when `findere` is used on the natural language dataset as well. With recommended z values ($z = 2$ to 5), compared to the query time of the Bloom filter, the `findere` query time is divided by a factor 1.6 to 3. With the default $z = 3$ value, query time is divided by 2.2 compared to the raw Bloom filter query.

3.4   Limit of the `findere` Approach

The `findere` algorithm generates so-called "construction false-positives" that occur when a negative K-mer contains only true positive k-mers. With recommended parameters, those cFP are negligible as shown in Fig. 2. However, as cFP depends only on true-positive calls, its value does not depend on the FPR_{AMQ}. Hence theoretically, when FPR_{AMQ} tends toward zero, the cFP rate (and thus `findere` FPR as well) becomes higher than FPR_{AMQ}. This effect can be observed on the natural language results (Fig. S2, Supp. Mat.) with BF FPR below 0.02%. However, one should remind that, first, the difference is insignificant (0.008% difference FPR when using 26 Go space), and second, the practical usage of an AMQ is usually with FPR higher than 0.1% to prevent huge space requirements.

4   Conclusion

We propose a method filtering results of any approximate member query (AMQ) data structure when used for querying words of length K from a query. Despite its amazing simplicity, applied on metagenomics and natural text data, compared to the non-filtered results: `findere` 1/ makes queries two times faster, 2/ enables to decrease by two orders of magnitude the false-positive rate or enables to decrease the space allocated to each element by two orders of magnitudes, and 3/ has no drawback when used with recommended values.

We are expecting an important impact of the `findere` tool, for which we propose an implementation. Indeed, AMQ data-structure are essential for indexing large datasets. In particular their usage is fundamental for indexing the genomic sequencing data, for which `findere` offers a new scaling breakthrough.

Acknowledgments. This work used HPC resources from the GenOuest bioinformatics core facility (https://www.genouest.org). The work was funded by ANR SeqDigger (ANR-19-CE45-0008).

References

1. Amid, C., et al.: The European nucleotide archive in 2019. Nucleic Acids Res. **48**(D1), D70–D76 (2020)
2. Bender, M.A., et al.: Don't thrash: how to cache your hash on flash. Proc. VLDB Endow. **5**(11), 1627–1637 (2012)
3. Benoit, G., et al.: Multiple comparative metagenomics using multiset k-mer counting. PeerJ Comput. Sci. **2**, e94 (2016)
4. Bloom, B.H.: Space/time trade-offs in hash coding with allowable errors. Commun. ACM **13**(7), 422–426 (1970)
5. Bray, N.L., Pimentel, H., Melsted, P., Pachter, L.: Near-optimal probabilistic RNA-SEQ quantification. Nat. Biotechnol. **34**(5), 525–527 (2016)
6. Chikhi, R., Holub, J., Medvedev, P.: Data structures to represent a set of k -long DNA sequences. ACM Comput. Surv. **54**(1), 1–22 (2021)

7. Fan, B., Andersen, D.G., Kaminsky, M., Mitzenmacher, M.D.: Cuckoo filter: practically better than bloom. In: Proceedings of the 10th ACM International on Conference on emerging Networking Experiments and Technologies, pp. 75–88 (2014)
8. HMP Integrative, Proctor, L.M., et al.: The integrative human microbiome project. Nature **569**(7758), 641–648 (2019)
9. Marchet, C., Boucher, C., Puglisi, S.J., Medvedev, P., Salson, M., Chikhi, R.: Data structures based on k-mers for querying large collections of sequencing data sets. Genome Res. **31**(1), 1–12 (2021)
10. Marchet, C., Iqbal, Z., Gautheret, D., Salson, M., Chikhi, R.: REINDEER: efficient indexing of k-mer presence and abundance in sequencing datasets. Bioinformatics, **36**(Supplement_1), i177–i185 (2020)
11. Ondov, B.D., et al.: Mash: fast genome and metagenome distance estimation using MinHash. Genome Biol. **17**(1), 132 (2016)
12. Pellow, D., Filippova, D., Kingsford, C.: Improving bloom filter performance on sequence data using k -mer bloom filters. J. Comput. Biol. **24**(6), 547–557 (2017)
13. Stephens, Z.D., et al.: Big data: astronomical or genomical? PLOS Biol. **13**(7), e1002195 (2015)
14. Weaver, S.A., Ray, K.J., Marek, V.W., Mayer, A.J., Walker, A.K.: Satisfiability-based set membership filters. J. Satisf. Boolean Model. Comput. **8**(3–4), 129–148 (2014)
15. Wood, D.E., Jennifer, L., Langmead, B.: Improved metagenomic analysis with Kraken 2. Genome Biol. **20**(1), 257 (2019). https://doi.org/10.1186/s13059-019-1891-0
16. Zielezinski, A., Vinga, S., Almeida, J., Karlowski, W.M.: Alignment-free sequence comparison: benefits, applications, and tools. Genome Biol. **18**(1), 186 (2017). https://doi.org/10.1186/s13059-017-1319-7

Repeats

A Separation of γ and b via Thue–Morse Words

Hideo Bannai[1,4]($\boxtimes$) (iD), Mitsuru Funakoshi[2,5] (iD), Tomohiro I[3] (iD),
Dominik Köppl[1] (iD), Takuya Mieno[2,5] (iD), and Takaaki Nishimoto[4]

[1] M&D Data Science Center, Tokyo Medical and Dental University, Tokyo, Japan
{hdbn.dsc,koeppl.dsc}@tmd.ac.jp
[2] Department of Informatics, Kyushu University, Fukuoka, Japan
{mitsuru.funakoshi,takuya.mieno}@inf.kyushu-u.ac.jp
[3] Department of Artificial Intelligence, Kyushu Institute of Technology, Iizuka, Japan
tomohiro@ai.kyutech.ac.jp
[4] RIKEN Center for Advanced Intelligence Project, Tokyo, Japan
takaaki.nishimoto@riken.jp
[5] Japan Society for the Promotion of Science, Tokyo, Japan

Abstract. We prove that for $n \geq 2$, the size $b(t_n)$ of the smallest bidirectional scheme for the nth Thue–Morse word t_n is $n + 2$. Since Kutsukake et al. [SPIRE 2020] show that the size $\gamma(t_n)$ of the smallest string attractor for t_n is 4 for $n \geq 4$, this shows for the first time that there is a separation between the size of the smallest string attractor γ and the size of the smallest bidirectional scheme b, i.e., there exist string families such that $\gamma = o(b)$.

1 Introduction

Repetitiveness measures for strings is an important topic in the field of string compression and indexing. Compared to traditional entropy-based measures, measures based on dictionary compression are known to better capture the repetitiveness in highly repetitive string collections [12]. Some well known examples of dictionary-compression-based measures are: the size r of the run-length Burrows–Wheeler transform [2] (RLBWT), the size z of the Lempel-Ziv 77 factorization [17], the size b of the smallest bidirectional (or macro) scheme [15].

Kempa and Prezza introduced the notion of *string attractors* [4], which gave a unifying view of dictionary-compression-based measures. A string attractor of a string is a set of positions such that any substring of the string has at least one occurrence which contains a position in the set. The size γ of the smallest string attractor of a word is a lower bound on the size of all known dictionary compression measures, but is NP-hard to compute. Kociumaka et al. [5,6] introduced another measure $\delta \leq \gamma$ that is computable in linear time, defined as the maximum over all integers k, the number of distinct substrings of length k in the string divided by k.

The landscape of the relations between these measures has been a focus of attention. For example, since z is a special case of a bidirectional scheme,

© Springer Nature Switzerland AG 2021
T. Lecroq and H. Touzet (Eds.): SPIRE 2021, LNCS 12944, pp. 167–178, 2021.
https://doi.org/10.1007/978-3-030-86692-1_14

$b \leq z$. Also, $b \leq 2r$ [13] and $r = O(z \log^2 N)$ [3] hold, where N is the length of the string. Notice that a string can be represented in space (with an extra factor of $\log N$ for bits) proportional to b, r, or z. Interestingly, while δ and γ do not give a direct representation of the string, it is known that the string can be represented in $O(\delta \log \frac{N}{\delta})$ or $O(\gamma \log \frac{N}{\gamma})$ space, respectively [4–6]. On the other hand, Kociumaka et al. [5,6] showed that for every length N and integer $\delta \in [2, N]$, there exists a family of length-N strings having the same measure δ, that requires $\Omega(\delta \log \frac{N}{\delta} \log N)$ bits to be encoded. Analogous results for γ are not yet known [5,6,12]. The bidirectional scheme is the most powerful among the dictionary-compression-based measures. The size b of the smallest bidirectional scheme is also known to satisfy $b = O(\gamma \log \frac{N}{\gamma})$, but again, the tightness of this bound was not known [12].

Following Mantaci et al. [8,9], Kutsukake et al. [7] investigated repetitiveness measures on Thue–Morse words [11,14,16] and showed that the size of the smallest string attractor for the n-th Thue–Morse word is 4, for any $n \geq 4$. They also conjectured that the size of the smallest bidirectional scheme for the n-th Thue–Morse word (which has length $N = 2^n$) is $\Theta(\log N)$, which would imply a separation between γ and b. Possibly due to the difficulty (NP-hardness) of computing the size of the smallest bidirectional scheme of a string [15], tight bounds for b have only been discovered for a very limited family of strings, most notably standard Sturmian words [10]. This was shown from the fact that the size r of the RLBWT of every standard Sturmian word is 2, therefore implying a constant upper bound on the smallest bidirectional scheme.

In this paper, we prove Kutsukake et al.'s conjecture by showing that for any $n \geq 2$, the size $b(t_n)$ of the smallest bidirectional scheme for t_n is exactly $n + 2$. For any value of $\gamma \geq 4$, we can construct a family of strings such that $b = \Theta(\gamma \log \frac{N}{\gamma})$ and N is the length of the string. Our result shows for the first time the separation between γ and b, i.e., there are string families such that $\gamma = o(b)$.

2   Preliminaries

We consider the alphabet $\Sigma = \{\mathsf{a}, \mathsf{b}\}$. A string is an element of Σ^*. For any string $w \in \Sigma^*$, let $|w|$ denote its length, and let $w = w[0] \cdots w[|w| - 1]$. Also, for any $0 \leq i \leq j < |w|$, let $w[i..j] = w[i] \cdots w[j]$.

A *string morphism* μ is a function mapping strings to strings such that each character is replaced by a single string (deterministically), i.e., $\mu(w) = \mu(w[0]) \cdots \mu(w[|w| - 1])$ for any string w. Let $\mu^0(w) = w$, and for any integer $n \geq 1$, let $\mu^n(w) = \mu(\mu^{n-1}(w))$. Now let μ be the morphism on the binary alphabet determined by $\mu(\mathsf{a}) = \mathsf{ab}$ and $\mu(\mathsf{b}) = \mathsf{ba}$. Then the n-th *Thue–Morse word* t_n is $\mu^n(\mathsf{a})$, and its length is $|t_n| = 2^n$.

A list of strings $b_1, \ldots, b_k$ is called a *parsing* of a string S, if $S = b_1 \cdots b_k$. Each b_i $(i = 1, \ldots, k)$ is called a *phrase*. A sequence $B = ((b_1, s_1), \ldots, (b_k, s_k))$ is a *bidirectional scheme* for S, if $b_1, \ldots, b_k$, is a parsing of S and for all $i = 1, \ldots, k$, $s_i \in [0, |S| - 1] \cup \{\bot\}$, such that $s_i = \bot$ if $|b_i| = 1$, and $b_i = S[s_i..s_i + |b_i| - 1]$

otherwise. We denote the *size* k of the bidirectional scheme B by $|B|$. We call s_i the *source* of the phrase b_i.

If $|b_i| = 1$ then we stipulate that $s_i = \bot$, and call b_i a *ground* phrase. (Consequently, there are no phrases of length one that have a source being a text position.) We denote the number of ground phrases in B by $\#_g(B)$. For convenience, we denote the starting position of phrase b_i by p_i, i.e., $p_1 = 0$ and $p_i = |b_1 \cdots b_{i-1}|$ for all $i = 2, \ldots, k+1$, where p_{k+1} is defined for technical reasons.

A bidirectional scheme B for the string S defines a function $f_B : [0, |S| - 1] \cup \{\bot\} \rightarrow [0, |S| - 1] \cup \{\bot\}$ over positions of S, where

$$f_B(x) = \begin{cases} \bot & \text{if } x = \bot \text{ or if } x = p_i, s_i = \bot \text{ for some } i, \\ s_i + x - p_i & \text{otherwise, i.e., if } p_i \leq x < p_{i+1}, s_i \neq \bot \text{ for some } i. \end{cases}$$

Let $f_B^0(x) = x$, and for any $j \geq 1$, let $f_B^j(x) = f_B(f_B^{j-1}(x))$. It is clear that if $f_B(i) \neq \bot$, then it holds that $S[i] = S[f_B(i)]$. A bidirectional scheme for S is *valid*, if there is no $i \in [0, |S| - 1]$ such that the function f_B contains a cycle, that is, for every $i \in [0, |S| - 1]$, there exists a $j \geq 1$ such that $f_B^j(i) = \bot$. A valid bidirectional scheme B of size k for S implies an $O(k)$-word size (compressed) representation of S, namely, the sequence $((|b_1|, s_1'), \ldots, (|b_k|, s_k')) \subset ([1, |S|] \times ([0, |S| - 1] \cup \Sigma))^k$, where $s_i' = b_i$ if $s_i = \bot$, and $s_i' = s_i$ otherwise. Note that the string S can be reconstructed from this sequence if and only if B is valid. A parsing $b_1, \ldots, b_k$ of S is *valid* if there exists a list of phrase sources $s_1, \ldots, s_k$ such that $((b_1, s_1), \ldots, (b_k, s_k))$ is a valid bidirectional scheme for S. See Fig. 1 for examples of representations of valid bidirectional schemes of t_3 and t_4.

Informally, $f_B(x)$ gives the position (source) from where we want to copy the character that restores $S[x]$ when reconstructing S from the compressed representation, where $f_B(x) = \bot$ indicates that the character is stored as a ground phrase, i.e., as a literal.

It is easy to see that a valid bidirectional scheme must have at least as many ground phrases as there are different characters appearing in S (the number of ground phrases is at least $|\Sigma|$ if all characters of Σ appear in S).

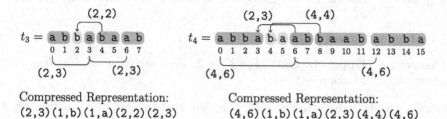

Fig. 1. Examples of valid compressed representations of t_3 and t_4.

3   Important Characteristics of Thue–Morse Words

Before proving our bounds, we first give some simple observations on Thue–Morse words that we will use later. Remember that the first index of t_n is 0, which is an even position.

Lemma 1. aa *and* bb *only occur at odd positions in* t_n.

Proof. The morphism μ implies that any substring of length 2 starting at an even position is either $\mu(a) = ab$ or $\mu(b) = ba$.   □

Lemma 2 *(Theorem 2.2.3 of [1]).* t_n *has no overlapping factors, i.e., two occurrences of the same string in* t_n *never share a common position.*

Lemma 3. abab *and* baba *only occur at even positions in* t_n.

Proof. Suppose to the contrary that there is an occurrence of abab that starts at an odd position. Then, Lemma 1 implies that b occurs immediately left of abab, i.e., there is an occurrence of the substring babab, thus contradicting Lemma 2 with the substring bab having two overlapping occurrences.   □

Let the *parity* of an integer i be $i \bmod 2 \in \{0,1\}$.

Lemma 4. *For any substring* $w \notin \{aba, bab, ab, ba, a, b\}$ *of* t_n, *the parities of all occurrences of* w *in* t_n *are the same.*

Proof. Any such substring w contains at least one of $\{aa, bb, abab, baba\}$ as a substring, and thus the result follows from Lemmas 1 and 3.   □

Further, we use that t_n is a prefix of t_{n+1} and $t_n[0..4] = abbab$ for $n \geq 3$.

4   Upper and Lower Bounds on b

We start with the upper bound on the smallest size of a (valid) bidirectional parsing by constructing such a parsing, and subsequently show that this bound is optimal by showing a lower bound whose proof is more involved.

4.1   Upper Bound

Theorem 1 (Upper bound). *For* $n \geq 2$, *there exists a valid bidirectional scheme for* t_n *of size* $n + 2$.

Proof. Proof by induction. For $n = 2$ it is clear that there is a valid bidirectional scheme of size 4.

Suppose that for some $n \geq 2$, there is a valid bidirectional scheme $B_n = ((b_1, s_1), \ldots, (b_k, s_k))$ of size k for t_n. We can assume that there are at least two ground phrases $b_{i_a} = t_n[p_{i_a}] = a$ and $b_{i_b} = t_n[p_{i_b}] = b$. Since $t_{n+1} = \mu(t_n)$, we first consider a bidirectional scheme B' for t_{n+1} where each phrase is constructed from phrases of B_n by applying μ, with the small exception for the two ground phrases. More precisely, the phrases of B' are $\mu(b_i)$ for $i \in [1,k] \setminus \{i_a, i_b\}$, and two ground phrases from each of $\mu(b_{i_a}) = $ ab and $\mu(b_{i_b}) = $ ba, resulting in a parsing of size $k+2$. For each non-ground phrase $\mu(b_i)$ in B', we can either choose the source to be (i) $2p_{i_a}$ or $2p_{i_b}$ if its length is 2, or (ii) $2s_i$ otherwise. The latter is because $\mu(b_i) = \mu(t_n[s_i..s_i+|b_i|-1]) = \mu(t_n)[2s_i..2s_i+2|b_i|-1] = t_{n+1}[2s_i..2s_i+2|b_i|-1]$. The validity of B' follows from the validity of B_n, and $f_{B'}$ has no cycles. It is easy to see that for any position i, the parities of i and $f_{B'}(i)$ are the same (unless $f_{B'}(i) = \perp$). Thus, noticing that $t_{n+1}[3..4] = $ ab, (1) the source of $t_{n+1}[3] = $ a at an odd position can eventually be traced to the ground phrase at position $2p_{i_b} + 1$, and (2) the source of $t_{n+1}[4] = $ b at an even position can eventually be traced to the ground phrase at position $2p_{i_b}$.

Next, we modify B' by combining the two consecutive ground phrases $(a, \perp)$ and $(b, \perp)$ corresponding to $\mu(b_{i_a})$, and replace them with a single $(ab, 3)$. This results in a bidirectional scheme B'' of size $k+1$. From the above observations (1) and (2), it is clear that B'' is still valid. Thus, $B_{n+1} = B''$ is a valid bidirectional scheme for t_{n+1} of size $k + 1$, thereby proving the theorem. $\qquad\square$

The bidirectional scheme of t_4 in Fig. 1 can be constructed by following the algorithmic instructions of the proof of Theorem 1.

4.2 Lower Bound

Theorem 2 (Lower Bound). *For $n \geq 2$, the smallest valid bidirectional scheme for t_n has size $n + 2$.*

To prove Theorem 2, we would like to, in essence, do the opposite of what we did in the proof of Theorem 1, and show that we can construct a bidirectional scheme for t_{n-1} of size $k-1$, given a bidirectional scheme for t_n of size k. However, the opposite direction involves halving the size of phrases, and thus does not work straightforwardly when there are phrases of odd length. Nevertheless, we will show that this can be done in an amortized way, and show the following.

Lemma 5. *For any $n \geq 5$, if there exists a valid bidirectional scheme of size k for t_n, then, for some $1 \leq i \leq 3$, there exists a valid bidirectional scheme of size at most $k - i$ for t_{n-i}.*

Since the size of the smallest bidirectional scheme for t_2, t_3, t_4 can be confirmed to be respectively $4, 5, 6$ by computer analysis, this with Lemma 5 implies Theorem 2.

In the rest of the section, we give an algorithm that, given a bidirectional scheme B_n for t_n, constructs a bidirectional scheme B_{n-1} for t_{n-1}, and claim that applying the algorithm repeatedly i times, for some $1 \leq i \leq 3$, we obtain a bidirectional scheme B_{n-i} for t_{n-i} such that $|B_{n-i}| \leq |B_n| - i$. The algorithm consists of 3 main steps:

1. Elimination of length-1 ground phrases.
2. Elimination of odd length phrases.
3. Application of the inverse morphism μ^{-1} on all phrases of the modified parsing.

The goal of Steps 1 and 2 is to modify the phrases of B_n to construct a bidirectional scheme B'_n so that all phrases in B'_n will be of even length. When modifying the phrases, we must take care in 1) defining the source of the phrase, and 2) ensuring that no cycles are introduced in the resulting bidirectional scheme B_{n-1}. To make this clear, we temporarily relax the definition for ground phrases in B'_n during the modification, so that the ground phrases of B'_n are phrases of length 2 that start at even positions. In this way, we can be sure that any position in a length-2 phrase starting at an even position in B'_n is not involved in a cycle. In Step 3, we create a new bidirectional scheme B_{n-1} of t_{n-1} by translating all phrase lengths and sources of B'_n according to the inverse morphism μ^{-1}, i.e., we map each non-ground phrase (b'_i, s'_i) of B'_n to the phrase $(\mu^{-1}(b'_i), s'_i/2)$ in t_{n-1}. The length-2 ground phrases in B'_n become length-1 ground phrases in B_{n-1}, and thus we obtain a valid bidirectional scheme B_{n-1} for t_{n-1}, without the relaxation, and the same size as B'_n.

Eliminating Length-1 Ground Phrases. The operation is done analogously and symmetrically for any length-1 ground phrase (a or b) that may occur at an even or odd position. We describe in detail the case for a ground phrase with character a that occurs at some odd position $2i + 1$.

For a consecutive pair of positions $2i, 2i + 1$, we call one a *partner* of the other. Let $i_b = 2i$ be the partner position of the length-1 ground phrase a, i.e., $t_n[i_b..i_b + 1] = ba$. The idea is to (re)move the phrase boundary that separates partner positions so that the ground phrase disappears. Since we are considering the case where the ground phrase is at an odd position, we extend the phrase (b_i, s_i) containing position i_b by one character, so that it includes the length-1 ground phrase $t_n[i_b + 1] = a$, thereby eliminating it. If possible, we would like to keep the source of the extended phrase the same, i.e., change (b_i, s_i) to $(b_i a, s_i)$, or equivalently, change $f_{B_n}(i_b + 1) = \bot$ to $f_{B_n}(i_b + 1) = s_i + |b_i|$. Note that if the parity of $f_{B_n}(i_b)$ is equal to that of i_b, this is always possible (i.e., $t_n[f_{B_n}(i_b) + 1] = a$ always holds). However, it may be that the position $i_b + 1$ gets involved in a cycle, due to this change. Notice that since we started from a valid (relaxed) bidirectional scheme, it is guaranteed that i_b is not involved in a cycle, i.e., $f_{B_n}^j(i_b) \neq i_b$ for any $j \geq 1$. Therefore, we further modify the phrase boundaries, if necessary, to ensure that the source of $t_n[i_b + 1] = a$ will belong in the same phrase as the source of $t_n[i_b] = b$. This is repeated until we are sure that all these changes made to eliminate the original length-1 ground phrase a do not introduce any cycles in the final bidirectional scheme. In other words, we ensure, for some *sufficiently large* j', $f_{B_n}^j(i_b + 1) = f_{B_n}^j(i_b) + 1$ for all $1 \leq j \leq j'$. Then, from the acyclicity of position i_b, the acyclicity of position $i_b + 1$ follows.

There are six cases where the process terminates, as shown in Fig. 2 (Case 3 is further divided into two sub-cases). As noted above, as long as the parity of $f_{B_n}^j(i_b)$ is the same as that of $f_{B_n}^{j-1}(i_b)$, the character of $f_{B_n}^j(i_b)$'s partner is always a, and we can ensure that $f_{B_n}^{j-1}(i_b)$ and $f_{B_n}^{j-1}(i_b+1)$ are in the same phrase by only (possibly) setting $f_{B_n}^j(i_b+1) = f_{B_n}^j(i_b) + 1$. Thus, we consider the cases where $j' \geq 1$ is the smallest integer such that the parities of $f_{B_n}^{j'-1}(i_b)$ and $f_{B_n}^{j'}(i_b)$ differ, in which case, Lemma 4 implies that $f_{B_n}^{j'-1}(i_b)$ is contained in a phrase in $\{\text{aba}, \text{bab}, \text{ab}, \text{ba}, \text{b}\}$. Each of the six cases corresponds to a distinct occurrence of b in the strings of this set. We show that in each case, we can modify the phrases so that both $f_{B_n}^{j'-1}(i_b)$ and $f_{B_n}^{j'-1}(i_b+1)$ are in the same length-2 phrase, i.e., a relaxed ground phrase, and be sure that $i_b + 1$ will not be involved in a cycle in the final bidirectional scheme. The details of each case are described in Fig. 2.

Although Cases 1, 2, 4 introduce a new length-1 ground phrase, the number of phrase boundaries that separate partner positions always decreases at the starting point, and never increases. Therefore the whole process terminates at some point, at which point, all length-1 ground phrases have been eliminated.

Eliminating Odd Length Phrases. In this step, we eliminate all remaining phrases with odd lengths. Since there are no more length-1 ground phrases, we first focus on removing phrases aba and bab of length 3. Below, we describe the operation for removing a phrase aba that starts at an odd position. The other cases are analogous or symmetric.

Starting with an occurrence of phrase aba that starts at an odd position $i_b + 1$, we know that this phrase is preceded by b. We move the phrase boundary that separates partner positions, so that the length-3 phrase shrinks to a length-2 phrase starting at an even position, i.e., a relaxed ground phrase, in this case, by expanding the phrase to its left. Since we have changed the source of the a at position $i_b + 1$, we ensure that for some sufficiently large j', $f_{B_n}^j(i_b+1) = f_{B_n}^j(i_b) + 1$ for all $1 \leq j \leq j'$, as we did for the elimination of length-1 ground phrases, so that $i_b + 1$ is not involved in a cycle.

There are five cases where the process terminates, as shown in Fig. 3. As noted previously, as long as the parity of $f_{B_n}^j(i_b)$ is the same as that of $f_{B_n}^{j-1}(i_b)$, then the character of $f_{B_n}^j(i_b)$'s partner is always a, and we can ensure that $f_{B_n}^{j-1}(i_b)$ and $f_{B_n}^{j-1}(i_b+1)$ are in the same phrase by only (possibly) setting $f_{B_n}^j(i_b+1) = f_{B_n}^j(i_b) + 1$. Thus, we consider the cases where $j' \geq 1$ is the smallest integer such that the parities of $f_{B_n}^{j'-1}(i_b)$ and $f_{B_n}^{j'}(i_b)$ differ, in which case, Lemma 4 and the previous step implies that $f_{B_n}^{j'-1}(i_b)$ is contained in a phrase in $\{\text{aba}, \text{bab}, \text{ab}, \text{ba}\}$. Each of the five cases corresponds to a distinct occurrence of b in strings of this set. The details of each case are described in Fig. 3.

After eliminating all phrases aba and bab of length 3, all remaining phrases are either of length 2 or do not belong to the set $\{\text{aba}, \text{bab}, \text{ab}, \text{ba}, \text{a}, \text{b}\}$. Therefore, we can move all phrase boundaries that separate partner positions to the right (or all of them to the left) and update the sources accordingly without intro-

i_b b | a	Starting point. We eliminate the length-1 ground phrase $t_n[i_b + 1] = $ a at an odd position $i_b + 1$, by including it in the same phrase as is partner $t_n[i_b] = $ b, in this case, on its left. We can do this by modifying the source of a to point to the position next to the source of b, i.e., setting $f_{B_n}(i_b + 1) = f_{B_n}(i_b) + 1$. This is done recursively at the source positions, until we reach one of the following cases, where the source of b no longer points to a position of the same parity.
↓ | a b a | ⇩ | a | b a |	Case 1: We introduce a new length-1 ground phrase, and modify the boundaries. We are done since both $t_n[f_{B_n}^{j'-1}(i_b)] = $ b and $t_n[f_{B_n}^{j'-1}(i_b) + 1] = $ a are in a length-2 phrase starting at an even position, i.e., a relaxed ground phrase. We are sure that $i_b + 1$ is not involved in a cycle. We recursively apply the procedure to the new length-1 ground phrase a at an odd position.
↓ | b a b | ⇩ | b a | b |	Case 2: Same as Case 1, with the exception that the new length-1 ground phrase is b at an even position.
↓ | b a b | a | ⇩ | b a | b a |	Case 3-1: This case is when there are no consecutive phrases of ba and ba in the current bidirectional scheme. We introduce a new length-2 phrase, and modify the boundaries. We are done since both $t_n[f_{B_n}^{j'-1}(i_b)] = $ b and $t_n[f_{B_n}^{j'-1}(i_b) + 1] = $ a are in a length-2 phrase starting at an even position, i.e., a relaxed ground phrase.
↓ | b a b | a | ⇩ | b a b a |	Case 3-2: This case is when there already are consecutive phrases of ba and ba in the current bidirectional scheme. We create the phrase baba and make its source be the consecutive phrases of ba and ba (possibly constructed in Case 3-1). No cycles are introduced since the new source are relaxed ground phrases.
↓ | a b | a | ⇩ | a | b a |	Case 4: Same as Case 1.
↓ | b a |	Case 5: There is nothing to do. We are done since both $t_n[f_{B_n}^{j'-1}(i_b)] = $ b and $t_n[f_{B_n}^{j'-1}(i_b) + 1] = $ a are in a length-2 phrase starting at an even position, i.e., a relaxed ground phrase.
↓ | b | a | ⇩ | b a |	Case 6: We expand the phrase to include its partner. We are done, since both $t_n[f_{B_n}^{j'-1}(i_b)] = $ b and $t_n[f_{B_n}^{j'-1}(i_b) + 1] = $ a are in a length-2 phrase starting at an even position, i.e., a relaxed ground phrase.

Fig. 2. Terminal cases for eliminating a length-1 ground phrase $t_n[i_b + 1] = $ a at an odd position $i_b + 1$ (see Sect. 4.2). The shaded squares are even positions. The vertical bars denote phrase boundaries. The black arrow points to the position $f_{B_n}^{j'-1}(i_b)$, where $j' \geq 1$ is the smallest integer such that the parities of $f_{B_n}^{j'-1}(i_b)$ and $f_{B_n}^{j'}(i_b)$ differ. The first line and second line of each case (except Case 5) respectively show the phrase boundaries before and after the modification.

ducing cycles, since length-2 phrases starting at odd positions become relaxed ground phrases, and the occurrences of each of the other phrases have the same parity due to Lemma 4. Thus, we now have a valid bidirectional scheme B'_n where all phrases are of even length, and length-2 phrases are considered to be relaxed ground phrases.

Analysis of the Number of Phrases. It is easy to see that Steps 2 and 3 do not increase the number of phrases. Also, Step 2 does not decrease the number of length-2 phrases that start at even positions, i.e., relaxed ground phrases, created in Step 1, which will become ground phrases in B_{n-1}. Thus, we focus on the analysis of Step 1.

Examining each case of Fig. 2, we can see that while at the start we eliminate a length-1 ground phrase and decrease the number of phrases, Cases 1, 2, 3-1, and 4 introduce a new phrase, thus do not change the total number of phrases. Also, notice that in Case 6, two ground phrases are eliminated, while the total number of phrases decreases only by one, since the second length-1 ground phrase is expanded. Case 3-1 can occur in total at most twice, once for consecutive phrases of ba and once for consecutive phrases of ab. Thus, we obtain the following inequality:

$$|B_{n-1}| \leq |B_n| - \lceil (\#_g(B_n) - 2)/2 \rceil. \tag{1}$$

If $|B_{n-1}| \leq |B_n| - 1$, then we can choose $i = 1$ for Lemma 5 and are done. Otherwise, $|B_{n-1}| = |B_n|$. This implies that $\#_g(B_n) = 2$, and also that Case 3-1 was applied twice. Thus, there exists at least 2 phrases of ab and ba each, which are converted by μ^{-1} to ground phrases in B_{n-1}, implying $\#_g(B_{n-1}) \geq 4$. Then, applying Eq. (1) for $n - 2$, we have

$$|B_{n-2}| \leq |B_{n-1}| - \lceil (\#_g(B_{n-1}) - 2)/2 \rceil$$
$$\leq |B_{n-1}| - 1 = |B_n| - 1.$$

If $|B_{n-2}| \leq |B_n| - 2$, then we can choose $i = 2$ for Lemma 5. Otherwise, $|B_{n-2}| = |B_n| - 1$. This implies that $\#_g(B_{n-1}) = 4$ and that Case 3-1 was applied twice, and Case 6 was applied once. Therefore, we get $\#_g(B_{n-2}) \geq 5$. Finally, applying Eq. (1) for $n - 3$, we have

$$|B_{n-3}| \leq |B_{n-2}| - \lceil (\#_g(B_{n-2}) - 2)/2 \rceil$$
$$\leq |B_{n-2}| - 2$$
$$= |B_n| - 3.$$

This proves Lemma 5, and thus Theorem 2.

i_b a b b │ a b a │ a	Starting point. We wish to eliminate the length-3 phrase **aba** starting at an odd position $i_b + 1$. We move the boundary so that the length-3 phrase shrinks to a length-2 phrase that starts at an even position. In this case, we extend the phrase on its left side to include $t_n[i_b + 1] = $ **a**. We can do this by modifying the source of **a** to point to the position next to the source of **b**, i.e., setting $f(i_b + 1) = f(i_b) + 1$. This is done recursively at the source positions, until we reach one of the following cases, where the source of **b** no longer points to a position of the same parity.
b │ a b a │ ⬇ │ b a b a │	Case 1: Noticing that the phrase is a substring of **baba**, we expand the phrase and make the source point to i_b. This is possible because $t_n[i_b..i_b + 3] = $ **baba**. Also, since each of the **ba**'s can finally be traced to the relaxed ground phrase created at the starting point, i.e., **ba** at position $i_b + 2$, we are done.
a b │ b a b │ a a	Case 2: We reach a different length-3 phrase **bab** that we wish to eliminate, that ends at an even position. We recursively apply the procedure to eliminate the phrase **bab**. In this case, that will shrink this length-3 phrase to a length-2 phrase by expanding the phrase to its right. Then, we are done with the elimination of the original length-3 phrase **aba**, since $t_n[f_{B_n}^{j'-1}(i_b)] = $ **b** and $t_n[f_{B_n}^{j'-1}(i_b) + 1] = $ **a** are in a length-2 phrase starting at an even position, i.e., a relaxed ground phrase.
│ b a b │ a ⬇ │ b a b a │	Case 3: Same as Case 1.
b │ a b │ a ⬇ │ b a b a │	Case 4: Same as Case 1.
│ b a │	Case 5: There is nothing to do. We are done, since both $t_n[f_{B_n}^{j'-1}(i_b)] = $ **b** and $t_n[f_{B_n}^{j'-1}(i_b) + 1] = $ **a** are in a length-2 phrase starting at an even position, i.e., a relaxed ground phrase.

Fig. 3. Terminal cases for eliminating a length-3 phrase **aba** that starts at an odd position $i_b + 1$ (see Sect. 4.2). The shaded squares are even positions. The vertical bars denote phrase boundaries. The black arrow points to the position $f_{B_n}^{j'-1}(i_b)$, where $j' \geq 1$ is the smallest integer such that the parities of $f_{B_n}^{j'-1}(i_b)$ and $f_{B_n}^{j'}(i_b)$ differ. The first and second lines in Cases 1, 3, 4 show the phrase boundaries before and after the modification. The characters outside the phrase considered for each case can be inferred from being a partner of a phrase, and also from Lemma 2

5   Conclusion

We have shown that for any $n \geq 2$, the size $b(t_n)$ of the smallest bidirectional scheme for the n-th Thue–Morse word t_n is exactly $n + 2$. From the result that the smallest string attractor of t_n is 4 for any $n \geq 4$ [7] and that $|t_n| = 2^n$, we have shown that Thue–Morse words are an example of a family of strings $\{S_n\}_{n \geq 1}$ in which each string S_n has $b(S_n) = \Theta(\gamma(S_n) \log \frac{|S_n|}{\gamma(S_n)})$ as the size of its smallest bidirectional parsing, where $\gamma(S_n)$ is the size of its smallest string attractor, and $|S_n| = 2^n$ is its length. Note that we can generalize this to hold for any $\gamma \geq 4$: Given a $\gamma \geq 4$, concatenate $k = \lfloor \gamma/4 \rfloor$ copies of t_n, each using distinct letters from a different binary alphabet. Finally, we add ($\gamma \bmod 4$) more distinct characters to make the smallest string attractor of the resulting string exactly γ. We thus can obtain a string of length $N = k \cdot 2^n + O(1)$ with $b = \Theta(kn) = \Theta(\gamma \log \frac{N}{\gamma})$. Whether this can be achieved for any γ by a family of binary strings is not yet known.

Our result shows for the first time the separation between γ and b, i.e., there are string families such that $\gamma = o(b)$. Although it is still open whether $O(\gamma \log N)$ bits is enough to represent any string of length N, it seems not possible by dictionary compression, i.e., copy/pasting within the string.

Acknowledgments. This work was supported by JSPS KAKENHI Grant Numbers JP20H04141 (HB), JP20J21147 (MF), JP19K20213 (TI), JP21K17701 (DK), JP20J11983 (TM).

References

1. Berstel, J., Reutenauer, C.: Square-free words and idempotent semigroups. In: Lothaire, M. (ed.) Combinatorics on Words, 2 edn, pp. 18–38. Cambridge Mathematical Library, Cambridge University Press (1997). https://doi.org/10.1017/CBO9780511566097.005

2. Burrows, M., Wheeler, D.J.: A block-sorting lossless data compression algorithm. Technical report (1994)

3. Kempa, D., Kociumaka, T.: Resolution of the burrows-wheeler transform conjecture. In: 61st IEEE Annual Symposium on Foundations of Computer Science, FOCS 2020, Durham, NC, USA, 16–19 November 2020, pp. 1002–1013. IEEE (2020). https://doi.org/10.1109/FOCS46700.2020.00097

4. Kempa, D., Prezza, N.: At the roots of dictionary compression: string attractors. In: Diakonikolas, I., Kempe, D., Henzinger, M. (eds.) Proceedings of the 50th Annual ACM SIGACT Symposium on Theory of Computing, STOC 2018, Los Angeles, CA, USA, 25–29 June 2018, pp. 827–840. ACM (2018). https://doi.org/10.1145/3188745.3188814

5. Kociumaka, T., Navarro, G., Prezza, N.: Towards a definitive measure of repetitiveness. In: Kohayakawa, Y., Miyazawa, F.K. (eds.) LATIN 2021. LNCS, vol. 12118, pp. 207–219. Springer, Cham (2020). https://doi.org/10.1007/978-3-030-61792-9_17

6. Kociumaka, T., Navarro, G., Prezza, N.: Towards a Definitive Measure of Repetitiveness. CoRR. **abs/1910.02151** (2019). http://arxiv.org/abs/1910.02151

7. Kutsukake, K., Matsumoto, T., Nakashima, Y., Inenaga, S., Bannai, H., Takeda, M.: On repetitiveness measures of Thue-Morse words. In: Boucher, C., Thankachan, S.V. (eds.) SPIRE 2020. LNCS, vol. 12303, pp. 213–220. Springer, Cham (2020). https://doi.org/10.1007/978-3-030-59212-7_15

8. Mantaci, S., Restivo, A., Romana, G., Rosone, G., Sciortino, M.: String attractors and combinatorics on words. In: Cherubini, A., Sabadini, N., Tini, S. (eds.) Proceedings of the 20th Italian Conference on Theoretical Computer Science, ICTCS 2019, Como, Italy, 9–11 September 2019. CEUR Workshop Proceedings, vol. 2504, pp. 57–71. CEUR-WS.org (2019). http://ceur-ws.org/Vol-2504/paper8.pdf

9. Mantaci, S., Restivo, A., Romana, G., Rosone, G., Sciortino, M.: A combinatorial view on string attractors. Theor. Comput. Sci. **850**, 236–248 (2021). https://doi.org/10.1016/j.tcs.2020.11.006

10. Mantaci, S., Restivo, A., Sciortino, M.: Burrows-Wheeler transform and Sturmian words. Inf. Process. Lett. **86**(5), 241–246 (2003). https://doi.org/10.1016/S0020-0190(02)00512-4

11. Morse, M.: Recurrent geodesics on a surface of negative curvature. Trans. Am. Math. Soc. **22**, 84–100 (1921)

12. Navarro, G.: Indexing highly repetitive string collections, part i: repetitiveness measures. ACM Comput. Surv. **54**(2), 1–31 (2021). https://doi.org/10.1145/3434399

13. Navarro, G., Ochoa, C., Prezza, N.: On the approximation ratio of ordered parsings. IEEE Trans. Inf. Theory **67**(2), 1008–1026 (2021). https://doi.org/10.1109/TIT.2020.3042746

14. Prouhet, E.: Mémoire sur quelques relations entre les puissances des nombres. C. R. Acad. Sci. Paris Sér **133**, 225 (1851)

15. Storer, J.A., Szymanski, T.G.: Data compression via textual substitution. J. ACM **29**(4), 928–951 (1982). https://doi.org/10.1145/322344.322346

16. Thue, A.: Über unendliche zeichenreihen. Norske vid. Selsk. Skr. Mat. Nat. Kl. **7**, 1–22 (1906)

17. Ziv, J., Lempel, A.: A universal algorithm for sequential data compression. IEEE Trans. Inf. Theory **23**(3), 337–343 (1977). https://doi.org/10.1109/TIT.1977.1055714

Lower Bounds for the Number of Repetitions in 2D Strings

Paweł Gawrychowski[1], Samah Ghazawi[2(✉)], and Gad M. Landau[2,3]

[1] Institute of Computer Science, University of Wrocław, Wrocław, Poland
[2] Department of Computer Science, University of Haifa, Haifa, Israel
idrees.samah@gmail.com
[3] NYU Tandon School of Engineering, New York University, Brooklyn, NY, USA

Abstract. A 2D string is simply a 2D array. We continue the study of the combinatorial properties of repetitions in such strings over the binary alphabet, namely the number of distinct tandems, distinct quartics, and runs. First, we construct an infinite family of $n \times n$ 2D strings with $\Omega(n^3)$ distinct tandems. Second, we construct an infinite family of $n \times n$ 2D strings with $\Omega(n^2 \log n)$ distinct quartics. Third, we construct an infinite family of $n \times n$ 2D strings with $\Omega(n^2 \log n)$ runs. This resolves an open question of Charalampopoulos, Radoszewski, Rytter, Waleń, and Zuba [ESA 2020], who asked if the number of distinct quartics and runs in an $n \times n$ 2D string is $\mathcal{O}(n^2)$.

Keywords: 2D strings · Quartics · Runs · Tandems

1 Introduction

The study of repetitions in strings goes back at least to the work of Thue from 1906 [36], who constructed an infinite square-free word over the ternary alphabet. Since then, multiple definitions of repetitions have been proposed and studied, with the basic question being focused on analyzing how many such repetitions a string of length n can contain. The most natural definition is perhaps that of palindromes, which are fragments that read the same either from left to right or right to left. Of course, any fragment of the string a^n is a palindrome, therefore we would like to count distinct palindromes. An elegant folklore argument shows that this is at most $n + 1$ for any string of length n [19], which is attained by a^n.

Another natural definition is that of squares, which are fragments of the form xx, where x is a string. Again, because of the string a^n we would like to count distinct squares. Using a combinatorial result of Crochemore and Rytter [16],

P. Gawrychowski—Partially supported by the Bekker programme of the Polish National Agency for Academic Exchange (PPN/BEK/2020/1/00444).

S. Ghazawi and G. M. Landau—Partially supported by the Israel Science Foundation grant 1475/18, and Grant No. 2018141 from the United States-Israel Binational Science Foundation (BSF).

T. Lecroq and H. Touzet (Eds.): SPIRE 2021, LNCS 12944, pp. 179–192, 2021.
https://doi.org/10.1007/978-3-030-86692-1_15

Fraenkel and Simpson [21] proved that a string of length n contains at most $2n$ distinct squares (see also [26] for a simpler proof, and [27] for an upper bound of $2n - \Theta(\log n)$). They also provided an infinite family of strings of length n with $n - o(n)$ distinct squares. It is conjectured that the right upper bound is actually n, however so far we only know that it is at most $11/6n$ [18]. Interestingly, a proof of the conjecture for the binary alphabet would imply it for any alphabet [30].

Perhaps a bit less natural, but with multiple interesting applications, is the definition of runs. A run is a maximal periodic fragment that is at least twice as long as its smallest period. Roughly speaking, runs capture all the repetitive structure of a string, making them particularly useful when constructing algorithms [15]. A well-known result by Kolpakov and Kucherov [29] is that a string of length n contains $\mathcal{O}(n)$ runs; they conjectured that it is actually at most n. After a series of improvements [12,32,33], with the help of an extensive computer search the upper bound was decreased to $1.029n$ [13,24]. Finally, in a remarkable breakthrough Bannai et al. [9] confirmed the conjecture. On the lower bound side, we current know an infinite family of strings with at least $0.944575712n$ runs [22,31,34]. Better bounds are known for the binary alphabet [20,25].

Given that we seem to have a reasonably good understanding of repetitions in strings, it is natural to consider repetitions in more complex structures, such as circular strings [7,17,35] or trees [14,23,28]. In this paper, we are interested in repetitions in 2D strings. Naturally, algorithms operating on 2D strings can be used for image processing, and combinatorial properties of such strings can be used for designing efficient pattern matching algorithms [1–4,11]. Therefore, we would like to fully understand what is a repetition in a 2D string, and what is the combinatorial structure of such repetition.

Apostolico and Brimkov [8] introduced the notions of tandems and quartics in 2D strings. Intuitively, a tandem consists of two occurrences of the same block W arranged in a 1×2 or 2×1 pattern, while a quartic consists of 4 occurrences of the same block W arranged in a 2×2 pattern. They considered tandems and quartics with a primitive W, meaning that it cannot be partitioned into multiple occurrences of the same W' (called primitively rooted in the subsequent work [10]), and obtained asymptotically tight bounds of $\Theta(n^3 \log n)$ and $\Theta(n^2 \log^2 n)$ for the number of such tandems and quartics in an $n \times n$ 2D string, respectively. Both tandems and quartics should be seen as an attempt to extend the notion of squares in a 1D string to 2D strings, and thus the natural next step is to consider distinct tandems and quartics (without restricting W to be primitive). Very recently, Charalampopoulos et al. [10] studied the number of distinct tandems and quartics in an $n \times n$ 2D string. For distinct tandems, they showed a tight bound of $\Theta(n^3)$ with the construction in the lower bound using an alphabet of size n. For distinct quartics, they showed an upper bound of $\mathcal{O}(n^2 \log^2 n)$ and conjectured that it is always $\mathcal{O}(n^2)$, similarly to the number of distinct squares in a 1D string of length n being $\mathcal{O}(n)$.

Amir et al. [5,6] introduced the notion of runs in 2D strings. Intuitively, a 2D run is a maximal subarray that is both horizontally and vertically periodic; we defer a formal definition to the next section. They proved that an $n \times n$ 2D string contains $\mathcal{O}(n^3)$ runs, showing an infinite family of $n \times n$ 2D strings with

$\Omega(n^2)$ runs. Later, Charalampopoulos et al. [10] significantly improved on this upper bound, showing that an $n \times n$ 2D string contains $\mathcal{O}(n^2 \log^2 n)$ runs, and conjectured that it is always $\mathcal{O}(n^2)$, similarly to the number of runs in a 1D string of length n being $\mathcal{O}(n)$.

Our Results. In this paper, we consider 2D strings and obtain improved lower bounds for the number of distinct tandems, distinct quartics, and runs. We start with the number of distinct tandems and extend the lower bound of Charalampopoulos et al. [10] to the binary alphabet in Sect. 3 by showing the following.

Theorem 1. *There exists an infinite family of $n \times n$ 2D strings over the binary alphabet containing $\Omega(n^3)$ distinct tandems.*

Then, we move to the number of distinct quartics in Sect. 4 and the number of runs in Sect. 5, and show the following.

Theorem 2. *There exists an infinite family of $n \times n$ 2D strings over the binary alphabet containing $\Omega(n^2 \log n)$ distinct quartics.*

Theorem 3. *There exists an infinite family of $n \times n$ 2D strings over the binary alphabet containing $\Omega(n^2 \log n)$ runs.*

By the above theorem, the algorithm of Amir et al. [6] for locating all 2D runs in $\mathcal{O}(n^2 \log n + \mathsf{output})$ time is worst-case optimal.

Our constructions exhibit a qualitative difference between distinct squares and runs in 1D strings and distinct quartics and runs in 2D strings. The number of the former is linear in the size of the input, while the number of the latter, surprisingly, is superlinear.

Our Techniques. For distinct tandems, our construction is similar to that of [10], except that we use distinct characters only in two columns. This allows us to replace them by their binary expansions, with some extra care as to not lose any counted tandems.

For both distinct quartics and runs, we construct an $n \times n$ 2D string recursively, but the high-level ideas behind both constructions are quite different.

For distinct quartics, our high-level idea is to consider subarrays with $\Theta(\log n)$ different aspect ratios. For each such aspect ratio, we want to obtain $\Omega(n^2)$ distinct quartics. To this end, we recursively define a family of rectangular arrays of the same width n but different heights, one for each aspect ratio, with the final array being the desired $n \times n$ 2D string. Each step of the recursion creates multiple new special characters, as to make the new quartics distinct; later we show how to implement this kind of approach with the binary alphabet.

For runs, we directly proceed with a construction for the binary alphabet, and build on the insight used by Charalampopoulos et al. [10] to show that the same quartic can be induced by $\Theta(n^2)$ runs. The construction is recursive, and allows us to obtain larger and larger 2D strings with many runs. In every step of the recursion, we compose multiple smaller 2D strings defined in the previous step. Then, we appropriately modify two of them to create many new runs. This needs to be carefully analyzed in order to lower bound the total number of runs.

2   Preliminaries

Let Σ be a fixed finite alphabet. A 2D string over Σ is an $m \times n$ array $A[0..m-1][0..n-1]$ with m rows and n columns, with every cell $A[i][j]$ containing an element of Σ. Furthermore, we use ϵ to denote an empty 2D string. A subarray $A[x_1..x_2][y_1..y_2]$ of $A[0..m-1][0..n-1]$ is an $(x_2-x_1+1) \times (y_2-y_1+1)$ array consisting of cells $A[i][j]$ with $i \in [x_1, x_2], j \in [y_1, y_2]$.

We consider three notions of repetitions in 2D strings.

Tandem. A subarray T of A is a tandem if it consists of 2×1 (or 1×2) subarrays $W \neq \epsilon$. Two tandems $T = \boxed{W|W}$ and $T' = \boxed{W'|W'}$ are distinct when $W \neq W'$.

Quartic. A subarray Q of A is a quartic if it consists of 2×2 subarrays $W \neq \epsilon$. Two quartics $Q = \dfrac{W|W}{W|W}$ and $Q' = \dfrac{W'|W'}{W'|W'}$ are distinct when $W \neq W'$.

Run. Consider an $r \times c$ subarray R of A. We define a positive integer p to be its horizontal period if the ith column of R is equal to the $(i+p)^{\text{th}}$ column of R, for all $i = 1, 2, \ldots, c - p$. The horizontal period of R is its smallest horizontal period, and we say that R is h-periodic when its horizontal period is at most $c/2$. Similarly, we define a vertical period, the vertical period, and a v-periodic subarray. An h-periodic and v-periodic R is called a run when extending R in any direction would result in a subarray with a larger horizontal or vertical period. Informally, for such R there exists a subarray W such that we can represent R as follows, with at least two repetitions of W in both directions, and we cannot extend R in any direction while maintaining this property.

$$R = \begin{array}{|c|c|c|c|} \hline W & \ldots & W & W' \\ \hline \ldots & \ldots & \ldots & \ldots \\ \hline W & \ldots & W & W' \\ \hline W'' & \ldots & W'' & W''' \\ \hline \end{array}$$

Where $W = \boxed{W'|U}$, $W = \dfrac{W''}{U'}$ and $W = \dfrac{W'''|V}{V'|V''}$, and any of the subarrays $W', W'', W''', U, U', V, V'$ and V'' may be ϵ.

3   Distinct Tandems

In this section, we show how to construct an $n \times n$ array A over the binary alphabet with $\Theta(n^3)$ distinct tandems, for any $\ell \geq 4$, where $n = 3 \cdot 2^\ell + 2(\ell + 2)$. The ith row of A is divided into 5 parts, where $0 \leq i < n$ (see Fig. 1). The first, third, and fifth part each consists of 2^ℓ cells, each containing the binary representation of 1. The second and fourth part each consists of $\ell + 2$ cells that contains the binary representation of the row number, i.e. the binary representation of i. Hence, all rows of A are different, see Fig. 2.

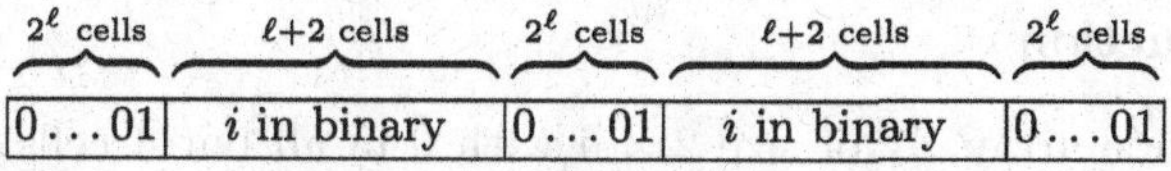

Fig. 1. The ith row of A, where $0 \le i < n$.

Proof. To lower bound the number of tandems in A, consider any $0 \le i \le j < n$ and $k \in \{1, 2, \dots, 2^\ell\}$. Then, let T be the subarray of width $2(2^\ell + \ell + 2)$ starting in the ith row and ending in the jth row with the top left cell of T being $A[i][k]$. We claim that for each choice of i, j, k we obtain a distinct tandem, making the number of distinct tandems in A is at least $n^2 \cdot 2^\ell = \Omega(n^3)$. It is clear that each such T is a tandem. To prove that all of them are distinct, consider any such T. The position of the leftmost 1 in its top row allows us to recover the value of k. Then, the next $\ell + 2$ cells contain the binary expansion of i, so we can recover i. Finally, the height of T together with i allows us to recover j. Thus, we can uniquely recover i, j, k from T, and all such tandems are distinct.   □

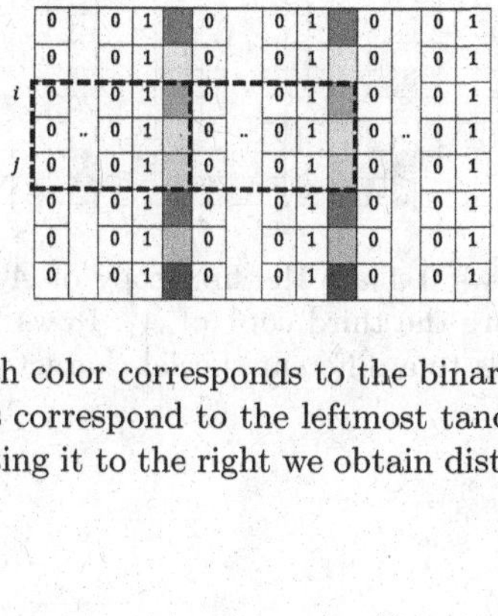

Fig. 2. Array A, where each color corresponds to the binary representation of the row number. The black borders correspond to the leftmost tandem of height $j - i + 1$ and width $2(2^\ell + \ell + 2)$; by shifting it to the right we obtain distinct tandems. (Color figure online)

4   Distinct Quartics

In this section, we show how to construct an $n \times n$ array A^ℓ_ℓ with $\Omega(n^2 \log n)$ distinct quartics, for any $\ell \ge 1$, where $n = 3^\ell - 1$. The construction is recursive, that is, we construct a series of arrays $A^\ell_1, A^\ell_2, \dots, A^\ell_\ell$, with A^ℓ_i being defined using A^ℓ_{i-1}. The number of columns of each array A^ℓ_i is the same and equal to n. The number of rows is increasing, starting with 2 rows in A^ℓ_1 and ending with n rows in the final array A^ℓ_ℓ. We provide the details of the construction in the next subsection, then analyze the number of distinct quartics in A^ℓ_ℓ in the subsequent subsection. Finally, in the last subsection we show how to use A^ℓ_ℓ to obtain a $n \times n$ binary array B^ℓ_ℓ with $\Omega(n^2 \log n)$ distinct quartics, where $n = \Theta(3^\ell \ell)$.

4.1   Construction

First, we provide array A_1^ℓ of size $2 \times n$ with 0 in all but 4 cells, namely, cells $A_1^\ell[0][\frac{n-2}{3}]$, $A_1^\ell[1][\frac{n-2}{3}]$, $A_1^\ell[0][\frac{2n-1}{3}]$ and $A_1^\ell[1][\frac{2n-1}{3}]$ containing the same special character. In particular, we are dividing the columns into 3 equal parts.

Second, we describe the general construction of an $M_i \times n$ array A_i^ℓ, for $i \geq 2$. We maintain the invariant that the columns of A_i^ℓ are partitioned into 3^i maximal ranges of N_i columns consisting of only 0s and separated with single columns, i.e., $N_1 = \frac{(n-2)}{3}$. To obtain A_i^ℓ, we first vertically concatenate 3 copies of A_{i-1}^ℓ, using different special characters in each copy, while adding a single separating row between the copies. Thus, $M_i = 3M_{i-1} + 2$. Initially, each separating row consists of only 0s. For each maximal range of columns in A_{i-1}^ℓ that consists of only 0s, we proceed as follows. We further partition the columns of the range into 3 sub-ranges of $\frac{N_{i-1}-2}{3}$ columns, separated by single columns. We create a new special character and insert its four copies at the intersection of each column separating the sub-ranges and each separating row. Overall, we create 3^{i-1} new special characters. See Fig. 3 for an illustration with $i = 2$.

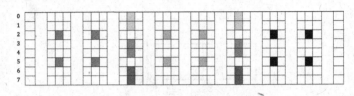

Fig. 3. Array A_2^ℓ, where rows 0–1 are the first copy of A_1^ℓ, rows 3–4 are the second copy of A_1^ℓ, and rows 6–7 are the third copy of A_1^ℓ. Rows 2 and 5 are the separating rows. Each color corresponds to a different special character. (Color figure online)

4.2   Analysis

Before we move to counting distinct quartics in each A_i^ℓ, we recall that the number of columns in each A_i^ℓ is the same and equal to n, while the number of rows M_i is described by the recurrence $M_1 = 2$ and $M_i = 3M_{i-1} + 2$ for $i \geq 2$, hence $M_i = 3^i - 1$. The size N_i of each maximal range of columns consisting only of 0s is described by the recurrence $N_1 = \frac{n-2}{3}$ and $N_i = \frac{N_{i-1}-2}{3}$ for $i \geq 2$, hence $N_i = \frac{n+1}{3^i} - 1$. For $n = 3^\ell - 1$ all these numbers are integers.

We now analyze the number of distinct quartics in each A_i^ℓ. We will be only counting quartics such that each quartic contains a single special character in each of the four copies that comprise it, and denote by Q_i the distinct quartics counted in the following argument that, similarly to the construction, considers first $i = 1$ and then the general case.

For $i = 1$, we count quartics of width $2(\frac{n-2}{3} + 1)$ and height 2 where each quartic contains a single special character in each of the four copies that comprise it. There are $Q_1 = \frac{n-2}{3} + 1 = \frac{n+1}{3}$ such quartics and all of them are distinct.

For the general case of $i \geq 2$, we consider two groups of distinct quartics. The first group consists of distinct quartics contained in the copies of A_{i-1}^ℓ. For each of the 3^{i-1} maximal ranges of N_{i-1} columns of A_{i-1}^ℓ consisting of 0s, the second group consists of all possible $2(M_{i-1}+1) \times 2(N_i+1)$ subarrays contained in that range. For each such range, we have $N_i + 1$ possible horizontal shifts and $M_{i-1}+1$ possible vertical shifts and for each of them we obtain a distinct quartic containing the new special character created for the range. As we use different special characters in every copy of A_{i-1}^ℓ and, for every range, in the separating rows of A_i^ℓ, overall we have at least $Q_i = 3Q_{i-1} + 3^{i-1}(N_i+1)(M_{i-1}+1)$ distinct quartics. See Fig. 4 for an illustration with $i = 2$.

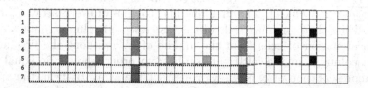

Fig. 4. Array A_2^ℓ, where the black border corresponds to the leftmost quartic in the third copy of A_1^ℓ, and by shifting it to the right we obtain distinct quartics. The red borders correspond to the leftmost quartic in each of the 3 maximal ranges of columns of A_1^ℓ consisting of 0s; by shifting each of them to the right and down we obtain distinct quartics. (Color figure online)

Substituting the formulas for N_i and M_{i-1}, we conclude that $Q_1 = \frac{n+1}{3}$ and $Q_i = 3Q_{i-1} + 3^{i-2}(n+1)$ for $i \geq 2$. Unwinding the recurrence, we obtain that $Q_i = 3^{i-1}Q_1 + (i-1)3^{i-2}(n+1) = 3^{i-1}\frac{n+1}{3} + (i-1)3^{i-2}(n+1)$. Therefore, $Q_i = 3^{i-2}i(n+1)$.

Theorem 4. *There exists an infinite family of $n \times n$ 2D strings containing $\Omega(n^2 \log n)$ distinct quartics.*

Proof. For each $\ell \geq 1$, we take $n = 3^\ell - 1$ and define arrays $A_1^\ell, A_2^\ell, \ldots, A_\ell^\ell$ as described above. The final array A_ℓ^ℓ consists of $M_\ell = n$ rows and n columns, and contains at least $Q_\ell = 3^{\ell-2}\ell(n+1)$ distinct quartics, which is $\Omega(n^2 \log n)$. □

While we were not concerned with the size of the alphabet in this construction, observe that the number of distinct special characters S_i in A_i^ℓ is described by the recurrences $S_1 = 1$ and $S_i = 3S_{i-1} + 3^{i-1}$ for $i \geq 2$. This is because we are using new special characters in each copy of A_{i-1}^ℓ and adding 3^{i-1} new special characters to divide the maximal ranges of N_{i-1} columns into 3 parts. The size of the alphabet used to construct A_ℓ^ℓ is thus $S_\ell + 1 = \frac{(n+1)\log_3(n+1)}{3} + 1 = 3^{\ell-1}\ell + 1$.

4.3   Reducing the Alphabet

In this subsection, we show how to modify the array A_ℓ^ℓ to obtain an array B_ℓ^ℓ over the binary alphabet, for any $\ell \geq 6$. Informally speaking, we will replace each character by a small gadget of size $k \times k$ encoding its binary representation. Both the width and height of the array increase by a factor of k, and therefore we need the number of distinct quartics increase by (roughly) a factor of k^2. Therefore we cannot consider only subarrays consisting of full gadgets, and need to adjust the gadgets so that sufficiently large subarrays with different horizontal or vertical offsets modulo k are certainly distinct.

Let $\Sigma = \{1, 2, \ldots, \sigma\}$ be the alphabet used to construct A_ℓ^ℓ, where $\sigma = 3^{\ell-1}\ell + 1$. We define arrays $C_1, C_2, \ldots, C_\sigma$ of the same size $k \times k$, where $k = \lceil * \rceil \sqrt{\log_2 \sigma} + 2 = \Theta(\sqrt{\ell})$. The first row and column of every array C_c contain only 0s, while the remaining cells of the last row and column contain only 1s. The concatenation of cells from the middle of C_c (without the first and last row and column), in the left-right top-bottom order, should be equal to the binary representation of c. Now, we construct the array B_ℓ^ℓ from the array A_ℓ^ℓ by repeating the recursive construction of arrays $A_1^\ell, A_2^\ell, \ldots, A_\ell^\ell$, but replacing a cell containing the character c with the array C_c. We denote the resulting arrays $B_1^\ell, B_2^\ell, \ldots, B_\ell^\ell$.

Let $n = (3^\ell - 1)k$. Each of the arrays B_i^ℓ consists of n columns and $M_i \cdot k$ rows, so the final array, B_ℓ^ℓ, is of size $n \times n$. We now analyze the number of distinct quartics in B_ℓ^ℓ. This will be done similarly as it was for A_i^ℓ, but we must be more careful about arguing quartics as being distinct, because we no longer have multiple distinct special characters. We first argue that, for all sufficiently wide and tall subarrays of B_i^ℓ, the horizontal and vertical shifts are uniquely defined modulo k, see Fig. 5.

Lemma 1. *Consider a subarray $R = B_i^\ell[x_1..x_2][y_1..y_2]$ with width and height at least k. Then $(x_1 \bmod k)$ and $(y_1 \bmod k)$ can be recovered from R.*

Proof. We only analyze how to recover $(y_1 \bmod k)$, recovering $(x_1 \bmod k)$ is symmetric. By construction of $C_1, C_2, \ldots, C_\sigma$, every k^{th} row of B_i^ℓ consists of only 0s, while in every other row there is at least one 1 in every block of k cells. Therefore, because the width of R is at least k, a row of R consists of 0s if and only if it is aligned with a row of B_i^ℓ that consists of 0s. Because the height of R is at least k such a row surely exists and allows us to recover $(y_1 \bmod k)$.   □

We argue that the number of distinct quartics in B_1^ℓ is at least $Q_1' = (\frac{n-2}{3})k + 1$. To show this, we consider subarrays spanning the whole height of B_1^ℓ and of width $2((\frac{n-2}{3})k + k)$. There are $(\frac{n-2}{3})k + 1$ such subarrays and each of them is a quartic that fully contains some C_c. Furthermore, subarrays starting in columns with different remainders modulo k are distinct by Lemma 1. Subarrays starting in columns with the same remainder modulo k are also distinct, as in such a case we can recover the special character from C_c fully contained in the subarray.

0	...	...	0	0	...	...	0	0	...	...	0
	...	1	...		1	...				1	
	...		...	...			...	...			...
0	1	...	1	0	1	...	1	0	1	...	1
0	...	...	0	0	...	...	0	0	...	...	0
	...	1	...		1	...				1	
	...		...	...			...	...			...
0	1	...	1	0	1	...	1	0	1	...	1

Fig. 5. Subarray of B_i^ℓ of size $2k \times 3k$, that is 2×3 concatenation of gadgets C_c. The red rectangle corresponds to R in Lemma 1. The yellow cells correspond to rows consisting of only 0s. The blue squares correspond to binary representations of symbols in Σ. (Color figure online)

For the general case, we claim that the number of distinct quartics in B_i^ℓ is at least $Q_i' = 3Q_{i-1}' + 3^{i-1}(N_i \cdot k + 1)(M_{i-1} \cdot k + 1)$ for $i \geq 2$. The argument proceeds as for A_i^ℓ; however, we must argue that the counted quartics are all distinct. By construction, each of them fully contains some C_c. Thus, quartics starting in columns with different remainders modulo k (and also in rows with different remainders modulo k) are distinct by Lemma 1. Now consider all counted quartics starting in columns with remainder y modulo k and rows with remainder x modulo k. For each of them, we can recover the special character from C_c fully contained in the quartic, so all of them are distinct.

Finally, we lower bound and solve the recurrence for Q_i' as follows.

$$
\begin{aligned}
Q_i' &= 3Q_{i-1}' + 3^{i-1}(N_i \cdot k + 1)(M_{i-1} \cdot k + 1) \\
&> 3Q_{i-1}' + 3^{i-1} \cdot k^2 \cdot N_i \cdot M_{i-1} \\
&= 3Q_{i-1}' + 3^{i-1} \cdot k^2 (\frac{n+1}{3^i} - 1)(3^{i-1} - 1) \\
&> 3Q_{i-1}' + 3^{i-1} \cdot k^2 (\frac{n}{3^i} - 1)3^{i-2} \\
&= 3Q_{i-1}' + 3^{2i-3} \cdot k^2 (\frac{n - 3^i}{3^i}) \\
&= 3Q_{i-1}' + 3^{i-3} \cdot k^2 (n - 3^i) \qquad\qquad \text{using } i \geq 2.
\end{aligned}
$$

Unwinding the recurrence, we obtain that $Q_i' > \sum_{j=2}^{i} 3^{i-j} \cdot 3^{j-3} \cdot k^2(n - 3^j) = 3^{i-3} \cdot k^2 \sum_{j=2}^{i}(n - 3^j) > 3^{i-3} \cdot k^2((i-1)n - \frac{3^{i+1}}{2})$.

Proof. For each $\ell \geq 6$, we take $n = (3^\ell - 1)k$ and define arrays $B_1^\ell, B_2^\ell, \ldots, B_\ell^\ell$ as described above. The final array B_ℓ^ℓ is over the binary alphabet by construction, consists of n rows and n columns, and contains at least Q_ℓ' distinct quartics.

Finally,

$$
\begin{aligned}
Q'_\ell &> 3^{\ell-3} \cdot k^2 \cdot ((\ell-1)n - \frac{3^{\ell+1}}{2}) \\
&= 3^{\ell-3} \cdot k^2 \cdot ((\ell-1)(3^\ell - 1) - \frac{3^{\ell+1}}{2}) \\
&> 3^{\ell-3} \cdot k^2 \cdot (\ell-4)(3^\ell - 1) \qquad \text{because } \frac{3^\ell}{2} < 3^\ell - 1 \\
&\geq 3^{\ell-3} \cdot k^2 \cdot \frac{\ell}{4} \cdot (3^\ell - 1) \qquad \text{because } \ell - 4 \geq \frac{\ell}{4}.
\end{aligned}
$$

Therefore, the number of runs in B_ℓ^ℓ is $\Omega(n^2 \cdot \ell)$. Because $k = \Theta(\sqrt{\ell})$ and $n = (3^\ell - 1)k$, this is $\Omega(n^2 \log n)$. □

5   Runs

In this section, we show how to construct an $n \times n$ array A_ℓ with $\Omega(n^2 \log n)$ runs, for any $\ell \geq 2$, where $n = 2 \cdot 4^\ell$. We note that here we do not restrict the runs to be distinct, that is, we count all occurrences. As in the previous section, the construction is recursive, i.e., we construct a series of arrays $A_1, A_2, \ldots, A_\ell$, with A_i being defined using A_{i-1}. Both the number of rows and columns in A_i is equal to $2 \cdot 4^i$, starting with 8 rows and columns in A_1. We describe the construction in the next subsection, then analyze the number of runs in A_ℓ in the subsequent subsection.

5.1   Construction

First, we provide array A_1 of size 8×8 with 1s in the cells $A_1[0][1]$, $A_1[1][0]$, $A_1[6][7]$ and $A_1[7][6]$, and 0s in the other cells, see Fig. 6 (left).

Second, we obtain array A_i by concatenating 4×4 copies of array A_{i-1} while using 1s to fill the antidiagonals in the upper left and bottom right copy of A_{i-1}, with A'_{i-1} denoting such modified copy of A_{i-1}, see Fig. 6 (right) for an illustration with $i = 2$.

The intuition behind the recursive construction is to duplicate the runs obtained in the previous arrays. For example, the array A_1 produces one run that does not touch the boundaries. This is duplicated 14 times in A_2, hence, A_1 contributes 14 runs to the counted runs produced by A_2. Moreover, the intuition behind filling the antidiagonals is to produce new runs such that the number of the new runs is equal to the size of the array up to some constant. As an example, A_2 produces 7^2 new runs between the antidiagonals of A'_1 such that the upper left and the bottom right corners of each run touch exactly two cells of the antidiagonals of the two copies of A'_1. Therefore, overall the counted runs produced by A_2 is $14 + 7^2 = 63$. The general case is analyzed in detail in the next subsection.

5.2   Analysis

The number N_i of rows and columns in A_i is described by the recurrence $N_1 = 8$ and $N_i = 4N_{i-1}$ for $i \geq 2$, so $N_i = 2 \cdot 4^i$. By straightforward induction, the antidiagonal of every A_i is filled with 0s.

We analyze the number of runs in A_i. We call a run new iff its upper left and the bottom right corners touch exactly two cells of the antidiagonals of the two copies of A'_{i-1}. We observe that a new run is not contained in any of the copies of A_{i-1} or A'_{i-1}. Let R_i denote the number of new runs in A_i.

Lemma 2. $R_i = (2 \cdot 4^{i-1} - 1)^2 = \frac{16^i}{4} - 4^i + 1.$

Proof. Consider any subarray R of A_i with the upper left and the bottom right corners touching exactly two cells of the antidiagonals of the two copies A'_{i-1}. It is easy to verify that N_{i-1} is a horizontal and a vertical period of R. Therefore, R is h-periodic and v-periodic. Now consider extending R in any direction, say by one column to the left. Then the topmost cell of the new column would contain a 1 from the antidiagonal of A'_{i-1}. For the horizontal period of the extended array to remain N_{i-1} we would need a 1 in the corresponding cell of the antidiagonal of A_{i-1}, but that cell contains a 0, a contradiction. Therefore, any such R is a run. The number of such subarrays is $(N_{i-1} - 1)^2 = (2 \cdot 4^{i-1} - 1)^2 = \frac{16^i}{4} - 4^i + 1$, because we have $(N_{i-1} - 1)$ possibilities for choosing the two corners.   □

Second, we have the runs contained in the 14 copies of A_{i-1}, hence A_{i-1} contributes $14R_{i-1}$ to the counted runs in A_i. Moreover, whenever A_{i-1} contains a copy of A_j, for some $j < i - 1$, all new runs of A_j are preserved in A_{i-1} and consequently in A_i. Additionally, we have the two copies of A'_{i-1}. Because we have filled their antidiagonals with 1s, we lose some of the runs. However, whenever A'_{i-1} contains a copy of A_j that does not intersect the antidiagonal, for some $j < i - 1$, all new runs of A_j are preserved in A'_{i-1} and consequently in A_i. For example, each copy of A_{i-1} contains 14 copies of A_{i-2} and each copy of A'_{i-1} contains 10 copies of A_{i-2} (5 above and 5 below the antidiagonal). Hence, A_i contains $14 \cdot 14 + 2 \cdot 10 = 216$ copies of A_{i-2}, thus A_{i-2} contributes $216 \cdot R_{i-2}$ to the counted runs in A_i. So, in order to bound the number of runs in the final array A_ℓ, we need to analyze how many copies of A_i are in A_ℓ, for $1 \leq i \leq \ell$.

Let X_i denote the number of copies of A_i in A_ℓ, and Y_i denote the number of copies of A'_i in A_ℓ. By construction, A_i consists of 14 copies of A_{i-1} and 2 copies of A'_{i-1}. Similarly, A'_i consists of 10 copies of A_{i-1} (5 above and 5 below the antidiagonal) and 6 copies of A'_{i-1} (4 intersecting the antidiagonal and the top left and bottom right copy). Consequently, we obtain the recurrences $X_\ell = 1$ and $X_i = 14X_{i+1} + 10Y_{i+1}$ for $i < \ell$, $Y_\ell = 0$ and $Y_i = 6Y_{i+1} + 2X_{i+1}$ for $i < \ell$. Instead of solving the recurrences, we show the following.

Lemma 3. $X_i \geq \frac{5}{6}16^{\ell-i}$

Proof. We first observe that $X_i + Y_i = 16(X_{i+1} + Y_{i+1})$, as A_{i+1} consists of the 4×4 smaller subarrays, each of them being A_i or A'_i. By unwinding the recurrence, $X_i + Y_i = 16^{\ell-i}(X_\ell + Y_\ell) = 16^{\ell-i}$. Furthermore, we argue that $X_i \geq 5Y_i$ for every $i < \ell$. This is proved by induction on i:

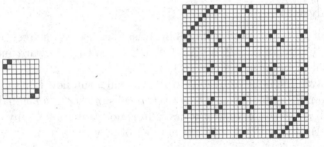

Fig. 6. Left: array A_1, where red cells contain 1s and white cells contain 0s. Right: Array A_2 consists of 14 copies of A_1 and 2 copies of A'_1. Red cells contain 1s and white cells contain 0s. Red is used to fill the antidiagonals of A'_1. (Color figure online)

$$i = \ell - 1 \quad X_{\ell-1} = 14X_\ell + 10Y_\ell = 14 \geq 5Y_{\ell-1} = 5(6Y_\ell + 2X_\ell) = 10.$$
$$i < \ell - 1 \quad \text{Assuming that } X_{i+1} \geq 5Y_{i+1}, \text{ we write } X_i = 14X_{i+1} + 10Y_{i+1} \geq 10X_{i+1} + 30Y_{i+1} \text{ and } 5Y_i = 30Y_{i+1} + 10X_{i+1}, \text{ so } X_i \geq 5Y_i.$$

Therefore, $16^{\ell-i} = X_i + Y_i \leq X_i + \frac{X_i}{5}$, so $X_i \geq \frac{5}{6}16^{\ell-i}$.   □

As explained earlier, whenever a copy of A_i occurs in A_ℓ, all of its new runs contribute to the counted runs in A_ℓ. Therefore, the total number of runs in A_ℓ is at least $\sum_{i=1}^{\ell} X_i \cdot R_i$.

Proof. For each $\ell \geq 2$, we take $n = 2 \cdot 4^\ell$ and construct the arrays $A_1, A_2, \ldots, A_\ell$ as described above. The final array A_ℓ is over the binary alphabet by construction, consists of n rows and columns and contains at least $\sum_{i=1}^{\ell} X_i \cdot R_i$ runs. By Lemma 2 and 3, this is at least

$$\sum_{i=1}^{\ell} \frac{5}{6} 16^{\ell-i} (2 \cdot 4^{i-1} - 1)^2 = \frac{5}{6} 16^\ell \sum_{i=1}^{\ell} 16^{-i} \left(\frac{16^i}{4} - 4^i + 1 \right)$$

$$= \frac{5}{6} 16^\ell \sum_{i=1}^{\ell} \left(\frac{1}{4} - \frac{1}{4^i} + \frac{1}{16^i} \right)$$

$$= \frac{5}{6} 16^\ell \left(\frac{\ell}{4} + \frac{1}{3 \cdot 4^\ell} - \frac{1}{15 \cdot 16^\ell} - \frac{4}{15} \right)$$

$$= \frac{5}{6 \cdot 4} \ell \cdot 16^\ell + \frac{5}{6 \cdot 3} 4^\ell - \frac{1}{6 \cdot 3} - \frac{4}{6 \cdot 3} 16^\ell$$

$$\geq \frac{5}{24} \ell \cdot 16^\ell - \frac{1}{18} - \frac{2}{9} 16^\ell$$

$$\geq \frac{1}{24} \ell \cdot 16^\ell = \Omega(n^2 \log n) \qquad \qquad \text{using } \ell \geq 2$$

□

References

1. Amir, A., Benson, G.: Efficient two-dimensional compressed matching. In: Data Compression Conference, pp. 279–288 (1992)
2. Amir, A., Benson, G.: Two-dimensional periodicity in rectangular arrays. SIAM J. Comput. **27**(1), 90–106 (1998)
3. Amir, A., Benson, G., Farach-Colton, M.: An alphabet independent approach to two dimensional pattern matching. SIAM J. Comput. **23**(2), 313–323 (1995)
4. Amir, A., Benson, G., Farach-Colton, M.: Optimal parallel two dimensional text searching on a CREW PRAM. Inf. Comput. **144**, 1–17 (1998)
5. Amir, A., Landau, G.M., Marcus, S., Sokol, D.: Two-dimensional maximal repetitions. In: 26th ESA, vol. 112, no. 2, pp. 1–14 (2018)
6. Amir, A., Landau, G.M., Marcus, S., Sokol, D.: Two-dimensional maximal repetitions. Theoret. Comput. Sci. **812**, 49–61 (2020)
7. Amit, M., Gawrychowski, P.: Distinct squares in circular words. In: Fici, G., Sciortino, M., Venturini, R. (eds.) SPIRE 2017. LNCS, vol. 10508, pp. 27–37. Springer, Cham (2017). https://doi.org/10.1007/978-3-319-67428-5_3
8. Apostolico, A., Brimkov, V.: Fibonacci arrays and their two-dimensional repetitions. Theoret. Comput. Sci. **237**(1–2), 263–273 (2000)
9. Bannai, H.I.T., Inenaga, S., Nakashima, Y., Takeda, M., Tsuruta, K.: The "runs" theorem. SIAM J. Comput. **46**(5), 1501–1514 (2017)
10. Charalampopoulos, P., Radoszewski, J., Rytter, W., Waleń, T., Zuba, W.: The number of repetitions in 2D-strings. In: 28th ESA, vol. 173, no. 32, pp. 1–18 (2020)
11. Cole, R., et al.: Optimally fast parallel algorithms for preprocessing and pattern matching in one and two dimensions. In: 34th FOCS, pp. 248–258 (1993)
12. Crochemore, M., Ilie, L.: Maximal repetitions in strings. J. Comput. Syst. Sci. **74**(5), 796–807 (2008)
13. Crochemore, M., Ilie, L., Tinta, L.: The "runs" conjecture. Theoret. Comput. Sci. **412**(27), 2931–2941 (2011)
14. Crochemore, M., et al.: The maximum number of squares in a tree. In: Kärkkäinen, J., Stoye, J. (eds.) CPM 2012. LNCS, vol. 7354, pp. 27–40. Springer, Heidelberg (2012). https://doi.org/10.1007/978-3-642-31265-6_3
15. Crochemore, M., Iliopoulos, C.S., Kubica, M., Radoszewski, J., Rytter, W., Waleń, T.: Extracting powers and periods in a word from its runs structure. Theoret. Comput. Sci. **521**, 29–41 (2014)
16. Crochemore, M., Rytter, W.: Squares, cubes, and time-space efficient string searching. Algorithmica **13**, 405–425 (1995). https://doi.org/10.1007/BF01190846
17. Currie, J.D., Fitzpatrick, D.S.: Circular words avoiding patterns. In: Ito, M., Toyama, M. (eds.) DLT 2002. LNCS, vol. 2450, pp. 319–325. Springer, Heidelberg (2003). https://doi.org/10.1007/3-540-45005-X_28
18. Deza, A., Franek, F., Thierry, A.: How many double squares can a string contain? Discret. Appl. Math. **180**, 52–69 (2015)
19. Droubay, X., Justin, J., Pirillo, G.: Episturmian words and some constructions of de Luca and Rauzy. Theoret. Comput. Sci. **255**(1), 539–553 (2001)
20. Fischer, J., Holub, Š, I, T., Lewenstein, M.: Beyond the runs theorem. In: Iliopoulos, C., Puglisi, S., Yilmaz, E. (eds.) SPIRE 2015. LNCS, vol. 9309, pp. 277–286. Springer, Cham (2015). https://doi.org/10.1007/978-3-319-23826-5_27
21. Fraenkel, A.S., Simpson, J.: How many squares can a string contain? J. Comb. Theory, Ser. A **82**(1), 112–120 (1998)

22. Franek, F., Yang, Q.: An asymptotic lower bound for the maximal number of runs in a string. Int. J. Found. Comput. Sci. **19**(1), 195–203 (2008)

23. Gawrychowski, P., Kociumaka, T., Rytter, W., Waleń, T.: Tight bound for the number of distinct palindromes in a tree. In: Iliopoulos, C., Puglisi, S., Yilmaz, E. (eds.) SPIRE 2015. LNCS, vol. 9309, pp. 270–276. Springer, Cham (2015). https://doi.org/10.1007/978-3-319-23826-5_26

24. Giraud, M.: Not so many runs in strings. In: Martín-Vide, C., Otto, F., Fernau, H. (eds.) LATA 2008. LNCS, vol. 5196, pp. 232–239. Springer, Heidelberg (2008). https://doi.org/10.1007/978-3-540-88282-4_22

25. Holub, S.: Prefix frequency of lost positions. Theoret. Comput. Sci. **684**, 43–52 (2017)

26. Ilie, L.: A simple proof that a word of length n has at most 2n distinct squares. J. Comb. Theory, Ser. A **112**(1), 163–164 (2005)

27. Ilie, L.: A note on the number of squares in a word. Theoret. Comput. Sci. **380**(3), 373–376 (2007)

28. Kociumaka, T., Radoszewski, J., Rytter, W., Walen, T.: String powers in trees. Algorithmica **79**(3), 814–834 (2017). https://doi.org/10.1007/s00453-016-0271-3

29. Kolpakov, R., Kucherov, G.: Finding maximal repetitions in a word in linear time. In: 40th FOCS, pp. 596–604. IEEE Computer Society (1999)

30. Manea, F., Seki, S.: Square-density increasing mappings. In: Manea, F., Nowotka, D. (eds.) WORDS 2015. LNCS, vol. 9304, pp. 160–169. Springer, Cham (2015). https://doi.org/10.1007/978-3-319-23660-5_14

31. Matsubara, W., Kusano, K., Ishino, A., Bannai, H., Shinohara, A.: New lower bounds for the maximum number of runs in a string. In: Proceedings of the Prague Stringology Conference 2008, pp. 140–145 (2008)

32. Puglisi, S.J., Simpson, J., Smyth, W.F.: How many runs can a string contain? Theoret. Comput. Sci. **401**(1–3), 165–171 (2008)

33. Rytter, W.: The number of runs in a string. Inf. Comput. **205**(9), 1459–1469 (2007)

34. Simpson, J.: Modified Padovan words and the maximum number of runs in a word. Australas. J. Combin. **46**, 129–146 (2010)

35. Simpson, J.: Palindromes in circular words. Theoret. Comput. Sci. **550**, 66–78 (2014)

36. Thue, A.: Über unendliche Zeichenreihen. Norske Vid Selsk. Skr. I Mat-Nat Kl. (Christiana) **7**, 1–22 (1906)

On Stricter Reachable Repetitiveness Measures

Gonzalo Navarro and Cristian Urbina[(✉)]

CeBiB—Center for Biotechnology and Bioengineering,
Department of Computer Science, University of Chile, Santiago, Chile

Abstract. The size b of the smallest bidirectional macro scheme, which is arguably the most general copy-paste scheme to generate a given sequence, is considered to be the strictest reachable measure of repetitiveness. It is strictly lower-bounded by measures like γ and δ, which are known or believed to be unreachable and to capture the entropy of repetitiveness. In this paper we study another sequence generation mechanism, namely compositions of a morphism. We show that these form another plausible mechanism to characterize repetitive sequences and define NU-systems, which combine such a mechanism with macro schemes. We show that the size $\nu \leq b$ of the smallest NU-system is reachable and can be $o(\delta)$ for some string families, thereby implying that the limit of compressibility of repetitive sequences can be even smaller than previously thought. We also derive several other results characterizing ν.

Keywords: Repetitiveness measures · Data compression · Combinatorics on words

1 Introduction

The study of repetitiveness measures, and of suitable measures of compressibility of repetitive sequences, has recently attracted interest thanks to the surge of repetitive text collections in areas like Bioinformatics, and versioned software and document collections. A recent survey [15] identifies a number of those measures, separating those that are *reachable* (i.e., any sequence can be represented within that space) from those that are not, which are still useful as lower bounds.

Reachable measures are, for example, the size g of the smallest context-free grammar that generates the sequence [8], the size c of the smallest *collage system* that generates the sequence [7] (which generalizes grammars), the number z of phrases of the Lempel-Ziv parse of the sequence [10], or the number b of phrases of a *bidirectional macro scheme* that represents the sequence [19]. Such a macro scheme cuts the sequence into phrases so that each phrase either is an explicit symbol or it can be copied from elsewhere in the sequence, in a way that no cyclic

Funded in part by Basal Funds FB0001, Fondecyt Grant 1-200038, and a Conicyt Doctoral Scholarship, ANID, Chile.

T. Lecroq and H. Touzet (Eds.): SPIRE 2021, LNCS 12944, pp. 193–206, 2021.
https://doi.org/10.1007/978-3-030-86692-1_16

dependencies are introduced. As such, macro schemes are the ultimate measure of what can be obtained by "copy-paste" mechanisms, which characterize repetitive sequences well.

Other measures are designed as lower bounds on the compressibility of repetitive sequences: γ is the size of the smallest *string attractor* for the sequence [6] and δ is a measure derived from the string complexity [3,17].

In asymptotic terms, it holds $\delta \leq \gamma \leq b \leq c \leq z \leq g$ and, except for $c \leq z$, there are string families where each measure is asymptotically smaller than the next. The recent result by Bannai et al. [2], showing that there exists a string family where $\gamma = o(b)$, establishes a clear separation between unreachable lower bounds (δ,γ) and reachable measures (b and the larger ones).

Concretely, Bannai et al. show that $b = \Theta(\log n)$ and $\gamma = O(1)$ for the Thue-Morse family, defined as $t_0 = 0$ and $t_{k+1} = t_k \overline{t_k}$, where $\overline{t_k}$ is t_k with 0s converted to 1s and vice versa. This family is a well-known example of the fixed point of a *morphism* ϕ, defined in this case by the rules $0 \to 01$ and $1 \to 10$. Then, t_k is simply $\phi^k(0)$. This representation of the words in the family is of size $O(1)$, and each word can be easily produced in optimal time by iterating the morphism.

Iterating a small morphism is arguably a relevant mechanism to define repetitive sequences. Intuitively, any short repetition $\alpha[1, k]$ that arises along the generation of a long string turns into a longer repetition $\phi^t(\alpha[1]) \cdots \phi^t(\alpha[k])$ in the final string, t steps later. More formally, if a morphism is k-uniform (i.e., all its rules are of fixed length k), then the resulting sequence is so-called k-*automatic* [1] and its prefixes have an attractor of size $\gamma = O(\log n)$ [18]. That is, many small morphisms lead to sequences with low measures of repetitiveness. Further, in the Thue-Morse family, morphisms lead to a reachable measure of repetitiveness that is $o(b)$, below what can be achieved with copy-paste mechanisms.

In this paper we further study this formalism. First, we define *macro systems*, a grammar-like extension that we prove equivalent to bidirectional macro schemes. We then study *deterministic Lindenmayer systems* [11,12], a grammar-like mechanism generating infinite strings via iterated morphisms; they are stopped at some level to produce a finite string. We combine both systems into what we call *NU-systems*. The size ν ("nu") of the smallest NU-system is always reachable and $O(b)$. Further, we show that there are string families where $\nu = o(\delta)$, thereby showing that δ is not anymore a lower bound for the compressibility of repetitive sequences if we include other plausible mechanisms to represent them. We present several other results that help characterize the new measure ν.

2   Basic Concepts

2.1   Terminology

Let Σ be a set of symbols, called the *alphabet*. A string w of length $|w| = n$ (also denoted as $w[1, n]$ when needed) is a concatenation of n symbols from Σ; in particular the string of length 0 is denoted by ε. The set of k-length concatenations of symbols from Σ is denoted Σ^k, and the set Σ^* of strings over Σ is defined as $\bigcup_{k \geq 0} \Sigma^k$; we also define $\Sigma^+ = \bigcup_{k \geq 1} \Sigma^k$. We juxtapose strings

(xy) or combine them with the dot operator $(x \cdot y)$ to denote their concatenation. A string x is a *prefix* of w if $w = xz$, a *suffix* of w if $w = yx$, and a *substring* of w if $w = yxz$, for some $y, z \in \Sigma^*$. Let $w[1, n]$ denote an n-length string. Then $w[i]$ is the i-th symbol of w, and $w[i, j]$ the substring $w[i]w[i + 1] \cdots w[j]$ if $1 \le i \le j \le n$, and ε if $j < i$.

2.2 Parsing Based Schemes

Probably the most popular measure of repetitiveness is the number z of *phrases* in the so-called *Lempel-Ziv parse* of a word $w[1, n]$ [10]. In such a parse, w is partitioned optimally into phrases $w = x_1 \cdots x_z$, so that every x_k is either of length 1 or it appears starting to the left in w (so the phrase x_k is copied from some *source* at its left). This parsing can be computed in $O(n)$ time.

Storer and Szymanski [19] introduced *bidirectional macro schemes*, which allow sources appear to the left or to the right of their phrases, as long as circular dependencies are avoided. We follow the definition by Bannai et al. [2].

Let $w[1, n]$ be a string. A bidirectional macro scheme of size k for w is a sequence $B = (x_1, s_1), \ldots, (x_k, s_k)$ satisfying $w = x_1 \cdots x_k$ and $x_i = w[s_i, s_i + |x_i| - 1]$ if $|x_i| > 1$, and $s_i = \bot$ if $|x_i| = 1$. We denote the starting position of x_i in w by $p_i = 1 + \sum_{j=1}^{i-1} |x_j|$. The function $f : [1, n] \to [1, n] \cup \{\bot\}$,

$$f(i) = \begin{cases} \bot & : \text{if } i = p_j, s_j = \bot \text{ for some } j \\ s_j + i - p_j : & \text{if } p_j \le i < p_{j+1} \text{ for some } j \text{ and } s_j \ne \bot \end{cases}$$

is induced by the macro scheme. For B to be a valid bidirectional macro scheme it must hold that, for each i, there exists some r satisfying $f^r(i) = \bot$. Therefore, it suffices with the values $|x_i|$ and s_i, plus x_i where $s_i = \bot$, to recover w.

We call $b \le z$ the number k of elements in the smallest bidirectional macro scheme generating a given string $w[1, n]$. There are string families where $b = o(z)$ [16]. While z is computed in linear time, computing b is NP-hard.

2.3 Grammars and Generalizations

The size g of the smallest context-free grammar generating (only) a word $w[1, n]$ [8] is a relevant measure of repetitiveness. Such a grammar has exactly one rule per nonterminal, and those can be sorted so that the right-hand sides mention only terminals and previously listed nonterminals. The size of the grammar is the sum of the lengths of the right-hand sides of the rules. The *expansion* of a nonterminal is the string of terminals it generates; the word defined by the grammar is the expansion of its last listed nonterminal.

More formally, a grammar over the alphabet of terminals Σ is a sequence of nonterminals $X_1, \ldots, X_r$, with a rule $X_k \to A_{k,1} \cdots A_{k,\ell_k}$, each $A_{k,r}$ being a terminal or some nonterminal in $X_1, \ldots, X_{k-1}$. The expansion of a terminal a is $exp(a) = a$, and that of a nonterminal X_k is $exp(X_k) = exp(A_{k,1}) \cdots exp(A_{k,\ell_k})$. The grammar represents the string $w = exp(X_r)$, and its size is $\sum_{k=1}^{r} \ell_k$.

Composition systems were introduced by Gasieniec et al. [5]. Those add the ability to reference any prefix or suffix of the expansion of a previous nonterminal (and, thus, substrings as prefixes of suffixes). Let us use the more general form, allowing terms $A_{k,r} = X_j[s,t]$ where $exp(X_j[s,t]) = exp(X_j)[s,t]$.

Kida et al. [7] extended composition systems with *run-length* terms of the form $(A_{k,r})^t$, so that $exp((A_{k,r})^t) = exp(A_{k,r})^t$, the expansion of $A_{k,r}$ concatenated t times. They called this extension a *collage system*. We call $c \leq g$ the size of the smallest collage system generating a word $w[1,n]$, and it always holds $b = O(c)^1$ and $c = O(z)$. There are string families where $b = o(c)$ [16], and where $z = o(g)$. Computing g (and, probably, c too) is NP-hard.

2.4   Lower Bounds

Kempa and Prezza introduced the concept of *string attractor* [6], which yields an abstract measure that lower-bounds all the previous reachable measures.

Let $w[1,n]$ be a string. A string attractor for w is a set of positions $A \subseteq [1,n]$ where for every substring $w[i,j]$ there exists a copy $w[i',j']$ (i.e., $w[i,j] = w[i',j']$) and a position $k \in A$ with $i' \leq k \leq j'$. The measure γ is defined as the cardinality of the smallest of such attractors for a given string $w[1,n]$, and it always holds that $\gamma = O(b)$. Further, a string family where $\gamma = o(b)$ exists [2].

Kociumaka et al. [3] used the *string complexity* of $w[1,n]$ to define a measure called δ. Let $S_w(k)$ be the number of distinct substrings of length k in w. Then $\delta = \max\{S_w(k)/k \mid 1 \leq k \leq n\}$. This measure is computed in $O(n)$ time and it always holds that $\delta = O(\gamma)$; there are string families where $\delta = o(\gamma)$ [9]. While δ is unreachable in some string families, any string can be represented in $O(\delta \log(n/\delta))$ space [9]. Measure δ has been proposed as a lower bound on the compressibility of repetitive strings, which we question in this paper.

2.5   Morphisms over Strings

We explain some general concepts about morphisms acting over strings [1,14]. A *monoid* $(M, *, e)$ is a set with an associative operation $*$ and a neutral element $e \in M$ satisfying $a * e = e * a = a$ for every $a \in M$. We write ab for $a * b$ and say that M is a monoid, instead of $(M, *, e)$. A *morphism of monoids* is a function $\phi : M_1 \to M_2$, where $(M_1, *_1, e_1)$ and $(M_2, *_2, e_2)$ are monoids, $\phi(a *_1 b) = \phi(a) *_2 \phi(b)$ for every $a, b \in M_1$, and $\phi(e_1) = e_2$.

Let Σ be a set of symbols, and $\cdot$ the concatenation of strings. Then $(\Sigma^*, \cdot, \varepsilon)$ is a monoid with string concatenation, called the *free monoid*. A *morphism of free monoids* $\phi : \Sigma^* \to \Delta^*$ is defined completely just by specifying ϕ on the symbols of Σ. If $\Sigma = \Delta$, then ϕ, is called an *automorphism*, and ϕ is *iterable*. We define the *n-iteration* (or composition) of ϕ over s as $\phi^n(s)$.

Let $\phi : \Sigma^* \to \Delta^*$ be a morphism of free monoids. We define $depth(\phi) = |\Sigma|$, $width(\phi) = \max_{a \in \Sigma} |\phi(a)|$, and $size(\phi) = \sum_{a \in \Sigma} |\phi(a)|$. We say ϕ is *expanding* if $|\phi(a)| > 1$, *non-erasing* if $|\phi(a)| > 0$, and *k-uniform* if $|\phi(a)| = k$, for every

---

[1] At least if the collage system is *internal*, that is, every $exp(X_k)$ appears in w.

$a \in \Sigma$. A *coding* is a 1-uniform morphism. We say ϕ is *prolongable* on $a \in \Sigma$ if $\phi(a) = as$ for a non-empty string s.

Let ϕ be an automorphism on Σ^*. Let ϕ be prolongable on a, so $\phi(a) = as$. Then, $w = as\phi(s)\phi^2(s) \cdots$ is the unique infinite *fixed point* of ϕ starting with a, that is, $\phi(w) = w$ [14]. Words built in this fashion are called *purely morphic words*. If we apply a coding to them, we obtain *morphic words*. A morphic word obtained from a k-uniform morphism is said to be *k-automatic* [1].

3 Macro Systems

Our first contribution is the definition of *macro systems*, a generalization of composition systems we prove to be as powerful as bidirectional macro schemes. That is, the smallest macro system generating a given string w is of size $O(b)$.

Definition 1. A *macro system* is a tuple $M = (V, \Sigma, R, S)$, where V is a finite set of symbols called the *variables*, Σ is a finite set of symbols disjoint from V called the *terminals*, R is the set of rules (exactly one per variable)

$$R : V \to (V \cup \Sigma \cup \{A[i,j] \mid A \in V, i, j \in \mathbb{N}\})^*,$$

and $S \in V$ is the *initial variable*. If $R(A) = \alpha$ is the rule for A, we also write $A \to \alpha$. The symbols $A[i,j]$ are called *extractions*. The rule $A \to \varepsilon$ is permitted only for $A = S$. The *size* of a macro system is the sum of the lengths of the right-hand sides of the rules, $size(M) = \sum_{A \in V} |R(A)|$.

We now define the string generated by a macro system as the expansion of its initial symbol, $exp(S)$. Such expansions are defined as follows.

Definition 2. Let $M = (V, \Sigma, R, S)$ be a macro system. The *expansion* of a symbol is a string over Σ^* defined inductively as follows:

- If $a \in \Sigma$ then $exp(a) = a$.
- If $S \to \varepsilon$, then $exp(S) = \varepsilon$.
- If $A \to B_1 \cdots B_k$ is a rule, then $exp(A) = exp(B_1) \cdots exp(B_k)$.
- $exp(A[i,j]) = exp(A)[i,j]$ (this second $[i,j]$ denotes substring).

We say that the macro system is *valid* if there is a single solution $w \in \Sigma^*$ for $exp(S)$. We say that the macro system *generates* the string w.

Note that a macro system looks very similar to a composition system, however, it does not impose an order so that each symbol references only previous ones. This algorithm determines the string generated by a macro system, if any:

1. Compute $|exp(A)|$ for every nonterminal A, using the rules:
   - If $a \in \Sigma$, then $|exp(a)| = 1$.
   - If $A \to B_1 \cdots B_k$, then $|exp(A)| = |exp(B_1)| + \cdots + |exp(B_k)|$.
   - $|exp(A[i,j])| = j - i + 1$.

This must generate a system of equations without loops (otherwise the macro system is invalid), which is then trivially solved.

2. Replace every symbol $A[i, j]$ by $A[i] \cdots A[j]$; we use $A[r]$ to denote $A[r, r]$.

3. Replace every $A[r]$, if $A \to B_1 \cdots B_k$, iterating until obtaining a terminal:
   - Let $p_i = 1 + \sum_{j=1}^{i-1} |exp(B_j)|$, for $1 \le i \le k + 1$.
   - Let s be such that $p_s \le r < p_{s+1}$.
   - If $B_s \in \Sigma$, replace $A[r]$ by B_s.
   - Otherwise replace $A[r]$ by $B_s[r - p_s + 1]$.

4. If the process to replace any $A[r]$ falls in a loop (i.e., we return to $A[r]$), then the system has no unique solution and thus it is invalid. Otherwise, we are left with a classical context-free grammar without extractions, and compute $w = exp(S)$ in the classical way.

Note that a rule like $A \to B \; A[1, (t-1)|exp(B)|]$ solves only for $exp(A) = exp(B)^t$, just like the run-length symbol B^t of collage systems. For example, $A \to ab$ and $S \to A \; S[1, 4]$ generates $ababab$ as follows:

$$A \; S[1] \; S[2] \; S[3] \; S[4]$$
$$A \; A[1] \; A[2] \; S[1] \; S[2]$$
$$A \; a \; b \; A[1] \; A[2]$$
$$A \; a \; b \; a \; b$$
$$a \; b \; a \; b \; a \; b$$

This shows that macro systems are at least as powerful as collage systems. But they can be asymptotically smaller. For example, the smallest collage system generating the Fibonacci string F_k (where $F_1 = b$, $F_2 = a$, and $F_{k+2} = F_{k+1}F_k$) is of size $\Theta(\log |F_k|)$ [16, Thm. 32]. Instead, we can mimic a bidirectional macro scheme of size 4 [16, Lem. 35] with a constant-sized macro system generating F_k: $S \to S[f_{k-2}+1, f_k-2] \; b \; a \; S[f_{k-2}+1, 2f_{k-2}]$ if k is odd and $S \to S[f_{k-2}+1, f_k-2] \; a \; b \; S[f_{k-2}+1, 2f_{k-2}]$ if k is even (where $f_k = |F_k|$). For example, for F_7 the system is $S \to S[6, 11] \; b \; a \; S[6, 10]$ and we extract $F_7 = exp(S)$ as follows, using that $F_7[1, 6] = F_7[6, 11]$, $F_7[7] = b$, $F_7[8] = a$, and $F_7[9, 13] = F_7[6, 10]$:

$$S[6] \; S[7] \; S[8] \; S[9] \; S[10] \; S[11] \; b \; a \; S[6] \; S[7] \; S[8] \; S[9] \; S[10]$$
$$S[11] \; b \; a \; S[6] \; b \; a \; b \; a \; S[11] \; b \; a \; S[6] \; b$$
$$a \; b \; a \; S[11] \; b \; a \; b \; a \; a \; b \; a \; S[11] \; b$$
$$a \; b \; a \; a \; b \; a \; b \; a \; a \; b \; a \; a \; b$$

In general, we can prove that a restricted class of our macro systems is equivalent to bidirectional macro schemes.

Definition 3. A macro system $M = (V, \Sigma, R, S)$ generating w is *internal* if $exp(A)$ appears in w for every $A \in V$. We use m to denote the size of the smallest internal macro system generating w.

Theorem 1. *It always holds that $m \le b$.*

Proof. Let $(x_1, s_1), \ldots, (x_b, s_b)$ be the smallest bidirectional macro scheme generating $w[1, n] = x_1 \cdots x_b$. We construct a macro system $M = (\{S\}, \Sigma, R, S)$ with a single rule $S \rightarrow A_1 \cdots A_b$, where A_i is the single terminal x_i if $s_i = \bot$, and the extraction symbol $S[s_i, s_i + |x_i| - 1]$ if not.

We now show that this macro system is valid. After we execute step 2 of our algorithm, the length of the resulting string (which we call W) is already n: it has only terminals and symbols of the form $W[i] = S[r]$. Note that this implies that $f(i) = r$ in the bidirectional macro scheme. In every step, we replace each such $S[r]$ by $W[r]$. Since the macro scheme is valid, for each i there is a finite k such that $f^k(i) = \bot$, and thus $W[i]$ becomes a terminal symbol after k steps. □

Theorem 2. *For every internal macro system of size m there is a bidirectional macro scheme of size $b \leq m$.*

Proof. An internal macro system $M = (V, \Sigma, R, S)$ generating $w[1, n]$ can always be transformed into one with a single rule for the initial symbol. Let $A \in V$ be such that $w[i, j] = exp(A)$. We can then replace every occurrence of A by $S[i, j]$, and every occurrence of $A[i', j']$ by $S[i' + i - 1, j' + i - 1]$, on the right-hand sides of all the rules. In particular, the rule defining S will now contain terminals and symbols of the form $S[i, j]$, and thus all the other nonterminals can be deleted.

From the resulting macro system $S \rightarrow A_1 \cdots A_{m'}$, where $m' \leq m$, we can derive a bidirectional macro scheme $(x_1, s_1), \ldots, (x_{m'}, s_{m'})$, as follows: if A_t is a terminal, then x_t is that terminal and $s_t = \bot$. Otherwise, A_t is of the form $S[i, j]$ and then $x_t = w[i, j]$ and $s_t = i$. The resulting scheme is valid, because our algorithm extracts any $S[i]$ after a finite number k of steps, which is then the k such that $f^{k+1}(i) = \bot$. □

That is, bidirectional macro schemes are equivalent to internal macro systems. General macro systems can be asymptotically smaller in principle, though we have not found an example where this happens.

4   Deterministic Lindenmayer Systems

In this section we study a mechanism for generating infinite sequences called *deterministic Lindenmayer Systems* [11,12], which build on morphisms. We adapt those systems to generate finite repetitive strings. Those systems are, in essence, grammars with only nonterminals, which typically generate longer and longer strings, in a levelwise fashion. For our purposes, we will also specify at which level d to stop the generation process and the length n of the string w to generate. The generated string $w[1, n]$ is then the n-length prefix of the sequence of nonterminals obtained at level d. We adapt, in particular, the variant called CD0L-systems, though we will use the generic name *L-systems* for simplicity.

Definition 4. An *L-system* is a tuple $L = (V, R, S, \tau, d, n)$, where V is a finite set of symbols called *variables*, $R : V \rightarrow V^+$ is the set of *rules*, $S \in V^*$ is a sequence of variables called the *axiom*, $\tau : V \rightarrow V$ is a coding, $d \in \mathbb{N}$ is the level where to stop (or depth), and $n \in \mathbb{N}$ is the length of the string to generate.

An L-system produces *levels* of strings $L_i \in V^*$, starting from $L_0 = S$ at level 0. Each level replaces every variable A from the previous level by $R(A)$, that is, $L_{i+1} = R(L_i)$ if we identify R with its homomorphic extension. The generated string is $w[1, n] = \tau(L_d[1, n]) \in V^*$, seeing τ as its homomorphic extension.

The *size* of an L-system is $|S| + \sum_{A \in V} |R(A)|$. We call ℓ the size of the smallest L-system generating a string w.

L-systems then represent strings by iterating a non-erasing automorphism. Somewhat surprisingly, we now exhibit a string family where $\delta = \Omega(\ell \log n)$, thus L-systems are a reachable mechanism to generate strings that can be asymptotically smaller than what was considered to be a stable lower bound.

Theorem 3. *There exist string families where* $\delta = \Omega(\ell \log n)$.

Proof. Consider the L-system $L = (V, R, S, \tau, d, n)$ where $V = \{0,1\}$, $S = 0$, $R(0) = 001$, $R(1) = 1$, $\tau(0) = 0$, $\tau(1) = 1$, and $n = 2^{d+1} - 1$. The family of strings is formed by all those generated by the systems L, where $d \in \mathbb{N}$. It is clear that all the strings in this family share the value $\ell = 5$.

The first strings of the family generated by this system (i.e., its levels L_i) are 0, 001, 0010011, 001001100100111, and so on. It is easy to see by induction that level i contains 2^i 0s and $2^i - 1$ 1s, so the string L_i is of length $2^{i+1} - 1$.

More importantly, one can see by induction that levels $i \geq 2$ start with 00 and contain all the strings of the form 01^j0 for $1 \leq j < i$. This is true for level 2. Then, in level $i+1$ the strings 01^j0 become 0011^j001, which contains $01^{j+1}0$, and the first 00 yields 001001, containing 010.

Consider now the number of d-length distinct substrings in L_d, for $d \geq 4$. Each distinct substring 01^j0, for $\lfloor d/2 \rfloor - 1 \leq j \leq d - 2$, yields at least $d - j - 1$ distinct d-length substrings (containing 01^j0 at different offsets; no single d-length substring may contain two of those). These add up to $d^2/8 + d/4$ distinct d-length substrings, and thus $\delta = \Omega(d) = \Omega(\log n)$ on the string $w = \tau(L_d)$.   □

On the other hand, L-systems are always reachable, which yields the immediate result that δ and ℓ are incomparable.

Theorem 4. *There exist string families where* $\ell = \Omega(\delta \log n)$.

Proof. Kociumaka et al. [9, Thm. 2] exhibit a string family of $2^{\Theta(\log^2 n)}$ elements with $\delta = O(1)$, so it needs $\Omega(\log^2 n) = \Omega(\delta \log^2 n)$ bits, to be represented with any method. On the other hand, an L-system of size ℓ is described with $O(\ell \log n)$ bits. Therefore $\ell = \Omega(\log n) = \Omega(\delta \log n)$ in this family.   □

Those strings are formed by n as, replacing them by bs at single arbitrary positions between $2 \cdot 4^{j-2} + 1$ and 4^{j-1} for every $j \geq 2$. While such a string is easily generated by a composition system of size $\Theta(\log n)$, we could only produce L-systems of size $\Theta(\log^2 n)$ generating it. We now prove bounds between L-systems and context-free grammars.

Theorem 5. *For any L-system $L = (V, R, S, \tau, d, n)$ of size ℓ generating w, there is a context-free grammar of size $(d+1)\ell$ generating w. If the morphism represented by R is expanding, then the grammar is of size $O(\ell \log n)$.*

Proof. Consider the derivation tree for w in L: the root children are $S = L_0$ at level 0, and if A is a node at level i, then the children of A are the elements in $R(A)$, at level $i + 1$. The nodes in each level i spell out L_i.

We create a grammar $G = (V', V, R', S')$ where V' contains the initial symbol S' and, for each variable $A \in V$ of the L-system, d nonterminals $A_0, \ldots, A_{d-1}$. The terminals of the grammar are the set of L-system variables, V. Then, for each L-system rule $A \to B_1 \cdots B_k$ appearing in level $0 \le i \le d - 2$, we add the grammar rule $A_i \to (B_1)_{i+1} \cdots (B_k)_{i+1}$. Further, for each rule $A \to B_1 \cdots B_k$ appearing in level $d - 1$, we add the grammar rule $A_{d-1} \to \tau(B_1) \cdots \tau(B_k)$. Finally, if $S = B_1 \cdots B_k$ is the L-system axiom, we add the grammar rule $S' \to (B_1)_0 \cdots (B_k)_0$ for its initial symbol.

It is clear that the grammar is of size at most $(d+1)\ell$ and it generates w. If every rule is of size larger than 1, and $d > \lg n$, then the prefix $w[1, n]$ of $\tau(L_d)$ is generated from the first symbol of $L_{d-\lceil \lg n \rceil}$, which can then be made the axiom and d reduced to $\lceil \lg n \rceil$. In this case, the grammar is of size $O(\ell \log n)$. $\square$

For example, consider our L-system $0 \to 001$ and $1 \to 1$. A grammar simulating a generation of $d = 3$ levels contains the rules $S' \to 0_0$, $0_0 \to 0_1 0_1 1_1$, $0_1 \to 0_2 0_2 1_2$, $1_1 \to 1_2$, $0_2 \to 001$, and $1_2 \to 1$. Note how the grammar uses the level subindices to control the point where the L-system should stop.

On the other hand, while we believe that composition systems can be smaller than L-systems, we can prove that L-systems are not larger than grammars.

Theorem 6. *It always holds that $\ell = O(g)$.*

Proof. Consider a grammar $G = (V, \Sigma, R, S)$ of height h generating $w[1, n]$. We define the L-system $L = (V \cup \Sigma, R', R(S), \tau, h, n)$, where R' contains all the rules in R except the one for S. We also include in R the rules $a \to a$ for all $a \in \Sigma$. The coding τ is the identity function.

It is clear that this L-system produces the same derivation tree of G, reaching terminals a at some level. Those remain intact up to the last level, h, thanks to the rules $a \to a$. At this point the L-system has derived $w[1, n]$.

The size of the L-system is that of G plus $|\Sigma|$, which is of the same order because every symbol $a \in \Sigma$ appears on some right-hand side (if not, we do not need to create the rule $a \to a$ for that symbol). $\square$

The following simple result characterizes a class of morphisms generating families with constant-sized L-systems.

Theorem 7. *Let $w \in \Sigma^*$, $\phi : \Sigma^* \to \Sigma^*$ be a non-erasing automorphism, and $\tau : \Sigma \to \Sigma$. Then $\ell = O(1)$ on the family $\{\tau(\phi^d(w)) \mid d > 0\}$.*

Proof. We can easily simulate ϕ on the L-system $L = (\Sigma, R, w, \tau, d, n)$ of fixed size, with $R(a) = \phi(a)$ and $n = |\phi^d(w)|$. The system generates $\tau(\phi^d(w))$ and, as d grows, it does not change its size. $\square$

This implies that $\ell = O(1)$ on families of n-iterations of the Thue-Morse morphism, the Fibonacci morphism, images of k-uniform morphisms (i.e., morphisms generating k-automatic words [1]), and standard Sturmian morphisms [4]. More generally, ℓ is $O(1)$ on the set of prefixes of any morphic word.

5   NU-Systems

We now define a mechanism that combines both macro systems and L-systems, yielding a computable measure that is reachable and strictly better than b.

Definition 5. A *NU-system* is a tuple $N = (V, R, S, \tau, d, n)$, which is understood in the same way as L-systems, except that we extend rules with *extractions*, that is, $R : V \to (V \cup E)^+$ and

$$E = \{A(l)[i, j] \mid A \in V, l, i, j \in \mathbb{N}\}.$$

The symbol $A(l)[i, j]$ means to expand variable A for l levels and then extract $\tau(A_l[i, j])$ from the string A_l at level l, recursively expanding extractions if necessary. This counts as a single expansion (one level) of a rule, that is, the levels L_i in the NU-system belong to V^*. We also use $A(l) = A(l)[1, |A_l|]$ to denote the whole level l of A. The *size* of the NU-system is $size(N) = |S| + \sum_{A \in V} |R(A)|$. We call ν the size of the smallest NU-system generating a string $w[1, n]$.

Just as macro systems, a NU-system is *valid* only if it does not introduce circular dependencies. Let $maxl$ be the maximum l value across every rule $A(l)[i, j]$ in the NU-system. The following algorithm determines the string generated by the system, if any:

1. Compute $|A_l|$ for every variable A and level $0 \le l \le maxl$, using the rules:
   - $|A_0| = 1$.
   - If $l > 0$ and $A \to B_1 \cdots B_k$, then $|A_l| = |(B_1)_{l-1}| + \cdots + |(B_k)_{l-1}|$.
   - Replace $|B(l)[i, j]| = j - i + 1$ on the previous summands $|(B_r)_{l-1}|$.
   This generates a system of equations without loops, which is trivially solved.
2. Replace every symbol $A(l)[i, j]$ in R by $A(l)[i] \cdots A(l)[j]$; we use $A(l)[r]$ to denote $A(l)[r, r]$.
3. Expand the rules, starting from the axiom, level by level as in L-systems. Handle the symbols $A(l)[r]$ as follows:
   (a) Replace every $A(0)[r]$ (so $r = 1$ if the NU-system is correct) by $\tau(A)$.
   (b) Replace every $A(l)[r]$, if $l > 0$ and $A \to B_1 \cdots B_k$, as follows:
       - Let $p_i = 1 + \sum_{j=1}^{i-1} |(B_j)_{l-1}|$, for $1 \le i \le k + 1$.
       - Let s be such that $p_s \le r < p_{s+1}$.
       - Replace $A(l)[r]$ by $B_s(l - 1)[r - p_s + 1]$.
   (c) Return to (a) until the extraction symbol disappears.

Note that the symbol B_s in step 3(b) can in turn be of the form $B_s = B(l')[r']$; we must then extract $B(l')[r']$ before continuing the extraction of $B_s(l-1)[r - p_s + 1]$. If, along the expansion, we return again to the original $A(l)[r]$, then the system has no unique solution and thus it is invalid. This is computable because the number of possible combinations $A(l)[r]$ is bounded by $|V| \cdot maxl \cdot n$.

We now show that NU-systems are at least as powerful as macro systems and L-systems.

Theorem 8. *It always holds that $\nu = O(\min(\ell, m))$.*

Proof. It always holds $\nu \leq \ell$ because L-systems are a particular case of NU-systems. With respect to m, let $M = (V, \Sigma, R, S)$ be a minimal macro system generating $w[1, n]$. Then we construct a NU-System $N = (V \cup \Sigma, R', S, \tau, d, n)$ where τ is the identity and $d = |V|$, which upper-bounds the height of the derivation tree. Each level of N will simulate the sequence of extractions that lead from each $A[r]$ to its corresponding terminal in the macro system. For each $a \in \Sigma$ we define the rule $a \to a$ in R'. For each rule $A \to B_1 \cdots B_m$ in R, we define the rule $A \to B'_1 \cdots B'_m$ in R', where $B'_i = B_i$ if $B_i \in V \cup \Sigma$, and $B'_i = A'(d)[j, k]$ if $B_i = A'[j, k]$. It is not hard to see that the NU-System N simulates the macro system M, and its size is $O(m)$. □

For example, consider our previous macro system $A \to ab$ and $S \to A\, S[1, 4]$. The corresponding NU-system would have the rules $a \to a$, $b \to b$, $A \to ab$, and $S \to A\, S(2)[1, 4]$. The derivation is then generated as follows:

$$L_0 = S \qquad \longrightarrow A\ S(2)[1]\ S(2)[2]\ S(2)[3]\ S(2)[4]$$
$$A\ A(1)[1]\ A(1)[2]\ (S(2)[1])(1)[1]\ (S(2)[2])(1)[1]$$
$$A\ a\ b\ (A(1)[1])(1)[1]\ (A(1)[2])(1)[1]$$
$$L_1 = A\ a\ b\ a\ b \quad \longleftarrow A\ a\ b\ a(1)[1]\ b(1)[1]$$
$$L_2 = a\ b\ a\ b\ a\ b.$$

Our new measure ν is then reachable, strictly better than b and incomparable with δ. It is likely, however, that computing ν (i.e., finding the smallest NU-system generating a given string $w[1, n]$) is NP-hard.

NU-systems easily allow us concatenating and composing automorphisms.

Theorem 9. *Let $N_1 = (V_1, R_1, S_1, \tau_1, d_1, n_1)$ and $N_2 = (V_2, R_2, S_2, \tau_2, d_2, n_2)$ be NU-systems generating w_1 and w_2, respectively. Then there are NU-systems of size $O(size(N_1) + size(N_2))$ that generate $w_1 \cdot w_2$ and the composition of w_1 and w_2, which is the string generated by N_2 with axiom w_1, $(V_2, R_2, w_1, \tau_2, d_2, n_2)$.*

Proof. Let $V'_1 = \{a_1 \mid a \in V_1\}$ and $V'_2 = \{a_2 \mid a \in V_2\}$ be disjoint copies of V_1 and V_2, respectively, and let R'_i, and S'_i be variants that operate on V'_i instead of V_i. We build a NU-system $N = (V, R, S, \tau, 1, n_1 + n_2)$ for $w_1 \cdot w_2$, where $V = V'_1 \cup V'_2 \cup V_1 \cup V_2 \cup \{Z_1, Z_2\}$, where Z_1 and Z_2 are new symbols. Let $R = R'_1 \cup R'_2 \cup \{Z_1 \to S'_1, Z_2 \to S'_2\}$, plus the rules $a \to a$ for $a \in V_1 \cup V_2$. The axiom is then $S = Z_1(d_1) \cdot Z_2(d_2)$. Finally, the mapping on V'_i is $\tau(a_i) = \tau_i(a)$, and $\tau(a) = a$ for $a \in V_1 \cup V_2$. It is easy to see that N generates $w_1 \cdot w_2$.

To generate the composition, V_2 should contain the image of V_1 by τ_1, but still V_1' is disjoint from V_2'. The axiom is $Z_1(d_1)$. The mapping is $\tau(a_1) = \tau_1(a)_2$ on V_1', $\tau(a_2) = \tau_2(a)$ on V_2', and $\tau(a) = a$ on $V_1 \cup V_2$. The depth is $d = 1 + d_2$. □

The theorem allows a family $\mathcal{F}$ to have $\nu = O(1)$, by finding a finite collection of families generated by fixed non-erasing automorphisms, and then joining them using a finite number of set unions, concatenations and morphism compositions.

6   Future Work

We leave a number of open questions. We know $\nu = O(m) = O(b) = O(\delta \log(n/\delta))$, but it is unknown if $\nu = O(\gamma)$; if so, then γ would be reachable. We know $\ell = O(g)$, but it is unknown if $\ell = O(c)$; we suspect it is not, but in general we lack mechanisms to prove lower bounds on ℓ or ν. We also know $m = O(b)$, but not if it can be strictly better. We also do not know if these measures are monotone, and if they are actually NP-hard to compute (they likely so).

We could prove that $g = O(\ell \log n)$, and thus the lower bound $\ell = \Omega(g/\log n)$, if every L-system could be made expanding, but this is also unknown. This, for example, would prove that the stretch $\ell = O(\delta/\log n)$ we found for a family of strings is the maximum possible.

7   Conclusions

Extending the study of repetitiveness measures, from parsing-based to morphism-based mechanisms, opens a wide number of possibilities for the study of repetitiveness. There is already a lot of theory behind morphisms, waiting to be exploited on the quest for a golden measure of repetitiveness.

We first generalized composition systems to macro systems, showing that a restriction of them, called internal macro systems, are equivalent to bidirectional macro schemes, the lowest reachable measure of repetitiveness considered in the literature. It is not yet known if general macro systems are more powerful.

We then showed how morphisms, and measures based on mechanisms capturing that concept called L-systems (and variations), can be strictly better than δ for some string families, thereby questioning the validity of δ as a lower bound for reachable repetitiveness measures. L-systems are never larger than context-free grammars, but probably not always as small as composition systems.

Finally, we proposed a novel mechanism of compression aimed at unifying parsing and morphisms as repetitiveness sources, called NU-systems, which builds on macro systems and L-systems. NU-systems can combine copy-paste, recurrences, and concatenations and compositions of morphisms. The size ν of the smallest NU-system generating a string is a relevant measure of repetitiveness because it is reachable, computable, always in $O(b)$ and sometimes $o(\delta)$.

A simple lower bound capturing the idea of recurrence on a string, and lower bounding ℓ, just like δ captures the idea of copy-paste and strictly lower bounds b, would be of great interest when studying morphism-based measures. For infinite strings, there exist concepts like *recurrence constant* and *appearance constant* [1], but an adaptation, or another definition, is needed for finite strings. Besides, like Lindenmayer systems, NU-systems could be used to model other repetitive structures beyond strings that appear in biology, like the growth of plants and fractals. In this sense, they can be compared with tree grammars; the relation between NU-systems and TSLPs [13], for example, deserves further study.

References

1. Allouche, J.P., Shallit, J.: Automatic Sequences: Theory, Applications, Generalizations. Cambridge University Press, Cambridge (2003)
2. Bannai, H., Funakoshi, M.I.T., Koeppl, D., Mieno, T., Nishimoto, T.: A separation of γ and b via Thue-Morse words. CoRR 2104.09985 (2021)
3. Christiansen, A.R., Ettienne, M.B., Kociumaka, T., Navarro, G., Prezza, N.: Optimal-time dictionary-compressed indexes. ACM Trans. Alg. **17**(1), Art. 8 (2020)
4. de Luca, A.: Standard Sturmian morphisms. Theor. Comput. Sci. **178**(1), 205–224 (1997)
5. Gasieniec, L., Karpinski, M., Plandowski, W., Rytter, W.: Efficient algorithms for Lempel-Ziv encoding. In: Karlsson, R., Lingas, A. (eds.) SWAT 1996. LNCS, vol. 1097, pp. 392–403. Springer, Heidelberg (1996). https://doi.org/10.1007/3-540-61422-2_148
6. Kempa, D., Prezza, N.: At the roots of dictionary compression: string attractors. In: Proceedings of 50th STOC, pp. 827–840 (2018)
7. Kida, T., Matsumoto, T., Shibata, Y., Takeda, M., Shinohara, A., Arikawa, S.: Collage system: a unifying framework for compressed pattern matching. Theor. Comp. Sci. **298**(1), 253–272 (2003)
8. Kieffer, J.C., Yang, E.H.: Grammar-based codes: a new class of universal lossless source codes. IEEE Trans. Inf. Theory **46**(3), 737–754 (2000)
9. Kociumaka, T., Navarro, G., Prezza, N.: Towards a definitive measure of repetitiveness. In: Kohayakawa, Y., Miyazawa, F.K. (eds.) LATIN 2021. LNCS, vol. 12118, pp. 207–219. Springer, Cham (2020). https://doi.org/10.1007/978-3-030-61792-9_17
10. Lempel, A., Ziv, J.: On the complexity of finite sequences. IEEE Trans. Inf. Theory **22**(1), 75–81 (1976)
11. Lindenmayer, A.: Mathematical models for cellular interactions in development I. Filaments with one-sided inputs. J. Theor. Biol. **18**(3), 280–299 (1968)
12. Lindenmayer, A.: Mathematical models for cellular interactions in development II. Simple and branching filaments with two-sided inputs. J. Theor. Biol. **18**(3), 300–315 (1968)
13. Lohrey, M.: Grammar-based tree compression. In: Potapov, I. (ed.) DLT 2015. LNCS, vol. 9168, pp. 46–57. Springer, Cham (2015). https://doi.org/10.1007/978-3-319-21500-6_3
14. Lothaire, M.: Algebraic Combinatorics on Words. Cambridge University Press, Cambridge (2002)
15. Navarro, G.: Indexing highly repetitive string collections, part I: repetitiveness measures. ACM Comput. Surv. **54**(2), Article 29 (2021)

16. Navarro, G., Ochoa, C., Prezza, N.: On the approximation ratio of ordered parsings. IEEE Trans. Inf. Theory **67**(2), 1008–1026 (2021)
17. Raskhodnikova, S., Ron, D., Rubinfeld, R., Smith, A.: Sublinear algorithms for approximating string compressibility. Algorithmica **65**(3), 685–709 (2013). https://doi.org/10.1007/s00453-012-9618-6
18. Shallit, J.: String attractors for automatic sequences. CoRR 2012.06840 (2020)
19. Storer, J.A., Szymanski, T.G.: Data compression via textual substitution. J. ACM **29**(4), 928–951 (1982)

Information Retrieval

Improved Topic Modeling in Twitter Through Community Pooling

Federico Albanese[1,2]([✉]) [iD] and Esteban Feuerstein[1,3] [iD]

[1] Instituto en Ciencias de la Computación, CONICET - Universidad de
Buenos Aires, Buenos Aires, Argentina
[2] Instituto de Cálculo, CONICET- Universidad de Buenos Aires,
Buenos Aires, Argentina
falbanese@dc.uba.ar
[3] Departamento de Computación,
Universidad de Buenos Aires, Buenos Aires, Argentina

Abstract. Social networks play a fundamental role in propagation
of information and news. Characterizing the content of the messages
becomes vital for different tasks, like breaking news detection, person-
alized message recommendation, fake users detection, information flow
characterization and others. However, Twitter posts are short and often
less coherent than other text documents, which makes it challenging to
apply text mining algorithms to these datasets efficiently. Tweet-pooling
(aggregating tweets into longer documents) has been shown to improve
automatic topic decomposition, but the performance achieved in this
task varies depending on the pooling method.

In this paper, we propose a new pooling scheme for topic modelling in
Twitter, which groups tweets whose authors belong to the same *community* (group of users who mainly interact with each other but not with
other groups) on a user interaction graph. We present a complete evaluation of this methodology, state of the art schemes and previous pooling
models in terms of the cluster quality, document retrieval tasks performance and supervised machine learning classification score. Results show
that our Community polling method outperformed other methods on the
majority of metrics in two heterogeneous datasets, while also reducing
the running time. This is useful when dealing with big amounts of noisy
and short user-generated social media texts. Overall, our findings contribute to an improved methodology for identifying the latent topics in
a Twitter dataset, without the need of modifying the basic machinery of
a topic decomposition model.

Keywords: Topic modelling · Community detection · Twitter · Text
mining · Text clustering

1 Introduction

Characterizing texts based on their content is an important task in machine
learning and natural language processing. Latent Dirichlet Allocation (LDA)

© Springer Nature Switzerland AG 2021
T. Lecroq and H. Touzet (Eds.): SPIRE 2021, LNCS 12944, pp. 209–216, 2021.
https://doi.org/10.1007/978-3-030-86692-1_17

is a generative model for unsupervised topic decomposition [6]. Documents are represented as random mixtures over topics with a Dirichlet distribution, and each topic is characterized by a distribution over words. LDA has been widely used for topic modeling in different areas such us medical science [17], political science [8], social computer science [19] and software engineering [9].

In practice, content analysis on microblogging services can be particularly challenging due to short and often vaguely coherent text [13,14]. Given the fact that Twitter has become a platform where a tremendous amount of content is generated, shared and consumed, this problem become of interest for the scientific community. Hong presented an intuitive solution to this problem: tweet pooling (making longer document by aggregating multiple tweets) [12]. Tweet-pooling has been shown to improve topic decomposition, but the performance varies depending on the pooling method [4,12–14,16]. For example, Mehrotra et al. [14] extended this idea by pooling all tweets that mention a given hashtag. More pooling techniques are described in detail in Sect. 2.

In this paper, we propose a novel pooling techniques based on community detection on graphs. Previous works stated that LDA has problems with sparse word co-occurrence matrix [13] and showed that users in a community tweet mostly about two or three particular topics [3]. Based on these issues, we propose a community pooling method which groups tweets whose authors belong to the same community on the retweet network, increasing the length of each document and reducing the total number of documents. We compare the schemes in terms of clustering quality, document retrieval, machine learning classification tasks and running time and we empirically show that this new scheme improves the performance over previous methods in two heterogeneous Twitter datasets.

The remainder of this work is organized as follows: In Sect. 2 we describe the different pooling schemes for topic models and propose a novel method. In Sect. 3 we describe the datasets that we used to test our method. In Sect. 4, we define the experiments and evaluation metrics that we use to measure the performance of all pooling schemes. In Sect. 5, we show the results of the experiments. Finally, we interpret the results in the Conclusions section.

2   Tweet Pooling for Topic Models

Microblog messages are very short texts. In particular Twitter posts are only 280 characters or shorter. Consequently, using each tweet as an individual document does not present adequate term co-occurrence data within documents [14]. This induced the idea that aggregating similar tweets gives place to larger documents and better LDA topic decomposition. In this section, we present a new pooling method for topic modeling based on community detection and describe five other methods proposed in the literature which were used for comparison.

Tweet-Pooling (Unpooled): The default approach which treats each tweet as a single document.

Author-Pooling: All tweets authored by a single user are aggregated in a single document. The number of documents is equal to the number of users. This pooling method outperforms the Unpooled scheme [12].

Hashtag Pooling: In this scheme, a document consists of all tweets that mention a given hashtag. A tweet that contains multiple hashtags appears in several documents. Tweets without hashtags are considered as individual documents. It has been shown that aggregating tweets this way outperforms the baseline scheme and user-pooling in some metrics and for some datasets [14].

Conversation Pooling: A document consists of all tweets in a conversation tree (i.e. a tweet, all the tweets written in reply to it, the replies to the replies, and so on). This schemes aggregate tweets form different authors and with multiple hashtags that belong to one conversation [4].

Network-Based Pooling: Twitter users are grouped together if they reply or are mentioned in a tweet or in replies to a tweet. Each single document consists of all tweets of a group of users. In contrast to Conversation pooling, only direct replies to an original tweet are considered since a conversation can shift its topic in time. This pooling scheme showed better results than the previous methods in most (not all) tasks and datasets [16].

Community Pooling: In this novel scheme, a retweet graph is defined in terms of $G = (N, E)$, where users are the nodes N, and retweets between them are edges E [5]. Since a user can retweet multiple times other user's tweets, the edges are weighted. A community in a social network is a group of users who mainly interact with each other but not with other groups. We determine these communities using the Louvain method for community detection [7], which seeks to maximize modularity by using a greedy optimization algorithm. Therefore, each community clusters users by their interactions. In our novel pooling method, we group in one document all the tweets authored by all users in each community. Therefore, there are as many documents as communities in the retweet network. Compared with the majority of the previous schemes, the number of words in a document is bigger and the number of documents is smaller, resulting in a denser word co-occurrence matrix, which is beneficial to LDA algorithm [4].

3   Twitter Dataset Construction

In order to evaluate the schemes in different scenarios and show the robustness of the methodology, we used two diverse datasets. Our experiments used data from Twitter Streaming API[1]. Similarly to previous works, we constructed two diverse datasets collecting tweets containing different queries and each tweet was labeled by the query that retrieved it [2,12,14,16]. We removed all tweets that were retrieved by more than one query, so as to preserve uniqueness of the tweet labels, which was important for our analysis. Besides, we prepossessed the

---

[1] https://developer.twitter.com/en.

tweets by lower-casing and removing stop-words. All tweets are in English. The two datasets are:

Generic Dataset: 115,359 tweets from December 15th to December 16th, 2020, concerning a wide range of themes and collected using the following queries (percentage of tweets retrieved by each query): music (36.78%), family (23.94%), health (17.21%), business (14.90%), movies (4.70%), sports (2.44%).

Event Dataset: 328,452 tweets from January 20th, 2021. A dataset composed of tweets belonging to a particular event: US president Biden inauguration day. We used the following queries: Biden (69.45%), joebiden (21.75%), kamalaharris (4.74%), inauguration2021 (4.04%).

4 Evaluation

As there is no standard way for evaluating topic models, previous works evaluated the proposed pooling methods using different metrics or tasks. In order to present a complete and exhaustive analysis, in this work we evaluate the schemes by the multiple metrics used in the different previous works: topic clustering metrics (Purity and Normalized Mutual Information) [1,4,11,14,16,20], a supervised machine learning classification task [10,12], a document retrieval task [2,4] and overall running time [4]. We briefly explain each of them.

Purity: We define each cluster as a topic and assign the tweets to their corresponding mixture topic of highest probability (a quantity estimated with LDA). The purity of a cluster measures the fraction of tweets in a cluster having the assigned cluster query label [21]. Formally, let T_i be the set of tweets in LDA topic cluster i and Q_j be the set of tweets with query label j. Let $T = \{T_1, T_2, ..., T_{|T|}\}$ be the set of size $|T|$ of all T_i and let $Q = \{Q_1, Q_2, ..., Q_{|Q|}\}$ be the set of size $|Q|$ of all Q_j. Then, the purity is defined as follows:

$$\text{Purity}(T, Q) = \frac{1}{|T|} \sum_{i \in \{1...|T|\}} max_{j \in \{1...|Q|\}} |T_i \cap Q_j| \tag{1}$$

A higher purity score reflects a better cluster representation and a better LDA decomposition.

Normalized Mutual Information (NMI): NMI measures the cluster quality using information theory and it is formally defined as follows:

$$\text{NMI}(T, Q) = \frac{2I(T, Q)}{H(T) + H(Q)} \tag{2}$$

where $I(\cdot, \cdot)$ is the mutual information and $H(\cdot)$ is the entropy, as defined in [21]. NMI's minimum and maximum values are resp. 0 when labels and clusters are independent sets and 1 when cluster results exactly match all labels.

Supervised Machine Learning Classifying Task: For the supervised machine learning task, we follow a basic machine learning classifying evaluation scheme [12]. We separate the dataset in two (train and test), train a classifier with the first one and evaluate on the second one. The first 80% of tweets (according to the time their were posted) were assigned to the train set and the other 20% to the test set. For this task, we train a naive Bayes classifier [15] and reported F-Measure (F1 score) on the test set.

Document Retrieval Task: We also evaluate the topic decomposition of the different pooling methods on a document retrieval task, using the same train-test split as the supervised classifier task. We use each tweet in the test set as a query and return the most similar tweets from the train set, according to their LDA topic decomposition. If the retrieved tweet has the same query label, we consider it relevant. More concretely, the methodology is as follows: we apply LDA using the different pooling techniques on the train set, for each tweet in the test set calculate its topic decomposition, compute the cosine similarity between its topic decomposition and the topic decomposition of all tweets in the train set and retrieve the top 10 most similar train tweets. Then, we calculate the F1 score in order to know if the categories of the retrieved tweets match the category of the test tweet. This task recreates a scenario of recommending content based on previous tweets.

Running Time: The measured time (in seconds) includes tweet pooling (aggregating the tweets in different documents) and the LDA topic modeling, which varies depending on the total number of documents of each pooling methods.

All experiments were run using the same hardware on a GTX 1080 NVIDIA graphic card.

5 Results

In this section we show and discuss the results of our evaluation. For each pooling scheme, we replicated the training workflow used in the literature and used an LDA model with 10 topics [14,16]. As we mentioned earlier, previous works showed that having denser co-occurrence matrix (fewer documents with more words each) is beneficial to LDA [4]. Table 1 reports the corpus characteristics and shows how our proposed model drastically reduced the number of documents and increased the number of words per document.

The results of the experiments can be seen in Table 2. The best performances are marked in bold. The table shows that Community pooling has the best performance of all examined methods in all metrics for the Generic Dataset, and in all metrics except the retrieval task for the Event Dataset.

Our methodology obtained the best cluster quality, having the highest Purity and NMI scores. Also our experiments showed that Community pooling outperformed the previous schemes in the supervised classification task, indicating that this topic decomposition was a good descriptor of the query label. Regarding the document retrieval task, this evaluation considers small changes in the

214 F. Albanese and E. Feuerstein

Table 1. Document characteristics for different pooling schemes and datasets.

Scheme	# of docs		Max # of words/doc		Mean # of words/doc	
	Generic	Event	Generic	Event	Generic	Event
Unpooled	115,359	328,452	783	1,023	137	128
Author	36,526	87,883	36,029	11,240	369	273
Hashtag	34,624	59,388	820,689	3,736,132	8295	173952
Conversation	35,484	67,276	12,480	41,024	141	130
Network-based	36,882	88,314	59,195	90,391	385	277
Community	24,657	31,303	2,077,085	5,284,617	874	1379

Table 2. Results for different pooling schemes and datasets.

Scheme	Purity		NMI		Classification		Retrieval		Running time	
	Generic	Event	Generic	Event	Generic	Event	Generic	Event	Generic	Event
Unpooled	0.664	0.733	0.436	0.110	0.814	0.843	0.837	0.893	137	388
Author	0.696	0.736	0.374	0.149	0.798	0.859	0.839	0.900	429	926
Hashtag	0.724	0.719	0.383	0.066	0.779	0.762	0.839	0.869	1,737	17,758
Conversation	0.658	0.733	0.436	0.110	0.814	0.843	0.835	0.908	738	1,569
Network-based	0.695	0.736	0.372	0.149	0.798	0.859	0.840	**0.910**	1131	2,841
Community	**0.780**	**0.779**	**0.439**	**0.310**	**0.827**	**0.889**	**0.843**	0.868	141	340

topic decomposition of a tweet, since it uses the cosine similarity between this decomposition instead of only taking into account the most likely topic as we did before with the clustering metrics. The results indicate that Community pooling had the best performance in a generic dataset where the topics of the labels ("family", "health" or "business") differentiate from each other. In contrast, we found that the Network-based method has a better score in this task for the event dataset, where the labels are closely related ("joebiden" and "kamalaharris"). Community pooling has better performance on all tasks and datasets, with the only exception of the retrieval task on the event dataset.

Finally, Community pooling had the best time performance among all pooling methods. From the fact that LDA time complexity depends on the number of documents [18] and Community pooling considerably reduced the number of documents by pooling together in a single document all the tweets posted by users of each community (see Table 1), it follows that our proposed method was faster than all other aggregation techniques (less than half the running time).

6 Conclusions

We presented a new way of pooling tweets in order to improve the quality of LDA topic modeling on Twitter, without requiring any modification of the underlying LDA algorithm. The proposed Community pooling uses the users' interaction information and aggregates into a single document all tweets of the users that belong to a community in the retweet network.

Our method was evaluated and compared with multiple pooling techniques on different task including clustering quality, a supervised classification problem and a retrieval tasks. The results on two heterogeneous datasets indicate that the novel Community based pooling outperforms all other pooling strategies in all tasks and metrics, with the only exception of the retrieval task on the event dataset. Also, the running time analysis shows that Community pooling has a significant improvement in time performance in comparison with previous pooling methods, due to its capacity of reducing the total number of documents. Future work includes further testing with other datasets from different social media.

References

1. Akhtar, N., Beg, M.: User graph topic model. J. Intell. Fuzzy Syst. **36**(3), 2229–2240 (2019)
2. Al-Sultany, G.A., Aleqabie, H.J.: Events tagging in twitter using twitter latent Dirichlet allocation. Int. J. Eng. Technol. **8**(1.5), 503–508 (2019)
3. Albanese, F., Lombardi, L., Feuerstein, E., Balenzuela, P.: Predicting shifting individuals using text mining and graph machine learning on twitter. arXiv preprint arXiv:2008.10749 (2020)
4. Alvarez-Melis, D., Saveski, M.: Topic modeling in twitter: aggregating tweets by conversations. In: Proceedings of the International AAAI Conference on Web and Social Media, vol. 10 (2016)
5. Aruguete, N., Calvo, E.: Time to # protest: selective exposure, cascading activation, and framing in social media. J. Commun. **68**(3), 480–502 (2018)
6. Blei, D.M., Ng, A.Y., Jordan, M.I.: Latent Dirichlet allocation. J. Mach. Learn. Res. **3**, 993–1022 (2003)
7. Blondel, V.D., Guillaume, J.L., Lambiotte, R., Lefebvre, E.: Fast unfolding of communities in large networks. J. Stat. Mech.: Theory Exp. **2008**(10), P10008 (2008)
8. Cohen, R., Ruths, D.: Classifying political orientation on twitter: it's not easy! In: Proceedings of the International AAAI Conference on Web and Social Media, vol. 7 (2013)
9. Gethers, M., Poshyvanyk, D.: Using relational topic models to capture coupling among classes in object-oriented software systems. In: 2010 IEEE International Conference on Software Maintenance, pp. 1–10. IEEE (2010)
10. Giorgi, S., Preotiuc-Pietro, D., Buffone, A., Rieman, D., Ungar, L.H., Schwartz, H.A.: The remarkable benefit of user-level aggregation for lexical-based population-level predictions. arXiv preprint arXiv:1808.09600 (2018)
11. Hajjem, M., Latiri, C.: Combining IR and LDA topic modeling for filtering microblogs. Procedia Comput. Sci. **112**, 761–770 (2017)
12. Hong, L., Davison, B.D.: Empirical study of topic modeling in twitter. In: Proceedings of the First Workshop on Social Media Analytics, pp. 80–88 (2010)
13. Ma, T., Li, J., Liang, X., Tian, Y., Al-Dhelaan, A., Al-Dhelaan, M.: A time-series based aggregation scheme for topic detection in Weibo short texts. Phys. A Stat. Mech. Appl. **536**, 120972 (2019)
14. Mehrotra, R., Sanner, S., Buntine, W., Xie, L.: Improving LDA topic models for microblogs via tweet pooling and automatic labeling. In: Proceedings of the 36th International ACM SIGIR Conference on Research and Development in Information Retrieval, pp. 889–892 (2013)

15. Müller, A.C., Guido, S.: Introduction to Machine Learning with Python: A Guide for Data Scientists. O'Reilly Media, Inc., Sebastopol (2016)
16. Ollagnier, A., Williams, H.: Network-based pooling for topic modeling on microblog content. In: Brisaboa, N.R., Puglisi, S.J. (eds.) SPIRE 2019. LNCS, vol. 11811, pp. 80–87. Springer, Cham (2019). https://doi.org/10.1007/978-3-030-32686-9_6
17. Paul, M., Dredze, M.: You are what you tweet: analyzing twitter for public health. In: Proceedings of the International AAAI Conference on Web and Social Media, vol. 5 (2011)
18. Pedregosa, F., et al.: Scikit-learn: machine learning in Python. J. Mach. Learn. Res. **12**, 2825–2830 (2011)
19. Pinto, S., Albanese, F., Dorso, C.O., Balenzuela, P.: Quantifying time-dependent media agenda and public opinion by topic modeling. Phys. A **524**, 614–624 (2019)
20. Quezada, M., Poblete, B.: A lightweight representation of news events on social media. In: Proceedings of the 42nd International ACM SIGIR Conference on Research and Development in Information Retrieval, pp. 1049–1052 (2019)
21. Schütze, H., Manning, C.D., Raghavan, P.: Introduction to Information Retrieval, vol. 39. Cambridge University Press, Cambridge (2008)

TSXor: A Simple Time Series Compression Algorithm

Andrea Bruno[1], Franco Maria Nardini[2], Giulio Ermanno Pibiri[2(✉)],
Roberto Trani[2], and Rossano Venturini[1,2]

[1] University of Pisa, Pisa, Italy
[2] ISTI-CNR, Pisa, Italy
giulio.pibiri@di.unipi.it

Abstract. Time series are ubiquitous in computing as a key ingredient of many machine learning analytics, ranging from classification to forecasting. Typically, the training of such machine learning algorithms on time series requires to access the data in temporal order for several times. Therefore, a compression algorithm providing good compression ratios and fast decompression speed is desirable. In this paper, we present TSXor, a simple yet effective lossless compressor for time series. The main idea is to exploit the redundancy/similarity between close-in-time values through a window that acts as a cache, as to improve the compression ratio and decompression speed. We show that TSXor achieves up to 3× better compression and up to 2× faster decompression than the state of the art on real-world datasets.

1 Introduction

In this paper, we focus on compressing *time series* that have become the de-facto data format for monitoring systems sharing content through the Internet [1]. As a result, time series are heavily used in several machine learning applications. In fact, machine learning algorithms learn analytics on time series data by accessing the data in temporal order and for several times during training. Fast and lossless decompression of time series is important to reduce training time without compromising the accuracy of the process.

We present TSXor, a simple yet effective encoder/decoder for time series that achieves high compression ratios and fast decompression speed. TSXor leverages on the similarity between values in a window. This permits to reference recently seen values using few bytes and, at the same time, to achieve fast decompression by using the window of decompressed values as a data cache. We measure the performance of TSXor in comparison to two state-of-the-art compression algorithms (Gorilla [2] by Facebook and FPC [3]) on seven public, real-world, time series datasets. Results show that TSXor achieves a compression ratio of up to 3× better compared to its competitors while decompressing up to 2× faster.

2 Background

A *uni-variate* time series is a collection of key-value pairs $\langle t_n, v_n \rangle$ for a single time-dependent variable, where the key t_n denotes the time at which the n-th

© Springer Nature Switzerland AG 2021
T. Lecroq and H. Touzet (Eds.): SPIRE 2021, LNCS 12944, pp. 217–223, 2021.
https://doi.org/10.1007/978-3-030-86692-1_18

Table 1. Cost in bits of a value Δ using the range-based encoding by Gorilla.

	Range	Value bits	Total bits
$\Delta \in [-0, 0]$	0	0	1
$\Delta \in [-2^6 + 1, 2^6]$	10	7	9
$\Delta \in [-2^8 + 1, 2^8]$	110	9	12
$\Delta \in [-2^{11} + 1, 2^{11}]$	1110	12	16
$\Delta \in [-2^{31} + 1, 2^{31}]$	1111	32	36

observation was made and v_n is the corresponding measured value. A *multivariate* time series has m time-dependent variables, hence each point can be regarded as a tuple $\langle t_n, [v_{n,1}, \ldots, v_{n,m}] \rangle$. In our experiments, in Sect. 4, we consider both types of time series. Refer to the book by Hamilton [4] for an introduction to time series.

FPC [3] is a lossless compression algorithm for double-precision floating-point data. FPC compresses sequences of IEEE 754 double-precision floating-point values by sequentially predicting each value. It uses variants of an FCM [5] and a DFCM [6] value predictor to predict the doubles. Both predictors are implemented using hash tables. The more accurate of the two predictions, i.e., that sharing the largest number of most significant bits with the true value, is XOR-ed with the true value. The XOR operation turns identical bits into zeros. Hence, if the binary representation of the predicted and that of the true value are similar, the result has many leading zeros. FPC then counts the number of leading zero bytes, encodes the count in a 3-bit value, and uses an extra bit to specify which of the two predictions was used. The resulting 4-bit code and the nonzero residual bytes are written to the output.

Gorilla [2] is an in-memory time-series database developed at Facebook. It uses compression techniques based on delta-encoding timestamps and values. The n-th timestamp t_n is turned into a "delta of a delta" as $\Delta = (t_n - t_{n-1}) - (t_{n-1} - t_{n-2})$ and encoded using the simple range-based encoding illustrated in Table 1: if Δ belongs to the k-th range $[\ell, r]$, first k is coded in unary, followed by the binary representation of Δ using $\lceil \log_2(r - \ell + 1) \rceil$ bits. Since most measurements occur at regular and constant intervals, this results in a very small difference between consecutive timestamps (with often $\Delta = 0$), thus achieving good compression effectiveness. Instead, the n-th value v_n is XOR-ed with the previous v_{n-1} and the result of the XOR, say x_n, is encoded as follows (the first value v_0 is written explicitly in 64 bits). If $x_n = 0$, then output a 0 bit. If $x_n \neq 0$, then output a 1 bit and calculate the number of leading and trailing zeros: if these quantities are the same as those of the previous XOR value x_{n-1}, then just output the different bits; otherwise store the number of leading zeros (in 5 bits), the number of different bits (in 6 bits), followed by the different bits themselves.

3   TSXor

Inspired by the XOR-based approach adopted by both FPC and Gorilla, we now present a novel lossless compressor, TSXor. We aim at improving the compression

Table 2. Examples of pairs of values and their corresponding IEEE 754 double-precision representation.

Value	Double-precision representation
11.3	0100000000100110100110011001100110011001100110011001100110011010
11.5	0100000000100111000000000000000000000000000000000000000000000000
−6.6	1100000000011010011001100110011001100110011001100110011001100110
−3.8	1100000000011100110011001100110011001100110011001100110011001100110
15.9	0100000000101111110011001100110011001100110011001100110011001101
12.4	0100000000101000110011001100110011001100110011001100110011001101

ratios of FPC and Gorilla, while achieving very fast decoding speed. In this preliminary version of the work we focus on compressing the values v_n, that are more challenging to compress effectively compared to the timestamps t_n.

Good compression has to necessarily exploit the empirical property of time series data in that close-in-time measurements are very similar if not exactly the same. To understand how to best exploit this property, we first study how the IEEE 754 double-precision binary representation of two values varies in comparison to their decimal representation. We contribute the following insight: *floating-point values that are very close in decimal format do not necessarily have a similar binary representation.* Table 2 illustrates some concrete examples. The first two rows are relative to 11.3 and 11.5 that are very close in decimal format but only share 16 bits out of 64 (25%). Instead, although the difference between −6.6 and −3.8 is larger than $11.5 − 11.3 = 0.2$, the binary format of −6.6 and −3.8 share 61 bits out of 64 (more than 95%).

As a result of this observation, it is not always effective to compress v_n relative to v_{n-1} (as Gorilla does). Better compression can instead be achieved by enlarging the number of values that should be compared to v_n as to select the one with most common bits. To achieve this, we compare v_n with its preceding $W \leq 127$ values, logically corresponding to the values seen in the time range $[t_{n-W}, t_{n-1}]$. Our goal is to compress v_n relative to this "window" containing the previous W values. We distinguish between 3 cases, namely *Reference*, *XOR*, and *Exception*.

Reference. If v_n is equal to a value in the window, just output its position p in the window. Since the window contains at most 127 values, 1 byte suffices to write the position with the most significant bit always equal to 0.

If the window does not contain v_n, then we search for the value u in the window such that $x = v_n \oplus u$ has the largest number of leading and trailing zeros bytes. Let p be the position of u in the window. We first write $p + 128$ using 1 byte. In this case the most significant bit will always be 1 because of sum, which allows us to distinguish this case from the Reference case. Let LZ and TZ indicate the number of leading and trailing zero bytes of x respectively.

XOR. If $LZ + TZ \geq 2$, we output a byte where 4 bits are dedicated to TZ and the other 4 bits to the length (in bytes) of the segment of x between the leading and trailing zero bytes. We then write such middle bytes.

Table 3. Basic statistics of the datasets: number of time series, size of each time series, and percentage of distinct values.

Dataset	Time series	Size	Distinct values
AMPds2 [7]	14 629 292	11	5.01%
Bar-Crawl [8]	14 057 564	4	12.45%
Max-Planck [9]	473 353	32	0.54%
Kinect [10]	733 432	80	41.07%
Oxford-Man [11]	143 397	19	79.85%
PAMAP [12]	3 127 602	44	0.38%
UCI-Gas [8]	2 841 954	18	0.63%

Exception. Otherwise, we output an exception code, i.e., the value 255 using 1 byte, followed by the plain double-precision representation of v_n using 8 bytes.

The decoding algorithm just reverts the encoding procedure. In particular, during decoding, the last W decoded values are cached in a separate data structure that represents the sliding window. If the Reference case occurs frequently, as we are going to show for several real-world datasets, decoding v_n defaults to an inexpensive lookup in the window, which is small and likely to be kept in the processor cache. Moreover, the encoding of v_n requires just 1 byte which is not possible with neither FPC nor Gorilla. The byte-level alignment maintained by the algorithm further contributes to keep the decoding process simple and efficient.

4   Experiments

In this section, we present the results of an experimental evaluation that compares the performance of TSXor, FPC, and Gorilla on seven public time series datasets. All experiments are carried out on a server machine equipped with Intel i7-7700 cores (@3.60 GHz), 64 GB of RAM, and running Ubuntu 18.04. The implementation of TSXor is written in C++ and available at https://github.com/andybbruno/TSXor. The code was compiled with gcc 9.1.0 with the -O3 optimization flag.

We test all algorithms on datasets belonging to different scientific fields so as to not introduce any bias in the results. The datasets comprehend uni-variate as well as multi-variate time series. We do not apply any normalization nor further pre-processing to the datasets. Table 3 reports some basic statistics.

Compression Effectiveness. TSXor achieves a higher compression ratio than FPC and Gorilla, compressing from 1.0 to 3.5× better than Gorilla and from 1.2 to 5.8× better than FPC (which is always outperformed by Gorilla). Furthermore, TSXor achieves a 6.4× compression ratio on the AMPds2 dataset, while the best competitor achieve only a 2.0× compression ratio on this dataset. Here,

Table 4. Performance of TSXor, FPC, and Gorilla. The best performance on each dataset is highlighted in bold.

	Compr. ratio			Decompr. speed (MB/s)			Compr. speed (MB/s)		
	TSXor	FPC	Gorilla	TSXor	FPC	Gorilla	TSXor	FPC	Gorilla
AMPds2	**6.39**×	1.10×	2.03×	**1174**	411	666	67	339	**704**
Bar-Crawl	**2.36**×	1.20×	1.44×	**710**	436	447	29	424	**466**
Max-Planck	**4.84**×	1.06×	2.97×	**1057**	355	859	52	313	**871**
Kinect	1.37×	1.09×	**1.41**×	**665**	287	636	17	166	**696**
Oxford-Man	**1.30**×	1.06×	1.28×	**604**	222	574	15	170	**630**
PAMAP	**4.85**×	1.01×	1.38×	**949**	224	487	45	182	**521**
UCI-Gas	**3.50**×	1.19×	1.23×	**642**	455	578	22	287	**654**

the strength of TSXor is the use of a single byte in 85% of the cases (see Table 5) to reference an identical 8-byte value that occurred in the sliding window.

On the dataset containing the highest percentage of distinct values, i.e., Oxford-Man, TSXor is still able to beat the other two algorithms. Interestingly enough, on this dataset only 23% of the values has been compressed using 9 bytes (see Table 5), thus spending an extra byte with respect to the 8 bytes needed by the uncompressed representation. In this case, our advantage comes from the 17% of the values that are compressed with only one byte (Reference case), while the remaining 59% of the values are compressed using 6.94 bytes on average.

Decompression Speed. Since the time series are compressed once but read several times, the most critical evaluation metric is decompression speed. There-fore, we start by analyzing the decompression speeds (reported in MB/s) of the different algorithms. Gorilla is from 1.0 to 2.6× faster than FPC. TSXor is the fastest algorithm, consistently on all datasets. In particular, TSXor is from 1.0 to 1.9× faster than Gorilla and from 1.4 to 4.2× faster than FPC. The byte granularity helps the algorithm to avoid bit shifts and costly functions calls. In particular, 92% of the times (see Table 5) we end up either in the Reference case or in the XOR case, by consuming only 2.62 bytes on average instead of the 8 bytes of the original representation. This means that TSXor heavily leverages on the window of cached values.

Compression Speed. Regarding the compression speeds (in MB/s), which is the less interesting case, Gorilla outperforms both FPC and TSXor. The reason lies in the simplicity of the algorithm. Indeed Facebook's approach requires nei-ther table lookups nor complicated calculations. The second fastest algorithm is FPC, which compresses the values exploiting two hash functions as predictors. TSXor trades compression speed for better compression and faster decoding speed. In fact, for each value to encode, the whole window is scanned.

Varying the Window Size. We now examine the performance achieved by TSXor when varying the window size W. We show this analysis in Fig. 1, which

Table 5. Percentage of TSXor cases (Reference, XOR, and Exception) over each dataset. For the XOR case, it is evident that TSXor spends less than 8 bytes for a double-precision value.

	Reference (1 byte)	XOR		Exception (9 bytes)
	%	%	bytes	%
AMPds2	84.87	14.87	3.19	0.26
Bar-Crawl	50.53	28.25	5.53	21.22
Max-Planck	77.93	21.94	4.15	0.13
Kinect	28.01	62.95	7.66	9.04
Oxford-Man	17.44	59.44	6.94	23.12
PAMAP	75.95	23.13	3.63	0.92
UCI-Gas	45.36	54.63	3.57	0.01
Average	54.30	37.89	4.95	7.81

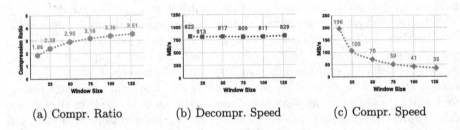

(a) Compr. Ratio   (b) Decompr. Speed   (c) Compr. Speed

Fig. 1. Compression ratio, decompression speed, and compression speed of TSXor by varying the size of the window. Each point represents the average of each metric over all datasets.

reports the average performance over all datasets. We did not observe noteworthy variations among the different datasets. Figure 1a shows that the compression ratio improves when increasing the window size. Not surprisingly, the larger the window, more compression opportunities are created for Reference and XOR cases. This improvement is balanced by the fact that the compression algorithm needs to look at more values when compressing. Indeed, Fig. 1c shows that the compression speed slows down when increasing the window size.

An interesting finding is that the window does *not* affect the decompression speed (Fig. 1b). The main reason is that during the decoding phase no searches are performed in the window, but only direct access to individual elements.

5   Conclusion and Future Work

In this short communication we introduced a lossless compression scheme for time series, TSXor, achieving good compression ratios and very fast sequential decoding. Despite its simplicity, TSXor provides very promising results: therefore, we think that room for improvement is possible with more sophisticated mechanisms. One defect of TSXor is certainly its encoding time, as it requires to scan the window for each value to encode. Future work will tackle this issue, e.g., by exploiting vectorized instructions. We will also explore the concrete applicability of TSXor to machine learning applications.

Acknowledgments. This work was partially supported by the projects: MobiData-Lab (EU H2020 RIA, grant agreement №101006879), OK-INSAID (MIUR-PON 2018, grant agreement №ARS01_00917), and "Algorithms, Data Structures and Combinatorics for Machine Learning" (MIUR-PRIN 2017).

References

1. Vashi, S., Ram, J., Modi, J., Verma, S., Prakash, C.: Internet of Things (IoT): a vision, architectural elements, and security issues. pp. 492–496, February 2017. https://doi.org/10.1109/I-SMAC.2017.8058399
2. Pelkonen, T., et al.: Gorilla: a fast, scalable, in-memory time series database. Proc. VLDB Endow. **8**(12), 1816–1827 (2015). https://doi.org/10.14778/2824032. 2824078. ISSN 2150-8097
3. Burtscher, M., Ratanaworabhan, P.: FPC: a high-speed compressor for double-precision floating-point data. IEEE Trans. Comput. **58**(1), 18–31 (2009). https://doi.org/10.1109/TC.2008.131. ISSN 0018-9340
4. Hamilton, J.D.: Time Series Analysis. Princeton University Press, Princeton (2020)
5. Sazeides, Y., Smith, J.E.: The predictability of data values. In: Proceedings of 30th Annual International Symposium on Microarchitecture, pp. 248–258, December 1997. https://doi.org/10.1109/MICRO.1997.645815. ISSN 1072-4451
6. Goeman, B., Vandierendonck, H., de Bosschere, K.: Differential FCM: increasing value prediction accuracy by improving table usage efficiency. In Proceedings HPCA Seventh International Symposium on High-Performance Computer Architecture, pp. 207–216, January 2001. https://doi.org/10.1109/HPCA.2001.903264
7. Makonin, S., Ellert, B., Bajić, I.V., Popowich, F.: Electricity, water, and natural gas consumption of a residential house in Canada from 2012 to 2014. Sci. Data **3**(1), 1–12 (2016)
8. Fonollosa, J., Sheik, S., Huerta, R., Marco, S.: Reservoir computing compensates slow response of chemosensor arrays exposed to fast varying gas concentrations in continuous monitoring. Sens. Actuators B Chem. **215**, 618–629 (2015)
9. Max Planck Institute for Biogeochemistry. Max-Planck-Institut fuer Biogeochemie - Wetterdaten (2019). https://www.bgc-jena.mpg.de/wetter/
10. Fothergill, S., Mentis, H., Kohli, P., Nowozin, S.: Instructing people for training gestural interactive systems. In: Proceedings of the SIGCHI Conference on Human Factors in Computing Systems, CHI 2012, New York, NY, USA, pp. 1737–1746 (2012). Association for Computing Machinery. https://doi.org/10.1145/2207676. 2208303. ISBN 9781450310154
11. Heber, G., Lunde, A., Shephard, N., Sheppard, K.: Oxford-Man Institute's Realized Library (2009)
12. Reiss, A., Stricker, D.: Introducing a new benchmarked dataset for activity monitoring. In: 2012 16th International Symposium on Wearable Computers, pp. 108–109. IEEE (2012)

Pattern Matching

Exploiting Pseudo-locality of Interchange Distance

Avivit Levy[(✉)]

Department of Software Engineering, Shenkar College, 52526 Ramat-Gan, Israel

Abstract. String metrics play an important role in various computational tasks. In this paper, we focus on a property shared by a subset of string rearrangement metrics, called *pseudo-locality*. This subset includes, for example, the swap, interchange and parallel-interchange distances. Intuitively, operators of *pseudo-local string metrics* have a bounded effect on the Hamming distance between the string prior to the operation and after it. The goal of this paper is to examine how to exploit pseudo-locality in order to transform tools derived for the Hamming distance in order to derive tools for other pseudo-local metrics. Specifically, we demonstrate such a way to exploit pseudo-locality of the interchange distance combined with additional techniques to derive:

1. The *first* efficient approximate nearest-neighbour (ANN) search data structure for the interchange distance, which is $\mathcal{NP}$-hard to compute for general strings even for binary alphabets.
2. The *first* linear time algorithm to compute approximate pattern matching with interchanges, which is a vast improvement from the $\Theta(nm)$ known algorithm.

In addition, we provide a highly accurate online space efficient construction of a histogram of a given pattern alphabet symbols in a text string as an infinite distance exclusion tool, which may be of independent interest.

1 Introduction

String metrics play an important role in various computational tasks, such as similarity search and analysis [9,11,15]), text editing [2,6,30], pattern matching [1,29,31] or comparative genomics [10,12,16,20].

Similarity search includes a range of mechanisms with the principle of searching (typically, very large) spaces of objects through computation of the similarity between any pair of objects. Various metrics are used for such similarity computations, where different measures may be preferred according to the applications. This is becoming increasingly important in large information repositories where the objects contained do not possess any natural order, for example large collections of images, sounds and other sophisticated digital objects. Nearest neighbor search (NN) is an important widely used subclass of similarity search [9,15].

String metrics are also used in text editing and pattern matching as a tool for approximate matching and error recovery. Edit distances find applications in natural language processing, where automatic spelling correction can determine

© Springer Nature Switzerland AG 2021
T. Lecroq and H. Touzet (Eds.): SPIRE 2021, LNCS 12944, pp. 227–240, 2021.
https://doi.org/10.1007/978-3-030-86692-1_19

candidate corrections for a misspelled word by selecting words from a dictionary that have a low distance to the word in question. Different models of errors express diverse possible applications needs as well as suggest different recovery algorithms (see e.g., [1,5,17,25]).

String rearrangement distances have gained interest in computational biology and were widely studied in order to establish phylogenetic information about the evolution of species. During the course of evolution, whole regions of genome may evolve through reversals, transpositions and translocations. Considering the genome as a string over the alphabet of genes, these cases represent a situation where the difference between the original string and the resulting one is in the locations of the different elements, where changes are caused by rearrangements operators. Many works have considered specific versions of this biological setting (*sorting by reversals* [12,20], *transpositions* [10], and *translocations* [19]).

In this paper, we focus on a property shared by a subset of string rearrangement metrics called *pseudo-locality* [3] (see formal definition in Sect. 2). This subset includes, for example, the swap, interchange and parallel-interchange distances (see formal definitions in Sect. 2). Intuitively, operators of *pseudo-local string metrics* have a bounded effect on the Hamming distance between the string prior to the operation and after it. The pseudo-locality was defined and observed in [3] to aid solving period recovery under this broaden set of string metrics using tools developed for the Hamming distance. Moreover, their techniques developed for pseudo-local metrics were applied even to the non-pseudo-local edit distance.

Our goal is to examine how pseudo-locality can be exploited beyond the context of the period recovery problem, in order to transform tools derived for the Hamming distance to other pseudo-local metrics. Specifically, we demonstrate such a way to exploit pseudo-locality of the interchange distance.

The Interchange Distance. The *interchange rearrangement problem* is the following: Given two strings x, y over alphabet Σ such that x, y have the same quantity of each symbol, the goal is to transform x (called the input string) to y (called the target string) using a succession of interchange operations. An *interchange* of two elements, a in position i and b in position j, puts element a in position j and element b in position i. The *interchange distance problem* is to find the minimum number of interchanges needed to transform the input string x to the target string y.

Note that, an interchange is the basic operation used in all comparison-based sorting algorithms. Indeed, the interchange distance problem is actually a classical problem mentioned back in 1849 by Cayley [13]. Cayley mainly studied permutation strings, in which all elements are distinct. In this case, strings can be viewed as permutations of $1, \ldots, m$, where m is the length of the string. This classical setting was well studied (e.g., [13,24]). Cayley [13] gives a characteristic theorem for the distance, from which a simple linear time algorithm for computing it on permutation strings can be immediately derived, as described in [1]. However, these results do not apply for the general strings case, posed as an open problem by Cayley. A generalization of the problem on permutations considering general input strings is studied by [4], thus solving the open problem of Cayley, while also examining various cost models.

The study of interchange distance pseudo-locality here focuses in the following key problems within the scope of similarity search and pattern matching.

The Scope Problems. We examine the following problems:

- **Approximate NN-Search (ANN).** Let $\mathcal{V}$ be some vector space of dimension d, and let *dist* be some distance function for $\mathcal{V}$. Given a database consisting of n vectors in $\mathcal{V}$, a slackness parameter $\epsilon > 0$, and a query vector q, a $C(\epsilon)$-approximate nearest neighbor of q is a database vector a such that for any other database vector b, $dist(q, a) \leq C(\epsilon) \cdot dist(q, b)$, where $C(\epsilon) > 1$ is some constant that depends on ϵ. It is assumed that d is considerably large, and that $n >> d$, i.e., n is much larger than d.[1]
- **Approximate Pattern Matching (PM).** Let *dist* be some distance function for strings. Given a pattern P of length m, a slackness parameter $\epsilon > 0$ and a text T of length $n > m$, the task is to output a $C(\epsilon)$-approximation of the distance *dist* between P and the m-length substring of the text T for each position i, $1 \leq i \leq n - m + 1$ (each position is called an *alignment*).

This Paper Contributions. The main contributions of this paper are:

- Demonstrating how the pseudo-locality condition can be used beyond the original period recovery problem, for which this property was first identified [3], to use tools derived for the Hamming distance in order to derive solutions for *pseudo-local metrics*.
- Obtaining the *first* efficient ANN-search data structure for the interchange distance, which is $\mathcal{NP}$-hard to compute for general strings even for binary alphabets [4].
- Providing the *first* linear algorithm to compute approximate PM with interchanges, which outperforms the complexity of the prior best algorithm while still providing only a slightly larger approximation ratio.
  While the former algorithm, which is implicit from [1] and [4], has complexity $\Theta(nm)$ with approximation ratio 1.5, the randomized algorithm given here is linear and provides approximation ratio of $2 + \epsilon$. For fixed-size alphabets, the algorithm is deterministic and its approximation ratio is 2.
- Providing a highly accurate online space efficient construction of the histogram of the pattern alphabet symbols appearing in every alignment of the text, which we call the *SEP-Vector*.

Paper Organization. The paper is organized as follows. Section 2 gives the preliminary definitions and lemmas regarding string metrics and pseudo-local metrics. Section 3 studies the approximate NN-search problem in high dimensions for pseudo-local metrics and applies it to derive an approximate NN-search

---

[1] The original definition of the problem required a $(1 + \epsilon)$-approximation, which is relaxed to require only a constant approximation depending on the slackness parameter.

for the interchange distance. Section 4 studies the approximate PM problem for pseudo-local metrics and applies it to derive an algorithm for approximate PM with interchanges, and also describes a highly accurate online space efficient construction of a histogram. Section 5 concludes with open questions.

2 Preliminaries

Consider a set Σ and let x and y be two n-long strings over Σ. [5] formally define the process of converting x to y through a sequence of operations. An *operator* ψ is a function $\psi : \Sigma^n \rightarrow \Sigma^{n'}$, with the intuitive meaning being that ψ converts n-long string x to n'-long string y with a cost associated to ψ. That cost is the distance between x and y. Formally,

Definition 1. [String Metric] [5]
Let $s = (\psi_1, \psi_2, \ldots, \psi_k)$ be a sequence of operators, and let $\psi_s = \psi_1 \circ \psi_2 \circ \cdots \circ \psi_k$ be the composition of the ψ_j's. We say that s converts x into y if $y = \psi_s(x)$.

Let Ψ be a set of rearrangement operators, we say that Ψ can convert x to y, if there exists a sequence s of operators from Ψ that converts x to y. Given a set Ψ of operators, we associate a non-negative cost with each sequence from Ψ, $cost : \Psi^* \rightarrow R^+$. The pair $(\Psi, cost)$ is called an edit system. Given two strings $x, y \in \Sigma^*$ and an edit system $\mathcal{R} = (\Psi, cost)$, the distance from x to y under $\mathcal{R}$ is defined to be:

$$d_{\mathcal{R}}(x,y) = \min\{cost(s)|s \text{ from } \mathcal{R} \text{ converts } x \text{ to } y\}$$

If there is no sequence that converts x to y then the distance is ∞.

It is easy to verify that $d_{\mathcal{R}}(x, y)$ is a metric, if there is also the inverse operation with equal cost for each operation in Ψ, thus, we have that $d_R(x, y) = d_R(y, x)$, and the cost function definition preserves the triangle inequality. Definition 2 gives examples of string metrics.

Definition 2. *1.* Hamming distance: $\Psi = \{\rho_{i,\sigma}^n | i, n \in \mathbb{N}, \ i \le n, \ \sigma \in \Sigma\}$, *where $\rho_{i,\sigma}^n(\alpha)$ substitutes the ith element of n-tuple α by symbol σ. The Hamming distance is denoted by H.*

2. Edit distance: *In addition to the substitution operators of the Hamming distance, Ψ also has insertion and deletion operators. The insertion operators are: $\{\iota_{i,\sigma}^n | i, n \in \mathbb{N}, \ i \le n, \ \sigma \in \Sigma\}$, where $\iota_{i,\sigma}^n(\alpha)$ adds the symbol σ following the ith element of n-tuple α, creating an $n+1$-tuple α'. The deletion operators are $\{\delta_i^n | i, n \in \mathbb{N}, \ i \le n\}$, where $\delta_i^n(\alpha)$ deletes the symbol at location i of n-tuple α, creating an $n-1$-tuple α'.*

3. Swap distance: $\Psi = \{\zeta_i^n | i, n \in \mathbb{N}, \ i < n\}$, *where $\zeta_i^n(\alpha)$ swaps the ith and $(i+1)$st elements of n-tuple α, creating an n-tuple α'. A valid sequence of operators in the Swap metric has the additional condition that if ζ_i^n and ζ_j^n are operators in a sequence then $i \ne j$, $i \ne j+1$, $i \ne j-1$.*

4. Interchange distance: $\Psi = \{\pi_{i,j}^n | i, n \in \mathbb{N}, \ i \le j \le n\}$, *where $\pi_{i,j}^n(\alpha)$ interchanges the ith and jth elements of n-tuple α, creating an n-tuple α'.*

5. Parallel-Interchange distance: $\Psi = \{\pi_{i,j}^n | i, n \in \mathbb{N}, \quad i \leq j \leq n\}$, where $\pi_{i,j}^n(\alpha)$ interchanges the ith and jth elements of n-tuple α, creating an n-tuple α'. A valid sequence of operators in the Parallel Interchange metric has the additional condition that if $\pi_{i,j}^n$ and $\pi_{i',j'}^n$ are operators in a sequence then $i \neq i'$, $j \neq j'$, $i \neq j'$ and $i' \neq j$.

Cost Models. Several known cost models are used for string metrics [5,25]. In the Unit-Cost Model (UCM) each operation is given a unit cost, so the problem is to transform a given sting x into a string y with a minimum number of operations. In the Length-Cost Model (LCM), the cost of an operation depends on its length characteristic. Other characteristics may be considered in the rearrangement problem. For example, some elements may be heavier than other elements. In such cases, moving light elements is preferable to moving heavy elements. This motivated researchers to explore the Element-Cost Model (ECM).

Definition 3. [Pseudo-local Metric] [3]
Let dist be a string metric under unit-cost model. dist is called a pseudo-local metric if there exists a constant $c \geq 1$ such that, for every two strings S_1, S_2, if $dist(S_1, S_2) = k$ then

$$k \leq H(S_1, S_2) \leq c \cdot k.$$

A metric that is pseudo-local with constant c is called a c-pseudo-local metric.

Note that pseudo-locality allows the resulted number of mismatches to be unboundedly far from each other (as may happen in an interchange) and therefore, a pseudo-local metric is not necessarily also local in the intuitive sense. Lemma 1 shows some interesting pseudo-local metrics and follows immediately from Definitions 2 and 3.

Lemma 1 [5]. *The following metrics are c-pseudo-local metrics:*

1. *Hamming distance (with $c = 1$).*
2. *Swap distance (with $c = 2$).*
3. *Interchange distance (with $c = 2$).*
4. *Parallel-Interchange distance (with $c = 2$).*

On the other hand, the Edit distance is not a pseudo-local metric, because a single deletion or insertion may cause an unbounded number of mismatches.

It will also be useful to make the following distinction between pseudo-local and *strong pseudo-local* metrics, defined as follows.

Definition 4. [Strong Pseudo-local Metric]
Let dist be a string metric under unit-cost model. dist is called a strong pseudo-local metric if there exists a constant $c \geq 1$ such that, for every two strings S_1, S_2, if $dist(S_1, S_2) = k$ then

$$H(S_1, S_2) = c \cdot k.$$

A metric that is strong pseudo-local with constant c is called a c-strong pseudo-local metric.

Lemma 2 immediately follows from Definitions 2 and 4.

Lemma 2. *The following metrics are c-strong pseudo-local metrics:*

1. *Hamming distance (with $c = 1$).*
2. *Swap distance (with $c = 2$).*
3. *Parallel-Interchange distance (with $c = 2$).*

Note, that the interchange distance is not a strong pseudo-local metric, since a sequence of interchanges may be performed on the same positions, thus the average effect of each interchange on the number of mismatches may be strictly less than 2. Consider, for example, the strings abc and bca, where the interchange distance is 2 and $H(abc, bca) = 3 < 2 \cdot 2 = 4$.

3   Approximate NN-Search in High Dimensional Spaces

In this section we show how pseudo-locality can be exploited to design a data structure enabling an efficient approximate nearest neighbor search. In fact, we demonstrate how the pseudo-locality condition enables to generalize the data structure of [28] for the Hamming distance to be used for the interchange distance. We follow [28] assuming, for simplicity of exposition, that the database set of points, DB, and query point q are in $\{0, 1\}^d$. As in [28], the results generalize to vector spaces over any finite field; thus we can handle a database of documents (strings) over any finite alphabet. When needed for generalizing the discussion, the distance is measured by a pseudo-local metric *dist*.

We begin by briefly describing the KOR data structure and search algorithm.

The KOR Test. The idea behind the KOR algorithm is to design a separate test for each distance ℓ. Given a query q, such a test either returns a database vector at distance at most $(1 + \epsilon)\ell$ from q, or informs that there is no database vector at distance ℓ or less from q. [28] define the β-test τ as follows. Pick a subset C of coordinates of the d-dimensional vectors by choosing each element in $\{1, 2, \ldots, d\}$ independently at random with probability β. For each of the chosen coordinates i, pick independently and uniformly at random $r_i \in \{0, 1\}$. For vector v in the database, define the value of τ at v, denoted $\tau(v)$ as follows:

$$\tau(v) = \sum_{i \in C} r_i \cdot v_i \pmod{2}.$$

Let q be a query, and let a, b be two database points with $H(q, a) \leq \ell$ and $H(q, b) > (1 + \epsilon)\ell$. For $\beta = \frac{1}{2\ell}$ this test distinguishes between a and b with constant probability [28].

The KOR Data Structure. Let $\epsilon > 0$, $\mu > 0$ be constants. [28] data structure $\mathcal{S}$ consists of d substructures $\mathcal{S}_1, \mathcal{S}_2, \ldots, \mathcal{S}_d$ (one for each possible distance except zero-distance, which can be easily verified separately). Fix $\ell \in \{1, 2, \ldots, d\}$, $\mathcal{S}_\ell$ consists of M structures $\mathcal{T}_1, \ldots, \mathcal{T}_M$, where $M = (d + \log d + \log \epsilon^{-1}) \log \frac{d}{\mu}$. Fix

$i \in \{1, 2, \ldots, M\}$, the structure $\mathcal{T}_i$ consists of a list of T $\frac{1}{2\ell}$-tests, where $T = O(\epsilon^{-2} \log \log \frac{d}{\mu})$, and a table of 2^T entries (one entry for each possible outcome of the sequence of T tests). Each entry of the table either contains a database point or is empty. For the structure $\mathcal{T}_i$, pick at random T $\frac{1}{2\ell}$-tests $t_1, \ldots, t_T$. For a database vector v, its *trace* is the vector $t(v) = t_1(v) \ldots t_T(v) \in \{0, 1\}^T$.

The KOR Search Algorithm. Given a query q, a binary search is performed in order to determine (approximately) the minimum distance ℓ to a database point. A step in the binary search consists of picking one of the structures $\mathcal{T}_i$ in $\mathcal{S}_\ell$ uniformly at random, computing the trace $t(q)$ of the list of tests in $\mathcal{T}_i$, and checking the table entry labeled $t(q)$. The binary search step succeeds if this entry contains a database point, and otherwise it fails. If the step fails, the search proceeds to larger ℓ values, otherwise, the search proceeds to smaller ℓ values. The search algorithm returns the database point contained in the last nonempty entry visited during the binary search.

Lemma 3, Lemma 4 and Theorem 1 summarize the properties of the KOR data structure that we need.

Lemma 3. [Lemma 2.6 in [28]]
For any query q, the probability that the binary search uses a structure $\mathcal{T}_i$ that fails at q is at most μ.

Lemma 4. [Lemma 2.7 in [28]]
If all the structures used by the binary search do not fail at q, then the distance from q to the database point a returned by the search algorithm is within a $(1+\epsilon)$-factor of the minimum distance from q to any database point.

Theorem 1. [KOR $(1 + \epsilon, \mu)$-ANN Data Structure [28]]
Let $\epsilon > 0$, $\mu > 0$ be constants. Given n d-dimensional binary database vectors, there exists a data structure S using $s = O(\epsilon^{-2} d^3 \log d(\log n + \log \log d) + d^2 \log d(n \log d)^{O(\epsilon^{-2})})$ space and constructed in $O(s \cdot n)$ time, which given a d-dimensional binary vector q finds $(1 + \epsilon)$-approximate nearest neighbor to q in the database with probability $1 - \mu$ in time $O(\epsilon^{-2} d(\log n + \log \log d + \log \frac{1}{\mu}) \log d)$.

ANN-Data Structure for Pseudo-local Metrics. Let *dist* be any c-pseudo-local metric and let $\epsilon > 0$, $\mu > 0$ be constants. The basic idea behind a construction of a $(c + \epsilon, \mu)$-ANN data structure for *dist* is to simply use a KOR $(1 + \epsilon', \mu)$-ANN data structure, where $\epsilon' = \frac{\epsilon}{c}$. We may try using the same query algorithm performed on the KOR $(1 + \epsilon', \mu)$-ANN data structure. Lemma 5 describes the guarantee we have on the database point returned to a query q, where the distance is measured by *dist*.

Lemma 5. [Pseudo-local Query Condition]
Let q be a query such that $\forall v \in DB$, $dist(q, v) < \infty$. Assume that all the structures used by the binary search do not fail at q, and let a be the point returned by the search algorithm. Let $\Delta = \min_{v \in DB} dist(q, v)$. Then,

$$dist(q, a) \le (c + \epsilon)\Delta.$$

Proof. Denote the minimum Hamming distance from q to any database point by Δ_H. Since $\forall v \in DB$, $dist(q, v) < \infty$, by Definition 3, we have that:

$$\Delta \leq \Delta_H \leq c \cdot \Delta.$$

Now, if $\ell < \Delta/(1 + \epsilon)$, then $\ell < \Delta/(1 + \epsilon') \leq \Delta_H/(1 + \epsilon')$. Then, no database point is within distance $(1 + \epsilon')\ell$ of q, and therefore, all the binary search steps that visit ℓ in this range fail. On the other hand, all binary search steps that visit ℓ in the range $\ell \geq \Delta_H$ succeed. Therefore, the binary search ends with a value $\Delta_H/(1+\epsilon') \leq \ell \leq \Delta_H$. By Lemma 4, for the point a returned by the KOR $(1 + \epsilon', \mu)$-approximate nearest neighbor data structure we have that:

$$H(q, a) \leq (1 + \epsilon')\ell \leq (1 + \epsilon')\Delta_H.$$

Therefore, by Definition 3 we have:

$$dist(q, a) \leq c(1 + \epsilon')\Delta = (c + \epsilon)\Delta.$$

$\square$

The condition that $\forall v \in DB$, $dist(q, v) < \infty$ is necessary for the guarantee of Lemma 5. To see this, consider the following example. Let $DB = \{a, b, c\}$, where $a = (0, 1, 0, 1, 0, 0)$, $b = (0, 0, 0, 1, 0, 1)$, $c = (0, 0, 0, 0, 1, 0)$, and let $q = (0, 0, 1, 0, 1, 0)$. We have that: $\Delta_{swap} = d_{swap}(q, a) = d_{swap}(q, b) = 2$ and $H(q, a) = H(q, b) = 4$, however, $\Delta_H = H(q, c) = 1$ and $d_{swap}(q, c) = \infty$. Thus, we get: $\Delta_H < \Delta_{swap}$.

This means that in the presence of vectors with infinite distance to a query, we have no guarantee on a point returned from a search based on the Hamming distance. In order to exploit Lemma 5, we must, therefore, monitor the vectors having infinite distance to query vectors. This monitoring should be done while constructing the database and is required, therefore, to be independent of the query. Fortunately, such a monitoring is possible for the interchange distance. To this end we need the following known definition and lemma.

Definition 5. [Parikh Vector]
Let $\Sigma = \{a_1, a_2, \dots, a_k\}$ be an alphabet. The Parikh vector of a word (string or vector) is the function $p : \Sigma^* \to \mathbb{N}^k$, given by: $p(w) = (|w|_{a_1}, |w|_{a_2}, \dots, |w|_{a_k})$, where $|w|_{a_i}$ denotes the number of occurrences of the letter a_i in the word w.

A Parikh vector of a binary vector should only specify the number of 1 bits.

Lemma 6. [Infinity Check for the Interchange Distance]
Let d_{int} be the interchange distance and let a, b be two vectors, then $d_{int}(a, b) < \infty$ if and only if $p(a) = p(b)$, where p is the Parikh vector function.

Monitoring Infinite Interchange Distances. Let $\epsilon > 0$, $\mu > 0$ be constants. The ANN-data structure for the interchange distance S consists of d^2 substructures $S_{1,1}, S_{1,2}, \dots, S_{d,d}$ (one for each possible distance and possible number of 1 bits). Note, that since there is only one vector with no 1 bits, it can be handled separately. If the query is the zeroes-vector, then the distance to the query is either 0, if the zeroes-vector is in the database, otherwise, the distance is ∞.

Fix a number of bits $\ell' \in \{1, 2, \ldots, d\}$ and a distance $\ell \in \{1, 2, \ldots, d\}$, $\mathcal{S}_{\ell,\ell'}$ consists of the same substructures as in [28], but with the additional requirement for a table entry to contain a point a of the database only if $p(a) = \ell'$. We also keep for each set of structures $\mathcal{S}_{\cdot,\ell'}$, a counter $n_{\ell'}$ of the total number of points with number of 1 bits equal to ℓ' that are stored in these structures.

The search algorithm for a query q is also refined to compute $p(q)$ and perform the binary search only on the structures $\mathcal{S}_{1,p(q)}, \ldots, \mathcal{S}_{d,p(q)}$. In addition, before the search we verify that the counter of points in the structures with number of 1 bits that equals $p(q)$ is not zero. If it is zero, then we do not perform the search and return that the distance to the query is ∞. By Lemma 6, this ensures that the search is done only over the database points that have finite distance from the query q, and thus Lemma 5 holds. Theorem 2 follows.

Theorem 2. [$(2 + \epsilon, \mu)$-ANN Data Structure for Interchange Distance]
Let $\epsilon > 0$, $\mu > 0$ be constants. Given n d-dimensional binary database vectors, there exists a data structure $\mathcal{S}$ using space $s = O((\frac{\epsilon}{2})^{-2}d^4 \log d(\log n + \log \log d) + d^3 \log d(n \log d)^{O((\frac{\epsilon}{2})^{-2})})$ that can constructed in $O(s \cdot n)$ time, which given a d-dimensional binary vector q finds $(2 + \epsilon)$-approximate nearest neighbor to q in the database with probability $1 - \mu$ in $O((\frac{\epsilon}{2})^{-2}d(\log n + \log \log d + \log \frac{1}{\mu}) \log d)$ time, where the distance is measured by the number of interchanges.

Remark. Note that Theorem 2 enables an efficient ANN-search under the interchange distance, which is $\mathcal{NP}$-hard to compute for strings having repeating symbols even for binary alphabet [4]. Using the linear time 1.5-approximation algorithm of [4] for computing approximate interchange distance between the query vector and each database vector would give a slightly better approximation (1.5 instead of $2 + \epsilon$) for the returned vector, however, much worse search time and space consumption when $n >> d$ as we assume in this problem.

Locally-Sensitive Hashing. To address the ANN problem, [23] proposed the Locality Sensitive Hashing scheme (LSH), which has since proved to be influential in theory and practice [7, 8, 21, 23]. In particular, LSH yields the best ANN data structures for the regime of sub-quadratic space and constant approximation factor, which turns out to be the most important regime from the practical perspective. The main idea is to hash the points such that the probability of collision is much higher for points which are close to each other than for those which are far apart.

It would be interesting to examine if some better data structures derived for the Hamming distance through LSH can also be adjusted for the interchange distance in order to improve our result.

4 Approximate Pattern Matching

In this section we show how pseudo-locality can be exploited to efficiently compute approximated distances between every alignment of the pattern to the text. Our study applies for computing approximate pattern matching with interchanges. We exploit known techniques for counting mismatches between every alignment of the pattern and the text of [18] and the randomized approximate pattern matching algorithm of [14] for Hamming distance.

Fixed-Size Alphabets Approximate PM for Hamming Distance. A wellknown method for computing the Hamming distances for binary alphabet is based on convolutions and gives the exact distances for every alignment of the pattern and the text in $O(n \log m)$ time [18]. This idea easily extends to fixedsize alphabets Σ in time $O(|\Sigma| n \log m)$, where $|\Sigma|$ is the alphabet size, thus providing an almost linear time algorithm for computing the Hamming distance in every pattern-text alignment. This is summarized in Theorem 3.

Theorem 3. [Approximate PM with Mismatches [18]]
For any fixed size alphabet, given a pattern P of length m and a text T of length n, there exists an algorithm for computing the Hamming distance between P and T for every alignment in time $O(n \log m)$.

Approximate PM Randomized Algorithm for Hamming Distance. For polynomial size alphabets, we use the randomized algorithm of [14] to approximately give the Hamming distances between T and P for every alignment.

Several efficient randomized (Monte-Carlo) algorithms for approximating all Hamming distances have been proposed. [26] obtained an $O(\epsilon^{-2} n \log n \log m)$-time algorithm, by randomly mapping the alphabet to $\{0, 1\}$, thereby reducing the problem to $O(\epsilon^{-2} \log n)$ binary alphabet instances. Each such instance can be solved in $O(n \log m)$ time by [18]. This algorithm can be de-randomized in $O(\epsilon^{-2} n \log^3 m)$ time, via ϵ-biased sample spaces or error-correcting codes. [22] solved the approximate decision problem for a fixed threshold in $O(\epsilon^{-3} n \log n)$ time, by using random sampling and performing $O(\epsilon^{-3} \log n)$ convolutions in $\mathbb{F}_2$, each in $O(n)$ time by a bit-packed version of FFT. The general problem can then be solved by examining logarithmically many thresholds, in $O(\epsilon^{-3} n \log n \log m)$ time. [27] obtained an $O(\epsilon^{-1} n \log n \log m)$-time algorithm, by randomly mapping the alphabet to $\{0, 1, \dots O(\epsilon^{-1})\}$, thereby reducing the problem to $O(\log n)$ instances with $O(\epsilon^{-1})$-size alphabet, each solved by $O(\epsilon^{-1})$ convolutions.

[14] is based on random sampling: the Hamming distance is estimated by checking mismatches at a random subset of positions. The algorithm picks a random prime p (of an appropriately chosen size) and a random offset b, and considers a subset of positions $\{b, b + p, b + 2p, \dots\}$. The structured nature of the subset enables more efficient computation. This result is summarized in Theorem 4.

Theorem 4. [$(1 + \epsilon)$-Approximate PM with Mismatches [14]]

Given a slackness parameter $\epsilon > 0$, a pattern P of length m and a text T of length n, there exists a randomized algorithm for computing a $(1+\epsilon)$-approximation of Hamming distance between P and T for every alignment in $O(\epsilon^{-2}n)$ time.

Pseudo-local Metric PM Algorithm. Let $dist$ be a c-pseudo-local metric and let $0 < \epsilon < 1$ be any slackness parameter. We may use exactly the same pattern matching algorithm of [14] with the slackness parameter ϵ/c and return the values computed by the algorithm as the approximated distances. For a c-strong pseudo-local metric we may use the same slackness parameter $\epsilon > 0$, but return the values computed by the algorithm divided by c as the approximated distances. Lemma 7 gives the approximation guarantees.

Lemma 7. *Let $d(\epsilon)$ be the value returned by [14]'s algorithm for some alignment of P and T for a slackness parameter $\epsilon > 0$, then:*

1. *If $dist$ is a c-pseudo local metric, and $\Delta_{dist} < \infty$ is the distance between P and T at some alignment, then for any $0 < \epsilon < 1$, $d(\epsilon/c)$ is a $(c+\epsilon)$-approximation of Δ_{dist}.*
2. *If $dist$ is a c-strong pseudo local metric, and $\Delta_{dist} < \infty$ is the distance between P and T at some alignment, then for any $0 < \epsilon < 1$, $d(\epsilon)/c$ is a $(1+\epsilon)$-approximation of Δ_{dist}.*

As in Sect. 3, in order to exploit Lemma 7, we need to exclude alignments of P and T where the distance is infinite. The following definition is needed.

Definition 6. [Online Infinite Distance Check]
Let $dist$ be any c-pseudo local metric. We say that $dist$ admits an online infinite distance check if, given P and a text T, there exists a $t(m)$-time per arriving symbol and $\tilde{O}(m)$ space algorithm to check for any alignment if the distance between P and T is infinite, where $t(m)$ is $O(m^\delta)$ function, for some $0 < \delta < 1$.

Theorem 5 then follows from Definition 6, Lemma 7 and Theorems 3 and 4.

Theorem 5. [Pseudo-local Metric Approximate PM]

1. *Let $dist$ be a c-pseudo local metric that admits an online infinite distance check in $O(t(m))$ time per arriving symbol, and let $\epsilon > 0$ be a slackness parameter. There is a randomized algorithm for the approximate dist PM problem that runs in $O(n((\frac{\epsilon}{c})^{-2} + t(m)))$ time. The algorithm gives $(c+\epsilon)$-approximation for the dist-distances.*
For fixed-size alphabet, there is a deterministic algorithm for the approximate dist PM problem that runs in $O(n(\log m + t(m)))$ time. The algorithm gives 2-approximation for the dist-distances.
2. *Let $dist$ be a c-strong pseudo local metric that admits an online infinite distance check in $O(t(m))$ time per arriving symbol, and let $\epsilon > 0$ be a slackness parameter. There is a randomized algorithm for the approximate dist PM problem that runs in $O(n(\epsilon^{-2} + t(m)))$ time. The algorithm gives $(1+\epsilon)$-approximation for the dist-distances.*

Lemma 8 is used for online infinite interchange distance exclusion.

Lemma 8. [Online Infinite Interchange Distance Check]
The interchange distance admits an online infinite distance check in $O(1)$ time per arriving symbol and:

1. $O(\log m)$-*bits space, for fixed-size alphabets.*
2. $O(m \log m)$-*bits space, for polynomial-size alphabets.*

Remark. The statement and proof of Lemma 8 assumes a word-size of $O(\log m)$ bits. If this assumption does not hold, only the time per arriving symbol changes to $O(\log m)$ instead of $O(1)$ and the time complexity for polynomial size alphabets has a $\log m$ factor.

Theorem 6 then follows from Theorem 5.1, Lemma 8 and Lemma 1.

Theorem 6. [Approximate PM with Interchanges]
Let $0 < \epsilon < 1$ be a slackness parameter. There is a randomized algorithm for the approximate PM with interchanges problem that runs in $O(n(\frac{\epsilon}{2})^{-2})$ time. The algorithm gives $(2 + \epsilon)$-approximation for the interchange distances.
For fixed-size alphabet, there is a deterministic algorithm for the approximate PM with interchanges problem that runs in $O(n \log m)$ time. The algorithm gives 2-approximation for the interchange distances.

Note that a direct use of the 1.5-approximation algorithm for the interchange distance of [4] for every alignment yields a 1.5-approximation for the approximate PM with interchanges problem in $\Theta(nm)$ time. On the other hand, there are known almost linear time algorithms for the approximate PM with the swap or the parallel-interchanges distances problems [1,2]. Thus, using their pseudo-locality property does not improve approximate PM algorithms for these metrics.

Space Efficient Online Infinite Interchange Distance Check. Due to space limitations this part is postponed to the paper full version.

5   Conclusion and Open Problems

This paper demonstrates how to exploit pseudo-locality for achieving an efficient ANN-search data structure and an approximate PM algorithm for the interchange distance. It is interesting to continue this line of research:

- Can better ANN data structures derived for the Hamming distance through LSH be adjusted for the interchange distance in order to improve the result presented in this paper?
- Can an infinite-distance check for swap and parallel-interchange distances be achieved to allow a use in ANN-search?
- Can pseudo-locality be exploited for deriving new solutions for pseudo-local metrics in problems other than ANN-search and PM?

Answering such questions will not only enrich our set of tools but also broaden our understanding of string metrics.

References

1. Amir, A., et al.: Pattern matching with address errors: rearrangement distances. J. Comput. Syst. Sci. **75**(6), 359–370 (2009)
2. Amir, A., Cole, R., Hariharan, R., Lewenstein, M., Porat, E.: Overlap matching. Inf. Comput. **181**(1), 57–74 (2003)
3. Amir, A., Eisenberg, E., Levy, A., Porat, E., Shapira, N.: Cycle detection and correction. ACM Trans. Algorithms (TALG) **9**(1), 13:1-13:20 (2012)
4. Amir, A., Hartman, T., Kapah, O., Levy, A., Porat, E.: On the cost of interchange rearrangement in strings. SIAM J. Comput. (SICOMP) **39**(4), 1444–1461 (2009)
5. Amir, A., Levy, A.: String rearrangement metrics: a survey. In: Elomaa, T., Mannila, H., Orponen, P. (eds.) Algorithms and Applications. LNCS, vol. 6060, pp. 1–33. Springer, Heidelberg (2010). https://doi.org/10.1007/978-3-642-12476-1_1
6. Amir, A., Lewenstein, M., Porat, E.: Approximate swapped matching. Inf. Process. Lett. **83**(1), 33–39 (2002)
7. Andoni, A., Indyk, P., Laarhoven, T., Razenshteyn, I., Schmidt, L.: Practical and optimal LSH for angular distance. In: Proceedings of the 28th International Conference on Neural Information Processing Systems, NIPS 2015, Cambridge, MA, USA, vol. 1, pp. 1225–1233. MIT Press (2015)
8. Andoni, A., Razenshteyn, I.: Optimal data-dependent hashing for approximate near neighbors. In: Proceedings of the Forty-Seventh Annual ACM Symposium on Theory of Computing, STOC 2015, New York, NY, USA, pp. 793–801. Association for Computing Machinery (2015)
9. Baeza-Yates, R., Ribeiro-Neto, B.: Modern Information Retrieval. Addison-Wesley, Reading (1999)
10. Bafna, V., Pevzner, P.A.: Sorting by transpositions. SIAM J. Discret. Math. **11**, 221–240 (1998)
11. Benson, G., Levy, A., Maimoni, S., Noifeld, D., Shalom, B.R.: LCSk: a refined similarity measure. Theoret. Comput. Sci. **638**, 11–26 (2016)
12. Caprara, A.: Sorting permutations by reversals and Eulerian cycle decompositions. SIAM J. Discret. Math. **12**(1), 91–110 (1999)
13. Cayley, A.: Note on the theory of permutations. Philos. Mag. **34**, 527–529 (1849)
14. Chan, T.M., Golan, S., Kociumaka, T., Kopelowitz, T., Porat, E.: Approximating text-to-pattern Hamming distances. In: Proceedings of the 52nd Annual ACM SIGACT Symposium on Theory of Computing, STOC 2020, New York, NY, USA, pp. 643–656. Association for Computing Machinery (2020)
15. Chavez, E., Navarro, G., Baeza-Yates, R.A., Marroquin, J.L.: Searching in metric spaces. ACM Comput. Surv. **33**(3), 273–321 (2001)
16. Christie, D.A.: Sorting by block-interchanges. Inf. Process. Lett. **60**, 165–169 (1996)
17. Clifford, R., Starikovskaya, T.: Approximate Hamming distance in a stream. In: Proceedings of the 43rd International Colloquium on Automata, Languages, and Programming ICALP, vol. 55, pp. 20:1–20:14 (2016)
18. Fischer, M., Paterson, M.: String matching and other products. In: Proceedings of the 7th SIAM-AMS Complexity of Computation, pp. 113–125 (1974)
19. Hannenhalli, S.: Polynomial algorithm for computing translocation distance between genomes. Discret. Appl. Math. **71**, 137–151 (1996)
20. Hannenhalli, S., Pevzner, P.A.: Transforming cabbage into turnip (polynomial algorithm for sorting signed permutations by reversals). In: Proceedings of the 27th Annual ACM Symposium on the Theory of Computing, pp. 178–187 (1995)

21. Har-Peled, S., Indyk, P., Motwani, R.: Approximate nearest neighbor: towards removing the curse of dimensionality. Theory Comput. **8**(1), 321–350 (2012)
22. Indyk, P.: Faster algorithms for string matching problems: matching the convolution bound. In: Proceedings of the 39th Annual Symposium on Foundations of Computer Science (USA), FOCS 1998, pp. 166–173. IEEE Computer Society (1998)
23. Indyk, P., Motwani, R.: Approximate nearest neighbors: towards removing the curse of dimensionality. In: Proceedings of the Thirtieth Annual ACM Symposium on Theory of Computing, STOC 1998, New York, NY, USA, pp. 604–613. Association for Computing Machinery (1998)
24. Jerrum, M.R.: The complexity of finding minimum-length generator sequences. Theoret. Comput. Sci. **36**, 265–289 (1985)
25. Kapah, O., Landau, G.M., Levy, A., Oz, N.: Interchange rearrangement: the element-cost model. Theoret. Comput. Sci. **410**(43), 4315–4326 (2009)
26. Karloff, H.: Fast algorithms for approximately counting mismatches. Inf. Process. Lett. (IPL) **48**(2), 53–60 (1993)
27. Kopelowitz, T., Porat, E.: A simple algorithm for approximating the text-to-pattern Hamming distance. In: Seidel, R. (ed.) 1st Symposium on Simplicity in Algorithms, SOSA 2018, OASICS, New Orleans, LA, USA, 7–10 January 2018, vol. 61, pp. 10:1-10:5. Schloss Dagstuhl - Leibniz-Zentrum für Informatik (2018)
28. Kushilevitz, E., Ostrovsky, R., Rabani, Y.: Efficient search for approximate nearest neighbor in high dimensional spaces. SIAM J. Comput. (SICOMP) **30**(2), 457–474 (2000)
29. Levenshtein, V.I.: Binary codes capable of correcting deletions, insertions and reversals. Sov. Phys. Dokl. **10**, 707–710 (1966)
30. Lowrance, R., Wagner, R.A.: An extension of the string-to-string correction problem. J. ACM **22**, 177–183 (1975)
31. Navarro, G.: A guided tour to approximate string matching. ACM Comput. Surv. **33**(1), 31–88 (2001)

Position Heaps for Cartesian-Tree Matching on Strings and Tries

Akio Nishimoto[1(✉)], Noriki Fujisato[1], Yuto Nakashima[1],
and Shunsuke Inenaga[1,2]

[1] Department of Informatics, Kyushu University, Fukuoka, Japan
{nishimoto.akio,noriki.fujisato,yuto.nakashima,
inenaga}@inf.kyushu-u.ac.jp
[2] PRESTO, Japan Science and Technology Agency, Kawaguchi, Japan

Abstract. The *Cartesian-tree pattern matching* is a recently introduced scheme of pattern matching that detects fragments in a sequential data stream which have a similar structure as a query pattern. Formally, Cartesian-tree pattern matching seeks all substrings S' of the text string S such that the Cartesian tree of S' and that of a query pattern P coincide. In this paper, we present a new indexing structure for this problem, called the *Cartesian-tree Position Heap* (*CPH*). Let n be the length of the input text string S, m the length of a query pattern P, and σ the alphabet size. We show that the CPH of S, denoted $\mathsf{CPH}(S)$, supports pattern matching queries in $O(m(\sigma + \log(\min\{h, m\})) + occ)$ time with $O(n)$ space, where h is the height of the CPH and occ is the number of pattern occurrences. We show how to build $\mathsf{CPH}(S)$ in $O(n \log \sigma)$ time with $O(n)$ working space. Further, we extend the problem to the case where the text is a labeled tree (i.e. a trie). Given a trie T with N nodes, we show that the CPH of T, denoted $\mathsf{CPH}(T)$, supports pattern matching queries on the trie in $O(m(\sigma^2 + \log(\min\{h, m\})) + occ)$ time with $O(N\sigma)$ space. We also show a construction algorithm for $\mathsf{CPH}(T)$ running in $O(N\sigma)$ time and $O(N\sigma)$ working space.

1 Introduction

If the Cartesian trees $\mathsf{CT}(X)$ and $\mathsf{CT}(Y)$ of two strings X and Y are equal, then we say that X and Y *Cartesian-tree match* (*ct-match*). The *Cartesian-tree pattern matching problem* (*ct-matching problem*) [18] is, given a text string S and a pattern P, to find all substrings S' of S that ct-match with P.

String equivalence with ct-matching belongs to the class of *substring-consistent equivalence relation* (*SCER*) [17], namely, the following holds: If two strings X and Y ct-match, then $X[i..j]$ and $Y[i..j]$ also ct-match for any $1 \le i \le j \le |X|$. Among other types of SCERs [3–5,14,15], ct-matching is the most related to order-peserving matching (*op-matching*) [7,9,16]. Two strings X and Y are said to op-match if the relative order of the characters in X and the relative order of the characters in Y are the same. It is known that with ct-matching one can detect some interesting occurrences of a pattern that cannot

T. Lecroq and H. Touzet (Eds.): SPIRE 2021, LNCS 12944, pp. 241–254, 2021.
https://doi.org/10.1007/978-3-030-86692-1_20

be captured with op-matching. More precisely, if two strings X and Y op-match, then X and Y also ct-match. However, the reverse is not true. With this property in hand, ct-matching is motivated for analysis of time series such as stock charts [11, 18].

This paper deals with the indexing version of the ct-matching problem. Park et al. [18] proposed the *Cartesian suffix tree* (*CST*) for a text string S that can be built in $O(n \log n)$ worst-case time or $O(n)$ expected time, where n is the length of the text string S. The $\log n$ factor in the worst-case complexity is due to the fact that the *parent-encoding*, a key concept for ct-matching introduced in [18], is a sequence of integers in range $[0..n-1]$. While it is not explicitly stated in Park et al.'s paper [18], our simple analysis (c.f. Lemma 10 in Sect. 5) reveals that the CST supports pattern matching queries in $O(m \log m + occ)$ time, where m is the pattern length and occ is the number of pattern occurrences.

In this paper, we present a new indexing structure for this problem, called the *Cartesian-tree Position Heap* (*CPH*). We show that the CPH of S, which occupies $O(n)$ space, can be built in $O(n \log \sigma)$ time with $O(n)$ working space and supports pattern matching queries in $O(m(\sigma + \log(\min\{h, m\})) + occ)$ time, where h is the height of the CPH. Compared to the afore-mentioned CST, our CPH is the *first* index for ct-matching that can be built in worst-case linear time for constant-size alphabets, while pattern matching queries with our CPH can be slower than with the CST when σ is large.

We then consider the case where the text is a labeled tree (i.e. a trie). Given a trie T with N nodes, we show that the CPH of T, which occupies $O(N\sigma)$ space, can be built in $O(N\sigma)$ time and $O(N\sigma)$ working space. We also show how to support pattern matching queries in $O(m(\sigma^2 + \log(\min\{h, m\})) + occ)$ time in the trie case. To our knowledge, our CPH is the first indexing structure for ct-matching on tries that uses linear space for constant-size alphabets.

Conceptually, our CPH is most related to the *parameterized position heap* (*PPH*) for a string [12] and for a trie [13], in that our CPHs and the PPHs are both constructed in an incremental manner where the suffixes of an input string and the suffixes of an input trie are processed in increasing order of their lengths. However, some new techniques are required in the construction of our CPH due to different nature of the *parent encoding* [18] of strings for ct-matching, from the *previous encoding* [3] of strings for parameterized matching.

2   Preliminaries

2.1   Strings and (Reversed) Tries

Let Σ be an ordered alphabet of size σ. An element of Σ is called a *character*. An element of Σ^* is called a *string*. For a string $S \in \Sigma^*$, let σ_S denote the number of distinct characters in S.

The empty string ε is a string of length 0, namely, $|\varepsilon| = 0$. For a string $S = XYZ$, X, Y and Z are called a *prefix*, *substring*, and *suffix* of S, respectively. The set of prefixes of a string S is denoted by Prefix(S). The i-th character of a string S is denoted by $S[i]$ for $1 \le i \le |S|$, and the substring of a string

S that begins at position i and ends at position j is denoted by $S[i..j]$ for $1 \le i \le j \le |S|$. For convenience, let $S[i..j] = \varepsilon$ if $j < i$. Also, let $S[i..] = S[i..|S|]$ for any $1 \le i \le |S| + 1$.

A *trie* is a rooted tree that represents a set of strings, where each edge is labeled with a character from Σ and the labels of the out-going edges of each node is mutually distinct. Tries are natural generalizations to strings in that tries can have branches while strings are sequences without branches.

Let $\mathbf{x}$ be any node of a given trie T, and let $\mathbf{r}$ denote the root of T. Let $\mathsf{depth}(\mathbf{x})$ denote the depth of $\mathbf{x}$. When $\mathbf{x} \ne \mathbf{r}$, let $\mathsf{parent}(\mathbf{x})$ denote the parent of $\mathbf{x}$. For any $0 \le j \le \mathsf{depth}(\mathbf{x})$, let $\mathsf{anc}(\mathbf{x}, j)$ denote the j-th ancestor of $\mathbf{x}$, namely, $\mathsf{anc}(\mathbf{x}, 0) = \mathbf{x}$ and $\mathsf{anc}(\mathbf{x}, j) = \mathsf{parent}(\mathsf{anc}(\mathbf{x}, j - 1))$ for $1 \le j \le \mathsf{depth}(\mathbf{x})$. It is known that after a linear-time processing on T, $\mathsf{anc}(\mathbf{x}, j)$ for any query node $\mathbf{x}$ and integer j can be answered in $O(1)$ time [6].

For the sake of convenience, in the case where our input is a trie T, then we consider its *reversed trie* where the path labels are read in the leaf-to-root direction. On the other hand, the trie-based data structures (namely position heaps) we build for input strings and reversed tries are usual tries where the path labels are read in the root-to-leaf direction.

For each (reversed) path $(\mathbf{x}, \mathbf{y})$ in T such that $\mathbf{y} = \mathsf{anc}(\mathbf{x}, j)$ with $j = |\mathsf{depth}(\mathbf{x})| - |\mathsf{depth}(\mathbf{y})|$, let $\mathsf{str}(\mathbf{x}, \mathbf{y})$ denote the string obtained by concatenating the labels of the edges from $\mathbf{x}$ to $\mathbf{y}$. For any node $\mathbf{x}$ of T, let $\mathsf{str}(\mathbf{x}) = \mathsf{str}(\mathbf{x}, \mathbf{r})$.

Let N be the number of nodes in T. We associate a unique *id* to each node of T. Here we use a bottom-up level-order traversal rank as the id of each node in T, and we sometimes identify each node with its id. For each node id i ($1 \le i \le N$) let $T[i..] = \mathsf{str}(i)$, i.e., $T[i..]$ is the path string from node i to the root $\mathbf{r}$.

2.2 Cartesian-Tree Pattern Matching

The *Cartesian tree* of a string S, denoted $\mathsf{CT}(S)$, is the rooted tree with $|S|$ nodes which is recursively defined as follows:

- If $|S| = 0$, then $\mathsf{CT}(S)$ is the empty tree.
- If $|S| \ge 1$, then $\mathsf{CT}(S)$ is the tree whose root r stores the left-most minimum value $S[i]$ in S, namely, $r = S[i]$ iff $S[i] \le S[j]$ for any $i \ne j$ and $S[h] > S[i]$ for any $h < i$. The left-child of r is $\mathsf{CT}(S[1..i - 1])$ and the right-child of r is $\mathsf{CT}(S[i + 1..|S|])$.

The *parent distance encoding* of a string S of length n, denoted $\mathsf{PD}(S)$, is a sequence of n integers over $[0..n - 1]$ such that

$$\mathsf{PD}(S)[i] = \begin{cases} i - \max_{1 \le j < i}\{j \mid S[j] \le S[i]\} & \text{if such } j \text{ exists,} \\ 0 & \text{otherwise.} \end{cases}$$

Namely, $\mathsf{PD}(S)[i]$ represents the distance to from position i to its nearest left-neighbor position j that stores a value that is less than or equal to $S[i]$.

A tight connection between CT and PD is known:

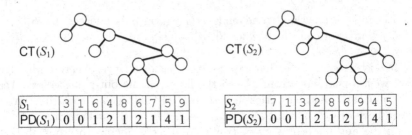

Fig. 1. Two strings $S_1 = 316486759$ and $S_2 = 713286945$ ct-match since $CT(S_1) = CT(S_2)$ and $PD(S_1) = PD(S_2)$.

Lemma 1 ([19]). *For any two strings S_1 and S_2 of equal length, $CT(S_1) = CT(S_2)$ iff $PD(S_1) = PD(S_2)$.*

For two strings S_1 and S_2, we write $S_1 \approx S_2$ iff $CT(S_1) = CT(S_2)$ (or equivalently $PD(S_1) = PD(S_2)$). We also say that S_1 and S_2 *ct-match* when $S_1 \approx S_2$. See Fig. 1 for a concrete example.

We consider the indexing problems for Cartesian-tree pattern matching on a text string and a text trie, which are respectively defined as follows:

Problem 1 (Cartesian-Tree Pattern Matching on Text String).

Preprocess: A text string S of length n.

Query: A pattern string P of length m.

Report: All text positions i such that $S[i..i+m-1] \approx P$.

Problem 2 (Cartesian-Tree Pattern Matching on Text Trie).

Preprocess: A text trie T with N nodes.

Query: A pattern string P of length m.

Report: All trie nodes i such that $(T[i..])[1..m] \approx P$.

2.3 Sequence Hash Trees

Let $W = \langle w_1, \ldots, w_k \rangle$ be a sequence of non-empty strings such that for any $1 < i \le k$, $w_i \notin \mathsf{Prefix}(w_j)$ for any $1 \le j < i$. The *sequence hash tree* [8] of a sequence $W = \langle w_1, \ldots, w_k \rangle$ of k strings, denoted $\mathsf{SHT}(W) = \mathsf{SHT}(W)^k$, is a trie structure that is incrementally built as follows:

1. $\mathsf{SHT}(W)^0 = \mathsf{SHT}(\langle \, \rangle)$ for the empty sequence $\langle \, \rangle$ is the tree only with the root.

2. For $i = 1, \ldots, k$, $\mathsf{SHT}(\mathcal{W})^i$ is obtained by inserting the shortest prefix u_i of w_i that does not exist in $\mathsf{SHT}(\mathcal{W})^{i-1}$. This is done by finding the longest prefix p_i of w_i that exists in $\mathsf{SHT}(\mathcal{W})^{i-1}$, and adding the new edge (p_i, c, u_i), where $c = w_i[|p_i| + 1]$ is the first character of w_i that could not be traversed in $\mathsf{SHT}(\mathcal{W})^{i-1}$.

Since we have assumed that each w_i in $\mathcal{W}$ is not a prefix of w_j for any $1 \leq j < i$, the new edge (p_i, c, u_i) is always created for each $1 \leq i \leq k$. This means that $\mathsf{SHT}(\mathcal{W})$ contains exactly $k + 1$ nodes (including the root).

To perform pattern matching queries efficiently, each node of $\mathsf{SHT}(\mathcal{W})$ is augmented with the *maximal reach pointer*. For each $1 \leq i \leq k$, let u_i be the newest node in $\mathsf{SHT}(\mathcal{W})^i$, namely, u_i is the shortest prefix of w_i which did not exist in $\mathsf{SHT}(\mathcal{W})^{i-1}$. Then, in the complete sequence hash tree $\mathsf{SHT}(\mathcal{W}) = \mathsf{SHT}(\mathcal{W})^k$, we set $\mathsf{mrp}(u_i) = u_j$ iff u_j is the deepest node in $\mathsf{SHT}(\mathcal{W})$ such that u_j is a prefix of w_i. Intuitively, $\mathsf{mrp}(u_i)$ represents the last visited node u_j when we traverse w_i from the root of the complete $\mathsf{SHT}(\mathcal{W})$. Note that $j \geq i$ always holds. When $j = i$ (i.e. when the maximal reach pointer is a self-loop), then we can omit it because it is not used in the pattern matching algorithm.

3 Cartesian-Tree Position Heaps for Strings

In this section, we introduce our new indexing structure for Problem 1. For a given text string S of length n, let $\mathcal{W}_S$ denote the sequence of the parent distance encodings of the non-empty suffixes of S which are sorted in increasing order of their lengths. Namely, $\mathcal{W}_S = \langle w_1, \ldots, w_n \rangle = \langle \mathsf{PD}(S[n..]), \ldots, \mathsf{PD}(S[1..]) \rangle$, where $w_{n-i+1} = \mathsf{PD}(S[i..])$. The *Cartesian-tree Position Heap* (*CPH*) of string S, denoted $\mathsf{CPH}(S)$, is the sequence hash tree of $\mathcal{W}_S$, that is, $\mathsf{CPH}(S) = \mathsf{SHT}(\mathcal{W}_S)$. Note that for each $1 \leq i \leq n + 1$, $\mathsf{CPH}(S[i..]) = \mathsf{SHT}(\mathcal{W}_S)^{n-i+1}$ holds.

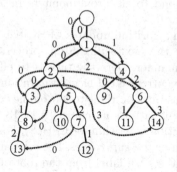

S	26427584365741
w_1	<u>0</u>
w_2	<u>00</u>
w_3	<u>000</u>
w_4	<u>0100</u>
w_5	<u>00100</u>
w_6	<u>012140</u>
w_7	<u>0012140</u>
w_8	<u>00012140</u>
w_9	<u>010012140</u>
w_{10}	<u>0010012140</u>
w_{11}	<u>01214512140</u>
w_{12}	<u>001214512140</u>
w_{13}	<u>0001214512140</u>
w_{14}	<u>01231214512140</u>

Fig. 2. $\mathsf{CPH}(S)$ for string $S = 26427584365741$. For each $w_i = \mathsf{PD}(S[n - i + 1..])$, the underlined prefix is the string that is represented by the node u_i in $\mathsf{CPH}(S)$. The dotted arcs are reversed suffix links (not all reversed suffix links are shown).

Our algorithm builds $\mathsf{CPH}(S[i..])$ for decreasing $i = n, \ldots, 1$, which means that we process the given text string S in a right-to-left online manner, by prepending the new character $S[i]$ to the current suffix $S[i + 1..]$.

For a sequence v of integers, let $\mathcal{Z}_v$ denote the sorted list of positions z in v such that $v[z] = 0$ iff $z \in \mathcal{Z}_v$. Clearly $|\mathcal{Z}_v|$ is equal to the number of 0's in v.

Lemma 2. *For any string S, $|\mathcal{Z}_{\mathsf{PD}(S)}| \leq \sigma_S$.*

Proof. Let $\mathcal{Z}_{\mathsf{PD}(S)} = z_1, \ldots, z_\ell$. We have that $S[z_1] > \cdots > S[z_\ell]$ since otherwise $\mathsf{PD}(S)[z_x] \neq 0$ for some z_x, a contradiction. Thus $|\mathcal{Z}_{\mathsf{PD}(S)}| \leq \sigma_S$ holds. □

Lemma 3. *For each $i = n, \ldots, 1$, $\mathsf{PD}(S[i..])$ can be computed from $\mathsf{PD}(S[i+1..])$ in an online manner, using a total of $O(n)$ time with $O(\sigma_S)$ working space.*

Proof. Given a new character $S[i]$, we check each position z in the list $\mathcal{Z}_{\mathsf{PD}(S[i+1..])}$ in increasing order. Let $\hat{z} = z + i$, i.e., $\hat{z}$ is the global position in S corresponding to z in $S[i+1..]$. If $S[i] \leq S[\hat{z}]$, then we set $\mathsf{PD}(S[i..])[z - i + 1] = z - i\ (> 0)$ and remove z from the list. Remark that these removed positions correspond to the front pointers in the next suffix $S[i..]$. We stop when we encounter the first z in the list such that $S[i] > S[\hat{z}]$. Finally we add the position i to the head of the remaining positions in the list. This gives us $\mathcal{Z}_{\mathsf{PD}(S[i..])}$ for the next suffix $S[i..]$.

It is clear that once a position in the PD encoding is assigned a non-zero value, then the value never changes whatever characters we prepend to the string. Therefore, we can compute $\mathsf{PD}(S[i..])$ from $\mathsf{PD}(S[i+1..])$ in a total of $O(n)$ time for every $1 \leq i \leq n$. The working space is $O(\sigma_S)$ due to Lemma 2. □

A position i in a sequence u of non-negative integers is said to be a *front pointer* in u if $i - u[i] = 1$ and $i \geq 2$. Let $\mathcal{F}_u$ denote the sorted list of front pointers in u. The positions of the suffix $S[i + 1..]$ which are removed from $\mathcal{Z}_{\mathsf{PD}(S[i+1..])}$ correspond to the front pointers in $\mathcal{F}_{\mathsf{PD}(S[i..])}$ for the next suffix $S[i..]$.

Our construction algorithm updates $\mathsf{CPH}(S[i+1..])$ to $\mathsf{CPH}(S[i..])$ by inserting a new node for the next suffix $S[i..]$, processing the given string S in a right-to-left online manner. Here the task is to efficiently locate the parent of the new node in the current CPH at each iteration.

As in the previous work on right-to-left online construction of indexing structures for other types of pattern matching [10,12,13,20], we use the *reversed suffix links* in our construction algorithm for $\mathsf{CPH}(S)$. For ease of explanation, we first introduce the notion of the *suffix links*. Let u be any non-root node of $\mathsf{CPH}(S)$. We identify u with the path label from the root of $\mathsf{CPH}(S)$ to u, so that u is a PD encoding of some substring of S. We define the suffix link of u, denoted $\mathsf{sl}(u)$, such that $\mathsf{sl}(u) = v$ iff v is obtained by (1) removing the first $0\ (= u[1])$, and (2) substituting 0 for the character $u[f]$ at every front pointer $f \in \mathcal{F}_u \subseteq [2..|u|]$ of u. The reversed suffix link of v with non-negative integer label a, denoted $\mathsf{rsl}(v, a)$, is defined such that $\mathsf{rsl}(v, a) = u$ iff $\mathsf{sl}(u) = v$ and $a = |\mathcal{F}_u|$. See also Fig. 2.

Lemma 4. *Let u, v be any nodes of $\mathsf{CPH}(S)$ such that $\mathsf{rsl}(v, a) = u$ with label a. Then $a \leq \sigma_S$.*

Proof. Since $|\mathcal{F}_u| \leq |\mathcal{Z}_v|$, using Lemma 2, we obtain $a = |\mathcal{F}_u| \leq |\mathcal{Z}_v| \leq \sigma_{S'} \leq \sigma_S$, where S' is a substring of S such that $\mathsf{PD}(S') = v$. □

The next lemma shows that the number of out-going reversed suffix links of each node v is bounded by the alphabet size.

Lemma 5. *For any node v in $\mathsf{CPH}(S)$, $|\{a \mid \mathsf{rsl}(v,a) = u$ for some node $u\}| \le \sigma_S + 1$.*

Proof. Let a be any integer such that $\mathsf{rsl}(v,a) = u$ exists for some node u. Since v is a node of $\mathsf{CPH}(S)$, $v = \mathsf{PD}(S')$ for some substring S' of S. Thus, by Lemma 4, we get $a \le \sigma_S$. Since $a \ge 0$, there can be at most $\sigma_S + 1$ different values for a such that $\mathsf{rsl}(v,a)$ is defined for any node v. □

Our CPH construction algorithm makes use of the following monotonicity of the labels of reversed suffix links:

Lemma 6. *Suppose that there exist two reversed suffix links $\mathsf{rsl}(v,a) = u$ and $\mathsf{rsl}(v',a') = u'$ such that $v' = \mathsf{parent}(v)$ and $u' = \mathsf{parent}(u)$. Then, $0 \le a - a' \le 1$.*

Proof. Immediately follows from $a = |\mathcal{F}_u|$, $a' = |\mathcal{F}_{u'}|$, and $u' = u[1..|u| - 1]$. □

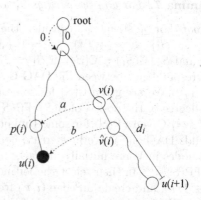

Fig. 3. We climb up the path from $u(i+1)$ and find the parent $p(i)$ of the new node $u(i)$ (in black). The label a of the reversed suffix link we traverse from $v(i)$ is equal to the number of front pointers in $p(i)$.

We are ready to design our right-to-left online construction algorithm for the CPH of a given string S. Since $\mathsf{PD}(S[i..])$ is the $(n - i + 1)$-th string w_{n-i+1} of the input sequence $\mathcal{W}_S$, for ease of explanation, we will use the convention that $u(i) = u_{n-i+1}$ and $p(i) = p_{n-i-1}$, where the new node $u(i)$ for $w_{n-i+1} = \mathsf{PD}(S[i..])$ is inserted as a child of $p(i)$. See Fig. 3.

Algorithm 1: Right-to-Left Online Construction of $\mathsf{CPH}(S)$

$i = n$ **(base case):** We begin with $\mathsf{CPH}(S[n..])$ which consists of the root $r = u(n + 1)$ and the node $u(n)$ for the first (i.e. shortest) suffix $S[n..]$ of S. Since $w_1 = \mathsf{PD}(S[n..]) = \mathsf{PD}(S[n]) = 0$, the edge from r to $u(n)$ is labeled 0. Also, we set the reversed suffix link $\mathsf{rsl}(r, 0) = u(n)$.

$i = n - 1, \ldots, 1$ **(iteration):** Given $\mathsf{CPH}(S[i+1..])$ which consists of the nodes $u(i + 1), \ldots, u(n)$, which respectively represent some prefixes of the already processed strings $w_{n-i}, \ldots, w_1 = \mathsf{PD}(S[i + 1..]), \ldots, \mathsf{PD}(S[n..])$, together with their reversed suffix links. We find the parent $p(i)$ of the new node $u(i)$ for $\mathsf{PD}(S[i..])$, as follows: We start from the last-created node $u(i + 1)$ for the previous $\mathsf{PD}(S[i + 1..])$, and climb up the path towards the root r. Let $d_i \in [1..|u(i + 1)|]$ be the smallest integer such that the d_i-th ancestor $v(i) = \mathsf{anc}(u(i+1), d_i)$ of $u(i+1)$ has the reversed suffix link $\mathsf{rsl}(v(i), a)$ with the label $a = |\mathcal{F}_{\mathsf{PD}(S[i..i+|v(i)|])}|$. We traverse the reversed suffix link from $v(i)$ and let $p(i) = \mathsf{rsl}(v(i), a)$. We then insert the new node $u(i)$ as the new child of $p(i)$, with the edge labeled $\mathsf{PD}(S[i..])[i+|u(i)|-1]$. Finally, we create a new reversed suffix link $\mathsf{rsl}(\hat{v}(i), b) = u(i)$, where $\hat{v}(i) = \mathsf{anc}(u(i+1), d_i - 1)$ and $\mathsf{parent}(\hat{v}) = v$. We set $b \leftarrow a+1$ if the position $i + |p(i)|$ is a front pointer of $\mathsf{PD}(S[i..])$, and $b \leftarrow a$ otherwise.

For computing the label $a = |\mathcal{F}_{\mathsf{PD}(S[i..i+|v(i)|])}|$ efficiently, we introduce a new encoding FP that is defined as follows: For any string S of length n, let $\mathsf{FP}(S)[i] = |\mathcal{F}_{\mathsf{PD}(S[i..n])}|$. The FP encoding preserves the ct-matching equivalence:

Lemma 7. *For any two strings S_1 and S_2, $S_1 \approx S_2$ iff $\mathsf{FP}(S_1) = \mathsf{FP}(S_2)$.*

Proof. For a string S, consider the DAG $\mathsf{G}(S) = (V, E)$ such that $V = \{1, \ldots, |S|\}$, $E = \{(j, i) \mid j = i - \mathsf{PD}(S)[i]\}$. By Lemma 1, for any strings S_1 and S_2, $\mathsf{G}(S_1) = \mathsf{G}(S_2)$ iff $S_1 \approx S_2$. Now, we will show there is a one-to-one correspondence between the DAG G and the FP encoding.

($\Rightarrow$) We are given $\mathsf{G}(S)$ for some (unknown) string S. Since $\mathsf{FP}(S)[i]$ is the in-degree of the node i of $\mathsf{G}(S)$, $\mathsf{FP}(S)$ is unique for the given DAG $\mathsf{G}(S)$.

($\Leftarrow$) Given $\mathsf{FP}(S)$ for some (unknown) string S, we show an algorithm that builds DAG $\mathsf{G}(S)$. We first create nodes $V = \{1, \ldots, |S|\}$ without edges, where all nodes in V are initially unmarked. For each $i = n, \ldots, 1$ in decreasing order, if $\mathsf{FP}(S)[i] > 0$, then select the leftmost $\mathsf{FP}(S)[i]$ unmarked nodes in the range $[i-1..n]$, and create an edge (i, i') from each selected node i' to i. We mark all these $\mathsf{FP}(S)[i]$ nodes at the end of this step, and proceed to the next node $i-1$. The resulting DAG $\mathsf{G}(S)$ is clearly unique for a given $\mathsf{PD}(S)$. □

For computing the label $a = |\mathcal{F}_{\mathsf{PD}(S[i..i+|v(i)|])}| = \mathsf{FP}(S[i..i + |v(i)|])[1]$ of the reversed suffix link in Algorithm 1, it is sufficient to maintain the induced graph $\mathsf{G}_{[i..j]}$ of DAG G for a variable-length sliding window $S[i..j]$ with the nodes $\{i, \ldots, j\}$. This can easily be maintained in $O(n)$ total time.

Theorem 1. *Algorithm 1 builds $\mathsf{CPH}(S[i..])$ for decreasing $i = n, \ldots, 1$ in a total of $O(n \log \sigma)$ time and $O(n)$ space, where σ is the alphabet size.*

Proof. **Correctness:** Consider the $(n - i + 1)$-th step in which we process $\mathsf{PD}(S[i..])$. By Lemma 6, the d_i-th ancestor $v(i) = \mathsf{anc}(u(i + 1), d_i)$ of $u(i + 1)$ can be found by simply walking up the path from the start node $u(i + 1)$. Note that there always exists such ancestor $v(i)$ of $u(i + 1)$ since the root r has the defined reversed suffix link $\mathsf{rsl}(r, 0) = 0$. By the definition of $v(i)$ and its reversed suffix link, $\mathsf{rsl}(v(i), a) = p(i)$ is the longest prefix of $\mathsf{PD}(S[i..])$ that is represented by $\mathsf{CPH}(S[i + 1..])$ (see also Fig. 3). Thus, $p(i)$ is the parent of the new node $u(i)$ for $\mathsf{PD}(S[i..])$. The correctness of the new reversed suffix link $\mathsf{rsl}(\hat{v}(i), b) = u(i)$ follows from the definition.

Complexity: The time complexity is proportional to the total number $\sum_{i=1}^{n} d_i$ of nodes that we visit for all $i = n, \ldots, 1$. Clearly $|u(i)| - |u(i+1)| = d_i - 2$. Thus, $\sum_{i=1}^{n} d_i = \sum_{i=1}^{n}(|u(i)| - |u(i+1)| + 2) = |u(1)| - |u(n)| + 2n \leq 3n = O(n)$. Using Lemma 5 and sliding-window FP, we can find the reversed suffix links in $O(\log \sigma_S)$ time at each of the $\sum_{i=1}^{n} d_i$ visited nodes. Thus the total time complexity is $O(n \log \sigma_S)$. Since the number of nodes in $\mathsf{CPH}(S)$ is $n+1$ and the number of reversed suffix links is n, the total space complexity is $O(n)$. □

Lemma 8. *There exists a string S of length n over a binary alphabet $\Sigma = \{1, 2\}$ such a node in $\mathsf{CPH}(S)$ has $\Omega(\sqrt{n})$ out-going edges.*

Proof. Consider string $S = 11211221 \cdots 12^k 1$. Then, for any $1 \leq \ell \leq k$, there exist nodes representing $01^{k-2}\ell$. Since $k = \Theta(\sqrt{n})$, the parent node 01^{k-2} has $\Omega(\sqrt{n})$ out-going edges. $\square$

Due to Lemma 8, if we maintain a sorted list of out-going edges for each node during our online construction of $\mathsf{CPH}(S[i..])$, it would require $O(n \log n)$ time even for a constant-size alphabet. Still, after $\mathsf{CPH}(S)$ has been constructed, we can sort all the edges offline, as follows:

Theorem 2. *For any string S over an integer alphabet $\Sigma = [1..\sigma]$ of size $\sigma = n^{O(1)}$, the edge-sorted $\mathsf{CPH}(S)$ together with the maximal reach pointers can be computed in $O(n \log \sigma_S)$ time and $O(n)$ space.*

Proof. We sort the edges of $\mathsf{CPH}(S)$ as follows: Let i be the id of each node $u(i)$. Then sort the pairs (i, x) of the ids and the edge labels. Since $i \in [0..n-1]$ and $x \in [1..n^{O(1)}]$, we can sort these pairs in $O(n)$ time by a radix sort. The maximal reach pointers can be computed in $O(n \log \sigma_S)$ time using the reversed suffix links, in a similar way to the position heaps for exact matching [10]. $\square$

4   Cartesian-Tree Position Heaps for Tries

Let T be the input text trie with N nodes. A naïve extension of our CPH to a trie would be to build the CPH for the sequence $\langle \mathsf{PD}(T[N..]), \ldots, \mathsf{PD}(T[1..]) \rangle$ of the parent encodings of all the path strings of T towards the root $\mathbf{r}$. However, this does not seem to work because the parent encodings are not consistent for suffixes. For instance, consider two strings 1432 and 4432. Their longest common suffix 432 is represented by a single path in a trie T. However, the longest common suffix of $\mathsf{PD}(1432) = 0123$ and $\mathsf{PD}(4432) = 0100$ is ε. Thus, in the worst case, we would have to consider all the path strings $T[N..], \ldots, T[1..]$ in T separately, but the total length of these path strings in T is $\Omega(N^2)$.

To overcome this difficulty, we reuse the FP encoding from Sect. 3. Since $\mathsf{FP}(S)[i]$ is determined merely by the suffix $S[i..]$, the FP encoding is suffix-consistent. For an input trie T, let the *FP-trie* T_{FP} be the reversed trie storing $\mathsf{FP}(T[i..])$ for all the original path strings $T[i..]$ towards the root. Let N' be the number of nodes in T_{FP}. Since FP is suffix-consistent, $N' \leq N$ always holds. Namely, FP is a linear-size representation of the equivalence relation of the nodes of T w.r.t. $\approx$. Each node v of T_{FP} stores the equivalence class $\mathcal{C}_v = \{i \mid T_{\mathsf{FP}}[v..] = \mathsf{FP}(T[i..])\}$ of the nodes i in T that correspond to v. We set $\min\{\mathcal{C}_v\}$ to be the representative of $\mathcal{C}_v$, as well as the id of node v.

250   A. Nishimoto et al.

Let Σ_T be the set of distinct characters (i.e. edge labels) in T and let
$\sigma_T = |\Sigma_T|$. The FP-trie T_{FP} can be
computed in $O(N\sigma_T)$ time and work-
ing space by a standard traversal on T,
where we store at most σ_T front pointers in each node of the current path in
T due to Lemma 3.

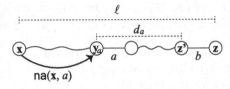

Fig. 4. Illustration for the data structure
of Lemma 9, where $(T[i..])[1..\ell] = \mathsf{str}(\mathbf{x},\mathbf{z})$.

Let $i_{N'}, \ldots, i_1$ be the node id's of
T_{FP} sorted in decreasing order. The
Cartesian-tree position heap for the input trie T is $\mathsf{CPH}(T) = \mathsf{SHT}(\mathcal{W}_T)$, where
$\mathcal{W}_T = \langle \mathsf{PD}(T[i_{N'}..]), \ldots, \mathsf{PD}(T[i_1..]) \rangle$.

As in the case of string inputs in Sect. 3, we insert the shortest prefix of
$\mathsf{PD}(T[i_k..])$ that does not exist in $\mathsf{CPH}(T[i_{k+1}..])$. To perform this insert operation, we use the following data structure for a random-access query on the PD
encoding of any path string in T:

Lemma 9. *There is a data structure of size $O(N\sigma_T)$ that can answer the following queries in $O(\sigma_T)$ time each.*

Query input: *The id i of a node in T and integer $\ell > 0$.*
Query output: *The ℓth (last) symbol $\mathsf{PD}((T[i..])[1..\ell])[\ell]$ in $\mathsf{PD}(T[i..])[1..\ell]$.*

Proof. Let $\mathbf{x}$ be the node with id i, and $\mathbf{z} = \mathsf{anc}(\mathbf{x}, \ell)$. Namely, $\mathsf{str}(\mathbf{x},\mathbf{z}) =
(T[j..])[1..\ell]$. For each character $a \in \Sigma_T$, let $\mathsf{na}(\mathbf{x}, a)$ denote the nearest ancestor
$\mathbf{y}_a$ of $\mathbf{x}$ such that the edge $(\mathsf{parent}(\mathbf{y}_a), \mathbf{y}_a)$ is labeled a. If such an ancestor does
not exist, then we set $\mathsf{na}(\mathbf{x}, a)$ to the root $\mathbf{r}$.

Let $\mathbf{z}' = \mathsf{anc}(\mathbf{x}, \ell-1)$, and b be the label of the edge $(\mathbf{z}, \mathbf{z}')$. Let D be an empty
set. For each character $a \in \Sigma_T$, we query $\mathsf{na}(\mathbf{x}, a) = \mathbf{y}_a$. If $d_a = |\mathbf{y}_a| - |\mathbf{z}'| > 0$
and $a \leq b$, then d_a is a candidate for $(\mathsf{PD}(T[j..])[1..\ell])[\ell]$ and add d_a to set D.
After testing all $a \in \Sigma_T$, we have that $(\mathsf{PD}(T[j..])[1..\ell])[\ell] = \min D$. See Fig. 4.

For all characters $a \in \Sigma_T$ and all nodes x in T, $\mathsf{na}(\mathbf{x}, a)$ can be pre-computed
in a total of $O(N\sigma_T)$ preprocessing time and space, by standard traversals on
T. Clearly each query is answered in $O(\sigma_T)$ time.   □

Theorem 3. *Let T be a given trie with N nodes whose edge labels are from an
integer alphabet of size $n^{O(1)}$. The edge-sorted $\mathsf{CPH}(T)$ with the maximal reach
pointers, which occupies $O(N\sigma_T)$ space, can be built in $O(N\sigma_T)$ time.*

Proof. The rest of the construction algorithm of $\mathsf{CPH}(T)$ is almost the same as
the case of the CPH for a string, except that the amortization argument in the
proof for Theorem 1 cannot be applied to the case where the input is a trie.
Instead, we use the nearest marked ancestor (NMA) data structure [2,21] that
supports queries and marking nodes in amortized $O(1)$ time each, using space
linear in the input tree. For each $a \in [0..\sigma_T]$, we create a copy $\mathsf{CPH}_a(T)$ of
$\mathsf{CPH}(T)$ and maintain the NMA data structure on $\mathsf{CPH}_a(T)$ so that every node
v that has defined reversed suffix link $\mathsf{rsl}(v, a)$ is marked, and any other nodes
are unmarked. The NMA query for a given node v with character a is denoted

by $\mathsf{nma}_a(v)$. If v itself is marked with a, then let $\mathsf{nma}_a(v) = v$. For any node $\mathbf{x}$ of T, let $\mathcal{I}_\mathbf{x}$ be the array of size at most σ_T s.t. $\mathcal{I}_\mathbf{x}[j] = h$ iff h is the jth smallest element of $\mathcal{F}_{\mathsf{PD}(\mathsf{str}(\mathbf{x}))}$.

We are ready to design our construction algorithm: Suppose that we have already built $\mathsf{CPH}(T[i_{k+1}..])$ and we are to update it to $\mathsf{CPH}(T[i_k..])$. Let $\mathbf{w}$ be the node in T with id i_k, and let $\mathbf{u} = \mathsf{parent}(\mathbf{w})$ in T_{FP}. Let u be the node of $\mathsf{CPH}(T[i_{k+1}..])$ that corresponds to $\mathbf{u}$. We initially set $v \leftarrow u$ and $a \leftarrow |\mathcal{F}_{\mathsf{PD}(T[i_k..i_k+|u|])}|$. Let $d(a) = \max\{|u| - \mathcal{I}_\mathbf{w}[a] + 1, 0\}$. We perform the following:

(1) Check whether $v' = \mathsf{anc}(u, d(a))$ is marked in $\mathsf{CPH}_a(T)$. If so, go to (2). Otherwise, update $v \leftarrow v'$, $a \leftarrow a - 1$, and repeat (1).
(2) Return $\mathsf{nma}(v, a)$.

By the definitions of $\mathcal{I}_\mathbf{w}[a]$ and $d(a)$, the node $v(i_k)$ from which we should take the reversed suffix link is in the path between v' and v, and it is the lowest ancestor of v that has the reversed suffix link with a. Thus, the above algorithm correctly computes the desired node. By Lemma 4, the number of queries in (1) for each of the N' nodes is $O(\sigma_T)$, and we use the dynamic level ancestor data structure on our CPH that allows for leaf insertions and level ancestor queries in $O(1)$ time each [1]. This gives us $O(N\sigma_T)$-time and space construction.

We will reuse the random access data structure of Lemma 9 for pattern matching (see Sect. 5.2). Thus $\mathsf{CPH}(T)$ requires $O(N\sigma_T)$ space. $\square$

5   Cartesian-Tree Pattern Matching with Position Heaps

5.1   Pattern Matching on Text String S with $\mathsf{CPH}(S)$

Given a pattern P of length m, we first compute the greedy factorization $\mathsf{f}(P) = P_0, P_1, \ldots, P_k$ of P such that $P_0 = \varepsilon$, and for $1 \leq l \leq k$, $P_l = P[\mathsf{lsum}(l-1) + 1..\mathsf{lsum}(l)]$ is the longest prefix of $P_l \cdots P_k$ that is represented by $\mathsf{CPH}(S)$, where $\mathsf{lsum}(l) = \sum_{j=0}^{l} |P_j|$. We consider such a factorization of P since the height h of $\mathsf{CPH}(S)$ can be smaller than the pattern length m.

Lemma 10. *Any node v in $\mathsf{CPH}(S)$ has at most $|v|$ out-going edges.*

Proof. Let (v, c, u) be any out-going edge of v. When $|u| - 1$ is a front pointer of u, then $c = u[|u|]$ and this is when c takes the maximum value. Since $u[|u|] \leq |u| - 1$, we have $c \leq |u| - 1$. Since the edge label of $\mathsf{CPH}(S)$ is non-negative, v can have at most $|u| - 1 = |v|$ out-going edges. $\square$

The next corollary immediately follows from Lemma 10.

Corollary 1. *Given a pattern P of length m, its factorization $\mathsf{f}(P)$ can be computed in $O(m \log(\min\{m, h\}))$ time, where h is the height of $\mathsf{CPH}(S)$.*

The next lemma is analogous to the position heap for exact matching [10].

Lemma 11. *Consider two nodes u and v in $\mathsf{CPH}(S)$ such that $u = \mathsf{PD}(P)$ the id of v is i. Then, $\mathsf{PD}(S[i..])[1..|u|] = u$ iff one of the following conditions holds: (a) v is a descendant of u; (b) $\mathsf{mrp}(v)$ is a descendant of u.*

We perform a standard traversal on $\mathsf{CPH}(S)$ so that one we check whether a node is a descendant of another node in $O(1)$ time.

When $k = 1$ (i.e. $\mathsf{f}(P) = P$), $\mathsf{PD}(P)$ is represented by some node u of $\mathsf{CPH}(S)$. Now a direct application of Lemma 11 gives us all the occ pattern occurrences in $O(m \log m + occ)$ time, where $\min\{m, h\} = m$ in this case. All we need here is to report the id of every descendant of u (Condition (a)) and the id of each node v that satisfies Condition (b). The number of such nodes v is less than m.

When $k \geq 2$ (i.e. $\mathsf{f}(P) \neq P$), there is no node that represents $\mathsf{PD}(P)$ for the whole pattern P. This happens only when $occ < m$, since otherwise there has to be a node representing $\mathsf{PD}(P)$ by the incremental construction of $\mathsf{CPH}(S)$, a contradiction. This implies that Condition (a) of Lemma 11 does apply when $k \geq 2$. Thus, the *candidates* for the pattern occurrences only come from Condition (b), which are restricted to the nodes v such that $\mathsf{mrp}(v) = u_1$, where $u_1 = \mathsf{PD}(P_1)$. We apply Condition (b) iteratively for the following $P_2, \ldots, P_k$, while keeping track of the position i that was associated to each node v such that $\mathsf{mrp}(v) = u_1$. This can be done by padding i with the off-set $\mathsf{lsum}(l - 1)$ when we process P_l. We keep such a position i if Condition (b) is satisfied for all the following pattern blocks $P_2, \ldots, P_k$, namely, if the maximal reach pointer of the node with id $i + \mathsf{lsum}(l-1)$ points to node $u_l = \mathsf{PD}(P_l)$ for increasing $l = 2, \ldots, k$. As soon as Condition (b) is not satisfied with some l, we discard position i.

Suppose that we have processed the all pattern blocks $P_1, \ldots, P_k$ in $\mathsf{f}(P)$. Now we have that $\mathsf{PD}(S[i..])[1..m] = \mathsf{PD}(P)$ (or equivalently $S[i..i+m-1] \approx P$) *only if* the position i has survived. Namely, position i is only a candidate of a pattern occurrence at this point, since the above algorithm only guarantees that $\mathsf{PD}(P_1) \cdots \mathsf{PD}(P_k) = \mathsf{PD}(S[i..])[1..m]$. Note also that, by Condition (b), the number of such survived positions i is bounded by $\min\{|P_1|, \ldots, |P_k|\} \leq m/k$.

For each survived position i, we verify whether $\mathsf{PD}(P) = \mathsf{PD}(S[i..])[1..m]$. This can be done by checking, for each increasing $l = 1, \ldots, k$, whether or not $\mathsf{PD}(S[i..])[\mathsf{lsum}(l - 1) + y] = \mathsf{PD}(P_1 \cdots P_l)[\mathsf{lsum}(l - 1) + y]$ for every position y ($1 \leq y \leq |P_l|$) such that $\mathsf{PD}(P_l)[y] = 0$. By the definition of PD, the number of such positions y is at most $\sigma_{P_l} \leq \sigma_P$. Thus, for each survived position i we have at most $k\sigma_P$ positions to verify. Since we have at most m/k survived positions, the verification takes a total of $O(\frac{m}{k} \cdot k\sigma_P) = O(m\sigma_P)$ time.

Theorem 4. *Let S be the text string of length n. Using $\mathsf{CPH}(S)$ of size $O(n)$ augmented with the maximal reach pointers, we can find all occ occurrences for a given pattern P in S in $O(m(\sigma_P + \log(\min\{m, h\})) + occ)$ time, where $m = |P|$ and h is the height of $\mathsf{CPH}(S)$.*

5.2 Pattern Matching on Text Trie T with $\mathsf{CPH}(T)$

In the text trie case, we can basically use the same matching algorithm as in the text string case of Sect. 5.1. However, recall that we cannot afford to store the PD encodings of the path strings in T as it requires $\Omega(n^2)$ space. Instead, we reuse the random-access data structure of Lemma 9 for the verification step. Since it takes $O(\sigma_T)$ time for each random-access query, and since the data structure occupies $O(N\sigma_T)$ space, we have the following complexity:

Theorem 5. *Let T be the text trie with N nodes. Using $\mathsf{CPH}(T)$ of size $O(N\sigma_T)$ augmented with the maximal reach pointers, we can find all occ occurrences for a given pattern P in T in $O(m(\sigma_P\sigma_T + \log(\min\{m,h\})) + occ)$ time, where $m = |P|$ and h is the height of $\mathsf{CPH}(T)$.*

Acknowledgments. This work was supported by JSPS KAKENHI Grant Numbers JP18K18002 (YN) and JP21K17705 (YN), and by JST PRESTO Grant Number JPMJPR1922 (SI).

References

1. Alstrup, S., Holm, J.: Improved algorithms for finding level ancestors in dynamic trees. In: Montanari, U., Rolim, J.D.P., Welzl, E. (eds.) ICALP 2000. LNCS, vol. 1853, pp. 73–84. Springer, Heidelberg (2000). https://doi.org/10.1007/3-540-45022-X_8
2. Amir, A., Farach, M., Idury, R.M., Poutré, J.A.L., Schäffer, A.A.: Improved dynamic dictionary matching. Inf. Comput. **119**(2), 258–282 (1995)
3. Baker, B.S.: A theory of parameterized pattern matching: algorithms and applications. In: STOC 1993, pp. 71–80 (1993)
4. Baker, B.S.: Parameterized pattern matching by Boyer-Moore type algorithms. In: Proceedings of 6th Annual ACM-SIAM Symposium on Discrete Algorithms, pp. 541–550 (1995)
5. Baker, B.S.: Parameterized pattern matching: algorithms and applications. J. Comput. Syst. Sci. **52**(1), 28–42 (1996)
6. Bender, M.A., Farach-Colton, M.: The level ancestor problem simplified. Theor. Comput. Sci. **321**(1), 5–12 (2004)
7. Cho, S., Na, J.C., Park, K., Sim, J.S.: A fast algorithm for order-preserving pattern matching. Inf. Process. Lett. **115**(2), 397–402 (2015)
8. Coffman, E., Eve, J.: File structures using hashing functions. Commun. ACM **13**, 427–432 (1970)
9. Crochemore, M., et al.: Order-preserving indexing. Theor. Comput. Sci. **638**, 122–135 (2016)
10. Ehrenfeucht, A., McConnell, R.M., Osheim, N., Woo, S.W.: Position heaps: a simple and dynamic text indexing data structure. J. Discrete Algorithms **9**(1), 100–121 (2011)
11. Fu, T., Chung, K.F., Luk, R.W.P., Ng, C.: Stock time series pattern matching: Template-based vs. rule-based approaches. Eng. Appl. Artif. Intell. **20**(3), 347–364 (2007)
12. Fujisato, N., Nakashima, Y., Inenaga, S., Bannai, H., Takeda, M.: Right-to-left online construction of parameterized position heaps. In: PSC 2018, pp. 91–102 (2018)
13. Fujisato, N., Nakashima, Y., Inenaga, S., Bannai, H., Takeda, M.: The parameterized position heap of a trie. In: Heggernes, P. (ed.) CIAC 2019. LNCS, vol. 11485, pp. 237–248. Springer, Cham (2019). https://doi.org/10.1007/978-3-030-17402-6_20
14. I., T., Inenaga, S., Takeda, M.: Palindrome pattern matching. In: Giancarlo, R., Manzini, G. (eds.) CPM 2011. LNCS, vol. 6661, pp. 232–245. Springer, Heidelberg (2011). https://doi.org/10.1007/978-3-642-21458-5_21

15. Kim, H., Han, Y.: OMPPM: online multiple palindrome pattern matching. Bioinformatics **32**(8), 1151–1157 (2016)
16. Kim, J., et al.: Order-preserving matching. Theor. Comput. Sci. **525**, 68–79 (2014)
17. Matsuoka, Y., Aoki, T., Inenaga, S., Bannai, H., Takeda, M.: Generalized pattern matching and periodicity under substring consistent equivalence relations. Theor. Comput. Sci. **656**, 225–233 (2016)
18. Park, S.G., Bataa, M., Amir, A., Landau, G.M., Park, K.: Finding patterns and periods in cartesian tree matching. Theor. Comput. Sci. **845**, 181–197 (2020)
19. Song, S., Gu, G., Ryu, C., Faro, S., Lecroq, T., Park, K.: Fast algorithms for single and multiple pattern Cartesian tree matching. Theor. Comput. Sci. **849**, 47–63 (2021)
20. Weiner, P.: Linear pattern-matching algorithms. In: Proceedings of 14th IEEE Annual Symposium on Switching and Automata Theory, pp. 1–11 (1973)
21. Westbrook, J.: Fast incremental planarity testing. In: Kuich, W. (ed.) ICALP 1992. LNCS, vol. 623, pp. 342–353. Springer, Heidelberg (1992). https://doi.org/10.1007/3-540-55719-9_86

Author Index

Printed in the United States
by Baker & Taylor Publisher Services